The Problem of Life

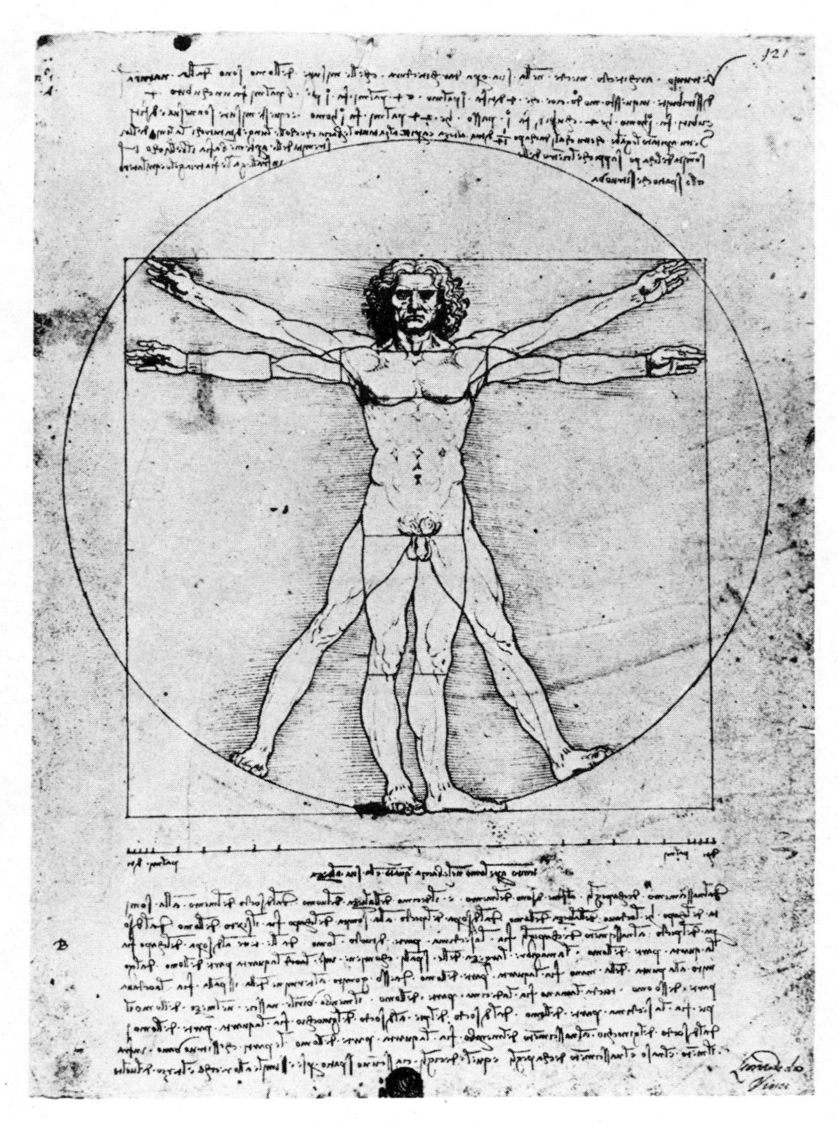

'VITRUVIAN' MAN

Leonardo da Vinci's famous drawing illustrates the theory of proportion worked out by the first-century Roman architect, Vitruvius. The theory proposed that a well-proportioned human body with feet together and arms outstretched fitted exactly into a square: a spread-eagled body, on the other hand, could be inscribed in a circle centred on the navel. Although Leonardo collected much data from living models, the reduction of human to geometrical proportion proved impossible. This Procrustean exercise has, however, been continued in a more abstract manner, and with greater success, by Leonardo's modern successors: the molecular biologists and biophysicists.

CE

THE PROBLEM OF LIFE

An Essay in the Origins of Biological Thought

C. U. M. SMITH

Department of Biological Sciences, University of Aston in Birmingham

M

First published 1976 by
THE MACMILLAN PRESS LTD
London and Basingstoke
Associated companies in New York Dublin Melbourne Johannesburg and Madras

SBN 333 17781 9

Printed in Great Britain by
Lowe and Brydone (Printers) Ltd., Thetford, Norfolk

Then, what is life? I cried.

Shelley: *The Triumph of Life*

As to the past, history continues to be the means by which we recognise what is new as well as what is not.

Frank Kermode: *Modern Essays*

. . . there is assuredly no more effective method of clearing up one's own mind on any subject than by talking it over, so to speak, with men of real power and grasp, who have considered it from a totally different point of view. The parallax of time helps us to the true position of a conception as the parallax of space helps us to that of a star.

T.H. Huxley: *On Animal Automatism*

PREFACE

This essay is of so prefatory a nature that it has no real need of any other preface. Writing this, as I am, surrounded by recently corrected galleys I am only too well aware of how many issues and how many periods have received all too cursory an examination. Yet to have attempted more would have risked obscuring the major themes in a fog of qualification and detail. I shall have succeeded if I have started sufficient intellectual hares to send the reader via the bibliography to the sources.

In the main I have concentrated on what appear to me to be salient episodes and seminal thinkers. These undoubtedly reflect my own biases and concerns. I must apologise in advance to readers interested in botany and microbiology. They will, I am afraid, find little about the origins of their subjects in the following pages. But it seems to me, as it seemed to Theophrastus the earliest of the historians of science, that the subject must have this partial (if not partisan) nature.

It remains to acknowledge, as in my previous books, my sense of indebtedness to the many men and women whose work and insight have illuminated the issues on which I have written. It remains also to acknowledge a sense of homage to the thinkers discussed in the following pages whom Claude Bernard likened to 'torches shining at long intervals to guide the advance of science. They light up their time . . .'

I am most grateful to the following sources for permission to reproduce illustrative material: Frontispiece, Venice Academy; Fig. 3.1 courtesy of Jonathan Cape from *The History of Man* by Carleton S. Coon (1955); Fig. 12.1 courtesy of the trustees of the British Museum; Fig. 13.2(a) courtesy of Anderson Giraudon; Fig. 13.2(b) courtesy of Staatsbibliothek Munchen Clm 23 638, fol. 11^{v}; Fig. 13.3 reproduced by gracious permission of Her Majesty Queen Elizabeth II; Fig. 14.2 courtesy of A. W. Singer Esq; Fig. 14.3 reproduced by gracious permission of Her Majesty Queen Elizabeth II; Fig. 16.3 courtesy of The Royal Society; Fig. 21.2 courtesy of J. C. Kendrew, F.R.S.; Fig. 21.3 courtesy of Mondadori Ltd. from *Biologia Molecolare* by C. U. M. Smith (1971); Fig. 22.1(a) courtesy A. W. Singer Esq.; Fig. 22.1(b) originally published by the University of California Press, reprinted by permission of the Regents of the University of California; Fig. 22.2 courtesy of the Hafner Press from *The Historical Development of Physiological*

Thought edited by Chandler McC. Brooks and P. C. Cranefield (1959); Fig. 22.3 originally published by the University of California Press, reprinted by permission of The Regents of the University of California.

I am most grateful to the following sources for permission to reproduce extracts: Regents of the University of California Press; Penguin Books Ltd.; Johns Hopkins Press; the literary trustees of Walter de la Mare and the Society of Authors as their representative.

Finally I would like once again to thank my wife for help in typing and proof reading and for being the most loyal of critics.

C.U.M.S.

CONTENTS

ILLUSTRATIONS

TABLES

INTRODUCTION

'What is life?' We have all asked ourselves Shelley's question—and probably had to be content with fairly dusty answers. This book tries to set the question in the perspective of history.

The story of the quest for an answer is largely the story of the growth of a science of biology. Largely, but not entirely; for the question 'What is life?' has concerned a circle far wider than that of the professional biologist. It has engaged the attention of poet and philosopher, theologian and physicist. This book, therefore, does not set out to be another academic study of the history of biology. It is not concerned so much to chronicle the minutiae of scientific advance as to investigate the historical and philosophical roots of our present understanding. Although I hope it will be of value to students of biology, I also hope it will be of interest to non-biologists.

Nowadays if we ask a professional biologist to say what 'life' is he would probably reply by describing the nature of self-replicating molecules, of energy-rich phosphate bonds and the mechanism of Darwinian evolution. If pressed he might agree that it was the name given to a peculiar state of matter existing at the interface between the planet and interplanetary space.

This is a very recent vision of the nature of 'life'. It has hardly yet impinged upon the general consciousness. Only a few generations back it would have seemed merely wild speculation, and in the remoter past simply blasphemous.

'Life' must surely be something more than this? Is there no meaning, no purpose, no 'privileged axis'? How can the scientist who is, after all, a sentient being like the rest of us seem to deny the possibility of consciousness? For it seems clear that 'matter in motion cannot think'. The biologist seems to give us Hamlet without the Prince of Denmark.

It is a major part of the impulse of this book to show how it is that we are arrived at this position. For it is too easily forgotten that the glittering spectacle of modern science and, in particular, of modern life science is but the surface of a historical process stretching back two millennia and more. In any account of contemporary science, whether it be of molecular biology or of neurobiology, we are bound to skim over the surface of countless unstated assumptions. To begin to answer Shelley's question we have, I believe, to investigate the hidden depth of our science. We have to use the perspectives of history to provide the parallax necessary to see ourselves and our science in the round.

It is clear that there are numerous ways of composing a book having the objectives outlined above. One might have taken isolated historical epochs, perhaps the Aristotelian biology of the fourth century BC, the Cartesian biology of the late seventeenth century and the biology of the nineteenth-century *Naturphilosophie*, and used them as triangulation points to determine the position and status of our present concepts. Or, perhaps, one might have analysed in depth the development of certain key biological concepts, epoch by epoch. Or one might have attempted to show how throughout the two millennia of our history social and economic forces have influenced biological theory and hence, by implication, suggest that we too are not immune from this bias. The approach finally adopted partakes to some extent of all three of those mentioned above. It tries, however, to set them in the frame of a continuous narrative and an unfolding plot. For, as I have already mentioned, I have tried to write a book which can appeal not only to students of biology but also to the general reader. For we are all interested in the answer, or the answers, to Shelley's question.

One way of viewing the historical process related in the ensuing pages is to see it as a progressive unravelling of a primitively inextricably intertwined knot of ideas. The most prominent threads in this tangled knot might be labelled the notions of 'objectivity' and 'subjectivity'. One of the book's major themes thus consists in following the gradual unravelling of these two attitudes towards events from their early entanglement. Another way of describing the same process is to say that one of the most important themes in the evolution of science in general and of biology in particular has been the separation of a teleological (animistic) from a non-teleological (mechanistic) account of the world.

Our subjective experience seems indisputably to be that of a 'tissue of purposes'. Few would quarrel with the existentialist's insistence that 'intentionality *is* consciousness'. Explanations of our own behaviour are invariably couched in terms of 'final' causes. We act to realise certain aims and ambitions. The scientific account of the world 'outside', however, disdains all recourse to teleological explanation. Final causes are anathema to the scientist. Such explanations seem to him to stifle further investigation. Scientific 'explanations' are invariably couched in terms of 'efficient' causes. The billiard ball does not progress across the table because it seeks the pocket but because it has received an impulse of appropriate strength and direction from the billiard cue. The heart does not beat with increased rapidity because it perceives a sudden danger and hence the need to escape but because of the influence of certain chemicals on its 'pacemaker' region. At the beginning of our history, however, we shall find these two types of explanation, the teleological and the non-teleological, systematically conflated.

The unravelling of objectivity and subjectivity, of final and efficient causation, has by no means been a continuous process. For long periods of time little advance was made. It seems, indeed, that 'progress' could only occur when the underlying social and technological conditions were appropriate. These conditions, to use an analogy due to Scheler (1), act as a valve controlling the development of scientific ideas. When social conditions are inappropriate the valve closes and prevents further scientific advance; when social conditions change the valve opens.

Now one of the crucial notions of an 'objective' or 'mechanistic' theory of nature is the notion of 'blindly running atoms'. If the phenomena of the world can be shown to be the outcome of the interaction of such submicroscopic 'billiard balls' then the case for teleology is very severely weakened. We shall see in the early chapters of this book that the intellectual origins of the atomic theory are to be found amongst the Eleatics and their Abderan successors in the fifth century BC. The theory, in fact, is the logical outcome of the early Greek way of looking at the world. But as we have just seen it has certain extrascientific implications. It implies that all the scents and sounds and colours of the world, all the strategems, ambitions and achievements, are but the outcome of 'chance and necessity' (2). The world emerges from the mere random collocation and dislocation of purposeless atoms. The theory implies that teleological explanations in terms of intention, whether human or divine, are illusory. These atheistic concomitants were unacceptable to societies which depended upon religious belief for their social cohesion. The atomic theory thus provides a classic instance of Scheler's 'social valving'. After a brief florescence in the ancient world it was outlawed until the seventeenth century AD.

Now the history of the atomic theory is closely bound up with the history of biological thought, for, as we have just noticed, it was from its inception profoundly non-teleological. It thus provided from the first a very considerable stumbling block for the biologist. We shall see, for example, the scornful reaction it produced in Aristotle, Galen and Harvey—three of biology's greatest names. Aristotle, the greatest of the three, was especially opposed. For not only was his biological intuition thoroughly outraged—how was an atomic theory of embryology possible?—but the teleological metaphysics which he had discussed at Plato's Academy for nearly twenty years was also totally antipathetic. This metaphysics had, of course, a social origin. It will be briefly discussed in chapter 7.

(1) M.Scheler, *Die Wissensformen und die Gesellschaft,* 1925. A useful account of Scheler's ideas is to be found in P.L.Berger and T.Luckmann, *The Social Construction of Reality,* Penguin, Harmondsworth (1967).

(2) A phrase from Democritus which Jacques Monod has used as the title for a celebrated book.

The evolution of the modern science of life is thus closely tied up with the evolution of the atomic theory of matter. Only when this had been achieved did it become possible to apply Occam's razor to the vital principles and entelechies which had formerly been conceived to occupy the bodies of living organisms. Contemporary molecular biology, perhaps the most significant of all advances towards an answer to the Shelleyan question, would have been impossible without the modern understanding of the nature of atoms and the complexes atoms form. Thus any consideration of the evolution of biological thought must include some account of the evolution of the atomic theory of matter.

It is interesting to notice (as Scheler would have predicted) that the post-seventeenth-century mechanistic biology, like its Aristotelian predecessor, had its sociological roots. For the atomic principle, in its widest sense, also lies at the source of Hobbes's bourgeois social philosophy. Society, the 'Leviathan', is seen as emerging from the interaction of innumerable 'blindly running' 'nasty, brutish and short' lives. The analogy is clear. Just as the expansions of gases are to be explained in terms of their invisible atomic constituents, so the behaviour of the state emerges from a summation of the multitudinous doings of its citizenry.

It is interesting, furthermore, to observe that the Hobbesian analysis also lies at the root of the empiricist doctrine of mind (chapter 22) and, through Godwin and Malthus, of the Darwinian theory of evolution (chapter 19). It was only when all these consequences had been fully worked through that it became clear that no natural event need escape the mechanistic interpretation. It is a great part of this book's object to trace the course of this dawning understanding in the realm of biology. For it is only in quite recent times that the full sharpness of the dichotomy between the 'subjective' and the 'objective' has become inescapable.

Our suggestion, then, is that the triumph of the atomic principle, of mechanism, in the modern world is connected at a deep level with the underlying evolution of society. It is the reflection of, or, better, it is allowed by, the triumph of a more manipulative, more operational, outlook. If one takes the next step and asks what it was that led to this triumph, then the obvious answer is the evolution of technology itself. Even in the ages of faith a slow but steady development of the means of production occurred in Europe (see chapter 13). And, going right back to the beginning of our history, we shall see, in chapter 3, that anthropologists are inclined to believe that it was man's early stone technologies which forced the development of his intelligence and his linguistic powers. It is part of the argument of the present book that technology has continued this role down to the present day. Only those things which man has created himself, which he has thoroughly learned to manipulate and control, does he fully comprehend.

The operational understanding gained from a familiarity with technology (in the widest sense of that term) can be used to gain an insight into the way naturally occurring entities, for example organisms, function. The lines of influence, of action and reaction, are thus complex and intertwined. The development of technology provides new explanatory paradigms. These, in turn, provide new ways of understanding and organising society. The spiral takes another turn when the perceived form of our society partly conditions the type of explanatory paradigm we prefer. The advance of scientific theory is thus partly controlled by social forces, partly by the practical techniques which society has mastered and last, but not least, partly by the sheer logic of ideas. The scientific 'objectivity' of which we are nowadays so proud is thus the expression of a whole complex of underlying socioeconomic and technological forces. Is it perhaps nothing more than the reflection of this 'reality' in our ineradicable subjectivity?

This 'complex of forces' lies behind the history outlined in this book. The 'open' societies of the Presocratics were succeeded by 'closed' societies whose theory was propounded at Athens and which spread in fact over all Europe in the centuries following the fall of Rome. The teleological biologies of Aristotle, Galen and Harvey reflect this social reality. The more 'open' societies which have succeeded the rebirth of science have, in turn, allowed the development of a mechanistic biology. The harbinger of this non-teleological biology was René Descartes. Descartes' mechanistic vision could, however, only be accepted in its full depth after the triumph of the Darwinian evolution theory in the early twentieth century. It is thus only in very recent times that we have become fully aware of the totally impersonal nature of the scientist's interpretation of the world. It is only in very recent times that the biologist is able to answer Shelley's question with conviction. 'What is life?' 'Read this text on biochemistry and then this on physical chemistry and this on physics'.

The book starts with a brief account of the part played by the human imagination in the formation of scientific theory. Albert Einstein considered scientific theories to be 'free creations of the human mind' and the first chapter seeks to emphasise this point of view. Such theories are the outcome of the mind's activity: its unceasing attempt to make sense of its surroundings. This drive seems to be basically the same in both the sciences and the arts. In both cases the data of experience are continually shuffled, consciously and/or unconsciously, until a meaningful pattern is obtained.

The next two chapters lay some of the essential groundwork for the subsequent discussion of biological theory in antiquity. In chapter 2 an attempt is made to show how alien to our present modes of thought were some of the predominant modes of thought in the primitive world. To understand the world-view of sympathetic magic requires an imaginative effort similar to those described in chapter 1. Instead of 'seeing' the world as a gigantic 'mechanism' the primitive mind is more inclined to see it as a personality, its parts interconnected not by 'external' forces but by mysterious indwelling sympathies. Chapter 3 takes up much the same major point showing that the words which we now apply at key points in our science—*action, energy, movement, nature, cause,* and so on—frequently had rather different connotations for the ancients. Their meanings seem, on the whole, to have been far richer: they contained psychical as well as physical reference. This is one more indication that in ancient times subjectivity and objectivity (the 'for-itself' and the 'in-itself', to use Sartre's terminology) were by no means as sharply distinguished as they have become today.

The history proper begins in chapter 4 with a discussion of early Greek science. It is hoped that the first three chapters have prepared the ground for this discussion of theories which may look absurd to twentieth-century eyes but which are in fact some of the most penetrating in existence. It is important to recognise that this early thought is almost equal parts poetry and science. It is partly for this reason that the discussion in chapter 1 dwelt on the theory of the imagination developed by the English Romantic poets.

The discussion of Presocratic Greek thought continues through chapters 5 and 6. In chapter 6 the important theme of atomism makes its first major appearance. As mentioned earlier in this introduction, the tension between this theory and the biologist's perpetual confrontation with 'fitness for purpose' is felt throughout our history.

Following chapter 6 come a group of chapters which consider the climax of classical biological thought first at Athens and finally in Alexandria and Rome. Here another of the book's themes makes its appearance: the effect of social conditions on scientific thought. Following Benjamin Farrington it is argued that Socrates' prime purpose was to save the Athenian democracy from the demagogues. The dialogues depend crucially on the doctrine that education consists in 'drawing out', in making explicit what is all the time implicit. Every slave has buried within him, unknown to him, the principles of geometry. Similarly every Athenian was possessed, in however hidden a fashion, of immutable forms of virtue, justice, statesmanship, and so on. If Athenians could only be brought to recognise the existence of these forms and that they had an incorrigibility similar to the truths of geometry, then all men could agree on what was best and on how the state should be governed.

But where were these hidden forms? Socrates believed that they were implanted before birth into the soul. The doctrine that the soul lies behind the body's appearances and governs its activities is taken up by Socrates' disciple : Plato. Plato makes great use of it throughout his work, and importantly for our purposes in the physiological psychology of the *Timaeus*. Plato's thought is thus thoroughly teleological: the ends govern the means. The contrast with the open Democritean vision is sharp and clear.

Plato's teleological metaphysics could scarcely have failed to influence his star pupil, Aristotle, who spent seventeen of his formative years at the Academy. Aristotle was moreover deeply involved in all aspects of biology. It need not therefore surprise us to find that not only the Stagirite's biology was deeply teleological but also his physics and his metaphysics. His perception that 'nature does nothing in vain' pervades his work. Aristotle, like the modern, saw no important dichotomy between the animate and the inanimate. There is, however, a vital distinction: whereas moderns seek to describe animate creation in terms derived from inanimate nature, Aristotle sought to do the opposite. His effort was so powerful, so well-articulated, so persuasive that his influence lasted two thousand years. Indeed it is still active today. It is for this reason that three chapters of the present book are devoted to his thought. He was, and is, as Darwin and many others have pointed out, the greatest of those who have attempted an answer to the question 'What is life'?

From Aristotle to Descartes there is a span of nearly two millennia. These nineteen hundred years of human endeavour are skimmed in four chapters. Such are the exigencies of one who would write a book in finite time and of finite size on the history of biological ideas. The first two of these four chapters discuss the aftermath of the great Athenians in the late classical and early mediaeval periods. We see how the habit of applying concepts derived from biological observation and psychological experience to inanimate nature impeded the growth of both physics and chemistry. In particular the great and long-enduring study of alchemy is seen as an instance of just this misapplication of ideas. In the last two of this set of four chapters the forces leading to an escape from this blind alley are considered. The slow evolution of technology is seen as the chief of these ultimately revolutionary forces. Once the Galilean insight had been accomplished, however, the way was clear for the development, or at least the promulgation, of a mechanistic biology.

René Descartes, in his justly famous division of being into the mental and the physical, permitted the development of a thoroughly mechanistic account of the physiology of the body. Descartes' psychophysiological speculations are considered in chapter 15. Speculations, however, they were bound to be, for in the seventeenth century the basic sciences of physics and chemistry just had not been developed far enough to allow an experimentally-based physicalist treatment of biology. The most important prerequisite was an understanding of the science of matter and of material change. Thus in chapter 16, after a preliminary look at the dead end into which post-Cartesian iatrochemists and iatrophysicists had worked themselves, the discussion turns to a brief review of the eighteenth-century chemical revolution.

Once this revolution had been accomplished the way was clear for the fulfilment of Descartes' vision. The mechanisation of physiology was largely the work of nineteenth-century workers, culminating in the syntheses of Claude Bernard and Charles Sherrington. 'What is life'? 'An intricate physicochemical mechanism'.

But late eighteenth and early nineteenth-century vitalists did not give in without a struggle. The differences between the crude automata of those days and the flexibility of living organisms was too glaring to be overlooked. Moreover a satisfactory answer to Shelley's question demanded some answer to the origin question. Engineering automata are, after all, built by engineers. Were living beings created by an Engineer in the skies? And if so did He perhaps imbue them with some special non-physical principle?

The origin problem was, of course, solved in essence by mid-nineteenth-century evolutionary theory. The emergence of the Darwinian theory is recounted in chapter 19. Darwinism, however, remained 'soft-centred' until a satisfactory understanding of the mechanism of heredity had been built up. This understanding required almost another century to accomplish. Thus it was not until the mid-point of the twentieth century had come and gone that the biologist could return a confident answer to Isaac Newton's query 'How came the bodies of animals to be contrived with so much art...'?

The last two chapters of the book carry the mechanistic analysis into the heartland of the teleological principle: firstly into embryology and then into neurobiology. Is it possible to show that physicalist interpretations obtain in these two fields also? It will be recalled that Aristotle had tried the exact opposite. He had attempted to use concepts suited to embryology and introspective psychology to describe physical phenomena. One may feel, however, that the mechanisation of embryology and neurobiology has been far more successful than the ancient attempt to vivify physics.

It has been more successful partly because of the evolution of matter-theory during the nineteenth and twentieth centuries. The present understanding of how atoms combine and recombine to form structures of ever greater complexity is a far cry from the discordant heaps and whirlpools envisaged by the Democriteans. The atom, moreover, has been shown to have a structure, to be splittable: a contradiction in terms to the ancients. It is thus the subatomic particle which truly corresponds to the concept of the founders of the theory. But here, of course, a radical difference obtrudes. The subatomic 'particle' since de Broglie and Schroedinger is understood to be *also* a wave. Thus are the ancient paradoxes resolved. And it is in modern molecular biology that the two-thousand-year tension between atom and organism at last finds a solution. The advent of X-ray diffraction and electron microscopy has shown there to be no discontinuity in fact, the evolving concept of the nature of the chemical bond has shown there to be no dissimilarity in concept, between the chemist's and the biologist's world.

But what of mind? Can this, too, be mechanised? The explosive development of computer science suggests an affirmative answer. A more careful analysis, however, suggests caution. In chapter 22 we trace the historical process whereby the 'mind', for long considered the 'active' principle of the organism and, indeed, the world, becomes transformed into a reflex mechanism: no longer a 'self-moved mover' but a contrivance reacting to external forces in a passive and ultimately predictable way. This revolutionary insight became possible immediately Descartes divided mind from matter. By the same token it remains only half the story. For our personal, subjective, experience remains as it was for the ancients: purposeful, goal-seeking. The book thus ends by drawing attention once more to this deep dichotomy in our understanding of things. The world we treat as if it were a mechanism drawing analogies from the technologies we have ourselves created; ourselves we understand teleologically. *What is life*? We still cannot answer. For our theories are man-generated, man-orientated; as Kant saw, they are in a way imposed upon the world. In one sense we can say that life, we ourselves, are at one with the Earth: are sprung from the planet's side and consequently fall within the province of natural science. In another and complementary sense we are embodiments of consciousness, of perception and understanding. We know what it is like to be the mechanism which the neurosurgeon dissects. 'What is life'? Both answers are valid. It is a physicochemical mechanism; it is the experience of intention and purposeful activity, of sadness and joy.

1 THE ACT OF IMAGINATION

The semantic imperative

Ludwig Wittgenstein was one of the seminal philosophers of the twentieth century. Among the many topics which he analysed was the topic of perception. This is, of course, one of the most fundamental in the whole field of philosophy. In order to enter the subject he made use of analogies drawn from visual perception. He was concerned to bring out the fact that we customarily use the verb 'to see' in two different senses (1). We are not only aware of the object in the visual field but we also attempt to interpret the sensation so that it fits into our 'pattern of expectations'. It is this second sense of 'see' which is of interest in the present context.

Students of perception have investigated many cases of this second use of the word 'see'. Some of these cases were touched upon in a previous book (2). Figure 1.1 is a picture used in that previous book. It is possible to 'see' it in two different ways. Some will see the face of a young woman, others that of an ancient crone. Most will be able to see both aspects, but not at the same time. The aspects alternate but the picture is always meaningful. This is the nub of the matter. The 'mind' insists that the image makes sense. A meaningless interpretation of the incoming information is rejected.

There are many other ambiguous figures of the same general type. Wittgenstein uses a 'goose–rabbit' as an example: the figure may be seen as a rabbit's head or a goose's (figure 1.2.). Other well-known examples are figures which may be seen as an overhanging cornice or a set of stairs, a pair of faces or the two sides of a jar, and so on.

The esemplastic power

Philosophers and psychologists have certainly not been the only individuals interested in the nature of perception. Shelley's question with which we started this book itself suggests a perceptual problem. Its resonances hint at a multitude of ways of framing an answer. It is perhaps a major part of the

(1) L. Wittgenstein, *Philosophical Investigations* (trans. G.E.M. Anscombe), Blackwell, London (1963), p.193.
(2) C.U.M. Smith, *The Brain: towards an understanding*, Faber and Faber, London (1970).

Figure 1.1. Wife or mother-in-law?

poet's trade to suggest the possibility of such alternative visions.

This, anyway, is very much what the English Romantic poets believed. William Blake was quite explicit: 'One power alone makes a poet, Imagination the Divine Vision'. Wordsworth and Coleridge, although

Figure 1.2. Goose or rabbit?

differing from Blake in many other respects, agreed with him about the centrality of the imagination in the poet's calling.

The *locus classicus* of the Romantic theory of the imagination is to be found in Coleridge's *Biographia Literaria*, chapter 13. A student of German philosophy, particularly of Kant, Coleridge refused to accept the contemporary empiricist concept of mind: a passive *tabula rasa* on which events were willy-nilly inscribed. On the contrary he believed that the mind was active and with its indwelling energy created for itself its so-called perceptions. 'The *primary* imagination', he writes, 'I hold to be the living power and prime Agent of all human Perception, and a repetition in the finite mind of the eternal act of creation in the infinite I AM.'

The instances of 'seeing-as' mentioned in the first section of this chapter give some hint of Coleridge's meaning. The individual *actively* sees either crone or girl in the same constellation of lines. Modern neurophysiology has also gone far towards showing that the brain receives a highly 'processed' image of the environment. The well-known experiments of Hubel and Wiesel have shown how the visual cortex of the brain is provided with highly filtered information from the retinae, information filtered in such a way that its biological significance is enhanced (3). Held and his co-workers have carried the analysis a step further (4). Matters were so arranged that over a period of

(3) An account of this work may be found in D.H. Hubel, The visual cortex of the brain, *Scientific American* **209**,5(1963),54, and also in C.U.M. Smith, *ibid.*
(4) R. Held and A. Hein, Movement-produced stimulation in the development of visually guided behaviour, *J. Comp. Physiol. Psychol.*, **56**(1963), 872–76.

a week or so two visually naive kittens were presented with exactly the same visual stimuli; but while one kitten was allowed to pad around its quarters, the other was carried around the same living space quite passively. It was shown that only the active kitten had learnt the ins and outs of its environment; its passive counterpart retained its original visual naivety. It is clear that modern investigations into the psychophysiology of perception tend to support Coleridge's position. Perception is an *active* process, a creative process, not a mere 'photographic' recording. It is a process in which the raw energies of the environment are actively shuffled and distorted until they fit into a preexistent conceptual frame.

The history of science is full of instances of this fitting of observations into preexistent conceptual frames, or, to use T S Kuhn's term, preexistent 'paradigms'. The history of science also shows, however, that occasionally the pressure of facts which do not quite fit the conceptual framework brings on a change in the paradigm itself. We shall examine one of these momentous events in chapters 13 and 14. In general, however, facts tend to be fitted into paradigms rather than paradigms rebuilt to fit the facts. In the words of Joseph Priestley (1774):

> We may take a maxim so strongly for granted that the plainest evidence of sense will not entirely change and often hardly modify our persuasions; and the more ingenious a man is, the more effectively he is entangled in his error, his ingenuity only helping him to deceive himself by evading the force of the truth (5).

But for Coleridge the active power of the primary imagination was only half the story, and the least important half. For, he says, it is the 'secondary' imagination which distinguishes the poet. It is this which

> dissolves, diffuses, dissipates in order to recreate ... it struggles to idealise and to unify. It is essentially vital as all objects (as objects) are essentially fixed and dead.

It is this power which, according to Coleridge, is the 'prime and loftiest faculty', a faculty found only in creative individuals. One of the most vivid and well-known descriptions of this power comes from the mathematician Henri Poincaré. He describes (6) several instances where solutions to his mathematical perplexities 'dawned' upon him while he was occupied with quite different matters. He recounts, for example, his experiences while composing his first mathematical memoirs. On several occasions the work seemed to reach an impasse and only when he had taken up something else, something quite different, did the correct treatment 'fall out'. This seems to have been the experience of many creative individuals. It is as if a subconscious part of the mind ceaselessly mulled over and rearranged matters while the conscious part was otherwise occupied. Coleridge suggested

(5) *Experiments and Observations on different kinds of Air*, London (1774), in *A Source Book in Chemistry* (ed. H.M. Leicester and H.S. Klickstein), McGraw-Hill, New York (1952), p.113.
(6) H. Poincaré, Mathematical creation (1908), in *The Creative Process* (ed. B. Ghiselin), University of California Press, Berkeley and Los Angeles (1955).

that the German word *Einsbildungskraft* best expresses this power: '... the power of co-adunation, the faculty which forms many into one—*In-eins-bildung*'.

Culture-dependent seeing

Coleridge's distinction between primary and secondary imagination seems to us today somewhat arbitrary. Perception of a particular form in a puzzle picture may well depend on the intellectual 'set' of the percipient. T S Kuhn quotes some work by Bruner and Postman (7) which showed that subjects shown unexpected playing cards, for example red spades, were subject to perceptual failure. In this case as in many others the 'set', the intellectual framework, depends on the training, experience and cultural background of the percipient. N R Hanson in his book *Patterns of Discovery* coins the term 'theory-laden seeing' to describe this tendency to see only what we expect to see. The term might be widened to 'culture-dependent seeing'. Theory-laden or culture-dependent seeing relies, of course, on memory. We are not born possessed of a particular culture, still less of a particular theory. Thus we can see that the Coleridgean distinction is far from being clear-cut. Our perceptions as much as our conceptions depend on memory. The distinction seems to be only temporal. Whereas the primary imagination requires only a fraction of a second, the secondary imagination may require hours, days, weeks or years. The distinction between the two is otherwise unreal. The primary and secondary imagination grade into each other, act and react upon each other and, like Coleridge's esemplastic power itself, dissolve, dissipate and diffuse into each other in an inextricable way.

The Coleridgean account of poets and poetry can of course be generalised. We have already noticed how the experience of the mathematician Henri Poincaré supports Coleridge's position. We may recall, also, that no less a scientist than Albert Einstein regarded scientific concepts as 'free creations of the human mind' (8). Indeed we may, with Shelley, read creative individual where Coleridge and Blake write poet. For in Shelley's famous phrases 'poets are the unacknowledged legislators of mankind' and 'all the authors of revolutions are ... necessarily poets as they are inventors'. Shelley, too, accepted the Romantic stress on the imagination as the central faculty of the poet. For him 'imaginative power is the ability to mark the before unapprehended relations of things', and this, as we have already remarked, is perhaps also the central faculty of the creative worker in science.

(7) J.S. Bruner and L. Postman, On the perception of incongruity: a paradigm, *J. of Personality,* **18** (1949), 206–23.

(8) A. Einstein and L. Infield, *The Evolution of Modern Physics,* Simon and Schuster, New York (1938), p. 33.

The Copernican 'saving of the appearances'

The most thoroughly discussed and researched instance of the power to 'lift the veil from the hidden beauty of the world and to make familiar objects seem as if they were not familiar' is that great seventeenth-century controversy which Galileo entitled 'Dialogue on the Two Great World Systems'. At a fairly superficial level this astronomical quarrel was about the 'saving of the appearances' (9). But clearly it had more emotional resonance than this. Otherwise Galileo would not have had narrowly to escape the fire or spend his declining years under house arrest. For, in fact, what the Copernican theory challenged was nothing less than the entire metaphysical and theological position of Aristotelian–Thomist orthodoxy. In place of an anthropocentric, intelligible world the Copernican theory led to a mere science of phenomena, ultimately unintelligible or, to use the term popularised by existentialist philosophers, absurd. But, to return to a more superficial level, Copernicus, on the face of it, was merely concerned to provide a new and somewhat simpler calculus to account for the movements of the heavenly bodies. Instead of an intricate system of 'wheel without wheel', of epicycle upon epicycle to describe and predict astronomical phenomena, Copernicus suggested the beginnings of a return to the simplicities of the Aristarchan heliocentric hypothesis. In this way, by assuming that the earth rotates on its axis every twenty-four hours and at the same time revolves about the sun, he was able, using the same mathematical devices as Ptolemy and his successors, to provide a somewhat simpler account of celestial kinematics:

> I hold it easier to concede this than to let the mind be distracted by an almost endless multitude of circles which those are obliged to do who detain the earth at the centre of the world. The wisdom of Nature is such that it produces nothing superfluous or useless but often produces many effects from one cause (10).

Copernicus was prudent enough to ensure that his masterpiece was published posthumously. The first edition was brought out in 1543, the year in which its author died. The full consequences of the new way of describing the movements of the stars and planets were not, however, felt until the next century—the century of Kepler, Galileo and Newton. It then became apparent, to paraphrase the words of Alexander Koyré (11), that man could no longer regard himself as the focus of a cosmic concern at the centre of a closed world, but rather as a stranger lost in an infinite universe. This very

(9) Bertrand Russell, for instance, writes that as all motion is relative it makes no difference whether we consider the earth to revolve about its own axis and the firmament to remain steady, or the heavens to revolve about a stationary earth. It is only that in the former case the mathematics is somewhat simpler. (*The ABC of Relativity,* Allen and Unwin, London (1969), p.13.)

(10) *De Revolutionibus Orbium Coelestium* (1543) quoted by A. C. Crombie in *Augustine to Galileo,* vol.2, Mercury Books, London (1961), p.170.

(11) *From the Closed World to the Infinite Universe,* The Johns Hopkins Press, Baltimore (1957).

thorough-going perceptual change did not of course occur suddenly except perhaps in a few individuals of genius. We may indeed suspect that it has not even yet worked its way completely through even Western science-based societies.

Biologies

But what of biology? Has the biologist's perception of the nature of life changed in the same way as the astronomer's perception of the nature of the universe?

Biology (12) has never been so dramatic a subject as physics or astronomy. Its concerns are not so abstract nor so wide-ranging. The Einsteinian concepts of space and time, of energy and mass escape all but the most highly gifted and highly trained minds. Everyone, however, is likely to have some opinion about how the heart works or how it is that offspring resemble their parents. Similar remarks may be made about the historical development of the two subjects. We read in histories of physics of how the substantial matter of Aristotle slowly vanishes over the millennia until in the twentieth century 'objective reality has evaporated' and quantum mechanics 'does not represent particles, but rather our knowledge, our observations, our consciousness of particles' (13). Or we may read how during the European Renaissance it was found necessary to 'strip space of its objectivity, of its substantial nature (*qua* Aristotle) and to rediscover it as a free ideal of a complex of lines' (14). Biologists are unused to such radical operations. The history of biology seems to have followed far more closely the form of its most famous theory: the theory of organic evolution. It is evolutionary both in the original and in the more modern sense of the term *evolutio*. It can be seen as an unfolding, as a florescence; it can also be seen as a process actuated by selective forces operating on random change.

That biologists saw their subject differently in the past cannot, however, be denied. Indeed in the following pages the outlines of perhaps four different biologies can be dimly discerned. These might be designated Aristotelian, Cartesian, Goethean and modern. Unlike the cases of physics and astronomy there is no very sharp line of demarcation between them. The development is continuous. But modern biology, in its fundamentals, differs as much from Aristotelian biology as modern astronomy differs from Ptolemy's. It is important to notice, nevertheless, that each of the biologies mentioned above

(12) The term 'biology' was not in fact introduced until the beginning of the nineteenth century (Burdoch, 1800; Lamarck and Treviranus, 1802). The subject for which the term stands has, of course, existed from the earliest times.

(13) K.R. Popper, Quantum mechanics without 'the observer', in *Quantum Theory and Reality* (ed. M. Bunge), Springer Verlag, Berlin, Heidelberg, New York (1967), p.7. The quoted statement represents, according to Popper, the ruling or Copenhagen interpretation of quantum mechanics.

(14) E. Cassirer, *Individual and Cosmos* (trans. M. Domandi), Oxford University Press, London (1963), p.182.

was in its time perfectly viable. They were no more lisping versions of modern biology than the foetus or larva is a lisping version of the adult, or than the coelacanth is a lisping version of an amphibian. They were viable because they were well adapted to the intellectual world of their time: they integrated the then known facts of biology in the most coherent manner then possible. In short, they made sense.

The Aristotelian biology had much the longest life span of the four mentioned above. It is still the biology of the non-biologist. It is fundamentally teleological: 'Nature does nothing in vain'. It survived for two thousand years, for it seemed to account for biological phenomena in a simple and self-evident fashion. Unfortunately, however, the Aristotelian physics in which it was grounded was from the first suspect. By the sixteenth century AD physicists had shown that its foundations were indeed rotten, and it was quickly swept away. In its place Galileo and Newton created a new and still valid structure.

The demolition of the Aristotelian physics left biology without a base. This was not immediately apparent to all. Indeed one of the very greatest of all biologists—William Harvey—remained a convinced Aristotelian until his death in 1657. To the mathematico-physical mind of Descartes it was, however, crystal clear. Accordingly he attempted to rebuild biology on what seemed to him the firmer foundations of the new physics. In the early seventeenth century this attempt, although brilliantly executed, was premature. Nevertheless the Cartesian automatism was a true insight into the sort of subject biology had to become in a post-Galilean world.

Reaction to the mechanising 'enlightenment' of the late seventeenth and early eighteenth centuries was not slow in coming. This reaction as it affected biology is perhaps best epitomised in Goethe's work. It is interesting to notice that Goethe, like Coleridge, had intellectual links with Kant—the philosophical herald of the Romantic movement. In one of Goethe's conversations with Eckermann the following passage occurs:

> Kant never took any notice of me, although independently I was following a course similar to his. I wrote my *Metamorphosis of Plants* before I knew anything of Kant and yet it is entirely in the spirit of his ideas (15).

Yet in many respects Goethe's approach to science was distinctly un-Kantian. In strong contrast to Kant, who, it will be remembered, regarded a subject as scientific in direct proportion to the quantity of mathematics it contained, Goethe remained to the end vehemently opposed to the use of mathematics. We may perhaps feel that Goethe's aversion was due to a misunderstanding of the nature of mathematics, for his deepest insight into the nature of the scientific endeavour was that it was a search for form. Particular animals and plants were only *instances* of an underlying

(15) *Conversations of Goethe with Eckermann and Soret* (trans. K. Oxenford), Bell, London (1892): 11 April, 1827.

archetypal unity of form. Goethe believed that the *Critique of Judgment* confirmed this view. In it, he believed, could be found the view that nature resembled art in that both possessed an 'inner life' which worked from 'within outward' (16).

We can see that the ancient notion of nature as craftsman is here struggling to be reborn. The excesses of the *Naturphilosophie* movement which Goethe did so much to originate are discussed in chapter 18.

Finally in quite modern times a fresh integration of biological facts has emerged. The work of nineteenth-century evolutionists and twentieth-century geneticists has made it clear that organic form and function can be explained in terms of unmysterious selective forces acting upon the chemical substances of the planet's crust. We may feel that three centuries after Descartes died his roboteer's vision has been vindicated. This development has been achieved as much by an evolution of the sciences of inorganic matter towards those of life as by a reduction of the life sciences to physics and chemistry.

Thus, although biology has not undergone the radical transitions of physics or astronomy, the way in which biologists 'see' their subject has greatly changed in two and a half millennia. We have, moreover, no reason to suppose that this evolution has reached its *terminus ad quem*. On the contrary, we have reason to suppose the opposite. Already many biologists are beginning to question the assumption made by Darwinists that nature 'red in tooth and claw' is the scene of endless war. In our post-Keynesian world this seems altogether too crude a vision. Concepts derived from cybernetics and computer science, concepts like feedback, information, homeostasis, begin to seem far more appropriate. Perhaps this presages a new phase in our understanding of biological phenomena and, in consequence, a new vision of man's place in nature.

(16) *Einwirkung der neueren Philosphie,* quoted by E. Cassirer in *Rousseau, Kant, Goethe* (trans. J. Gutman, P.O. Kristeller and J.A. Randall), Princeton, N J (1945), p.63.

2 BEGINNINGS

The 'interior' and the 'exterior' view

When we attempt to discern the origins of man we often feel ourselves seized by a species of double vision. It is an instance of that schizophrenia which attends all our attempts to describe man totally within the framework of natural science. In some respects it resembles the oscillatory perception we have of Wittgenstein's goose-rabbit or Hill's girl-drone. We shall see as we go on that this systematic duality forms one of the central strands of this book's argument.

On the one hand we are presented by the physical anthropologist and the primatologist with a mass of evidence indicating that man's lineage may be traced back to a population of Dryopithecine apes living in the African Miocene some fifteen million years ago. On the other hand we are shown by the cultural anthropologist and the archaeologist evidence—paintings in the Altamiran caves, the dwelling-places and utensils at Jarmo and Mount Carmel—which we instantly recognise as the work of people like ourselves. The first group of experts shows us the 'exterior', the second the 'interior'. The first shows us skulls, bones, teeth—comparative anatomy; the second shows us 'spirit'. It is the latter we instantly recognise as human. The upright post-arboreal Unguiculate might otherwise have been just another mammal.

In spite of this instant 'recognition' the mentality of the humans responsible for these persistent artefacts was probably vastly different from our own. We saw in the last chapter how culture-dependent are our perceptions. Although the world in which these dawn humans lived must in some sense have been the same as the world which surrounds us today, it must still have *seemed* very different: so different, indeed, that we may suspect that an individual of those far-off times transported by some miraculous time machine to our present age would not only be profoundly bewildered but might well, like some of the long-term blind suddenly given sight, die of it. It follows that we too, if we took up the offer of a return trip on the time machine, would be similarly bewildered. Our scientifically conditioned minds would find themselves in a deeply alien environment, an intellectual environment in which the objects of perception were organised according to principles profoundly different from those used by contemporary scientists.

Sympathetic magic

As a first hypothesis let us suppose that the major intellectual orientation prevalent in those distant times resembled, in its essentials, that still displayed by contemporary primitives. However, this is perhaps not quite the help which it at first sight seems. For it has often been pointed out that the range of intellectual 'sets' shown by modern primitives is surprisingly large. The striking difference, for example, between the solemn rituals of the Bantu and the mocking 'secularity' of the neighbouring forest Pygmies has been pointed out by Turnbull (1). Anthropologists following Durkheim have been inclined to see these different attitudes as reflections of differing social organisation and hence social experience. Whereas the Bantu, to return to Turnbull's example, live a reasonably settled and structured existence, the Pygmies have a far more fluid, wandering life-style. Individuals are always on the move from one Pygmy group to another.

Nevertheless, in spite of the considerable differences in social organisation, both Bantu and Pygmy share a common attitude to their environment, an attitude which differs radically from that customary among the citizens of contemporary industrial societies. For want of a better term this attitude may be described as magico-religious. With the Bantu, however, this outlook is more formalised, more ritualised, than it is among their peripatetic neighbours.

Can we suppose that this intellectual orientation was also characteristic in the Near East during preclassical times? For it is from the ancient societies established around the shores of the Eastern Mediterranean that the Greek intellectual revolution—science—took its origin.

The answer to this important question is not quite straightforward. Desmond Morris (2) and others have, for example, suggested that contemporary primitives should be seen as specialised cultural adaptations to difficult environments, trapped in dead ends off the main line of evolution. Some support for this view may be found in the profound and fascinating book (3) in which de Santillana and von Dechend suggest that perhaps millennia before the birth of science in Magna Graecia astronomical and hence, by implication, mathematical knowledge had developed to a very high level. Similar suggestions have been made by students of mesolithic stone circles such as Stonehenge. Could it be, then, that far from being fossilised remnants of the 'mainstream' from which the classical civilisations of the Mediterranean originated, contemporary primitives are in fact untypical, adapted only to fit unimportant ecological niches?

(1) C.M.Turnbull, *Wayward Servants: the two worlds of the African Pygmies,* Eyre and Spottiswoode, London (1965).
(2) D. Morris, *The Naked Ape,* Cape, London (1967), chapter 1.
(3) G. de Santillana and H. von Dechend, *Hamlet's Mill: an essay on myth and the frame of time,* Gambit, Boston (1969).

There may indeed be some truth in this suggestion. So far, however, as the 'magico-religious' attitude towards the world is concerned, there seems to be good evidence that this was quite general in ancient times and similar, in its essentials, to that which is found among modern primitives. According to Thorndike (4) the word 'magic' itself may be traced far back into the limbo between historical and prehistorical times. It may have originated in the Sumerian or Turanian word *imga*, meaning 'deep' or 'profound'. Furthermore if we examine the poems and hymns deciphered by the archaeologist we frequently find expressed, even by peoples advanced far beyond the palaeolithic, a philosophy which chimes well with the beliefs reported by anthropologists. Thus part of a hymn to the Babylonian fire-god, Nusku, runs as follows:

> Those who made images of me, reproducing my features,
> Who have taken away my breath, torn my hairs,
> Who have rent my clothes, have hindered my feet from treading the dust,
> May the fire-god, the strong one, break their charm (5).

A living world

From what has been said above it follows that one of the ways of gaining the beginnings of an insight into the thinking of those at the root of science is to examine, very briefly, the magico-religious attitude towards the world. In *The Golden Bough* Sir James Frazer has collected a great mass of detailed observations relating to this primal world-view. From this extensive field-work he believed that two sets of regularities—two laws—might be distilled. These serve the primitive much as Newton's laws serve the modern. Frazer named the first of these laws the law of similarity (6) and the second the law of contagion. According to the first like causes like, an effect resembles its cause—'fat pigs', for example, 'are driven by a fat steward', or a wife stumbling over a household box may cause her distant husband to stumble in front of his foe. According to the second, things that have once been in contact with each other continue to exert an influence on each other long after the contact has been severed. Many primitives, for example, are very careful not to let their personal possessions, their clothes or hair trimmings fall into the hands of their enemies. In both cases, both in the law of similarity and in the law of contagion, a mysterious sympathy is felt to exist between things.

This notion of a 'sympathy' running between things and connecting them

(4) L. Thorndike, *A History of Magic and Experimental Science,* vol. 1, Macmillan, London (1929–56), p.4.
(5) Quoted in G. Childe, *The Dawn of European Civilisation,* Kegan Paul, London (1927).
(6) 'The source and seed of all magical power is the attraction of like things and the repulsion of unlike things that takes place in nature.' Giambattista Porta, *Magiae naturalis, Libri viginti,* 1, 2. Quoted by E. Cassirer, *ibid.*

together is not restricted to the primitive mind. It has had a long and distinguished career in the history of medicine and physiology. In later chapters we shall see that as late as the eighteenth century AD physiologists were still accounting for the harmoniousness of the body's activities in terms of a 'sympathy' running between and amongst its parts. The cheek, for example, swells in sympathy with an aching tooth. Because for eighteenth-century physiologists all sympathy presupposed feeling, it came to be believed that sympathetic interactions were dependent on the nervous system (7). Thus in the twentieth century we still speak of sympathetic and parasympathetic nervous systems, of sympathetic ganglia, of sympathetico-mimetic drugs, and so on.

Among primitive peoples, however, sympathy is conceived to exist not only in the world of physiology but also in the world of physics. A concept which more properly functions as an explanation of the workings of man, the microcosm, is misapplied to account for the workings of the world of nature, the macrocosm. Indeed it has often been pointed out that the world to the primitive mind appears to be, precisely, an organism (8). The harmonious interaction of its parts is comparable to the interdependent activity of the parts of a living body, his own body. Even in our society we are sometimes confronted by the question, usually apropos some pest or parasite, 'but what is it for?', as if each part of nature had a function which contributed to the well-being of the whole, as does each part of a living organism. The sympathetic influence of one part of the world on another is only an instance, though a very important instance, of this misanalogy of the world to an organism. We shall find as we go on through this book that this misanalogy between microcosm and macrocosm has been extremely influential.

Returning, however, to *The Golden Bough* and Sir James Frazer we find that he derives the laws of sympathetic magic not from an explicit mistaking of the world for a living organism but from the cognate confusion of psychological with physical reality. Psychoanalysis has today made widely familiar older notions (see chapter 22) of the associationist school of psychology. Similar ideas, that is, ideas with similar connotations, seem to recall each other to consciousness. Ideas which, although dissimilar, have been entertained in the mind together over an extended period of time also tend to recall each other. It is clear that we have here in the psychological realm an exact counterpart of Frazer's laws of sympathetic magic.

If sympathetic magic is indeed partly to be accounted for by a primitive conflation of physical and psychological reality, it is of interest to note what the founder of psychoanalysis himself had to say about magic and magicians.

(7) See R.K.French, *Robert Whytt: the soul and medicine*, Wellcome Institute of the History of Medicine (1969), pp.32–36.

(8) H.Frankfort *et al.*, *Before Philosophy: the intellectual adventure of ancient man*, Penguin, Harmondsworth (1949), p.14: 'Primitive man simply does not know an inanimate world .... The world appears to primitive man neither inanimate nor empty but redundant with life... '

Freud (9) writes of the magician's belief in 'the omnipotence of thought' and suggests that the magician greatly overvalues 'psychical acts': indeed, that he seems to have no sure grasp of the commonly recognised boundary between subjective and objective, between self and not-self. The magician, he writes, believes that to have thought a thing is tantamount to having done it. Dawson in his study of magic and the origins of medicine in ancient Egypt is also impressed by the magician's conviction of his own omnipotence. 'The mere word', he writes, 'is often alone considered as marking the demon's defeat' (10). It is interesting to notice how reminiscent this primitive blurring of the line between subjective and objective is of the findings of certain child psychologists. According to Piaget, for example, the small infant does not think of his mother as a separate person with feelings of her own. He is aware only of his own emotions. He seems to live 'in a world in which to have a fantasy about doing a thing may be indistinguishable from thinking he has actually done it ...'.

We suggested above that some magico-religious practices may be seen as expressions of underlying tensions in tightly knit social organisations. It may thus be that we can account for the behaviour of magicians as ritualised expressions of these underlying tensions and desires in total ignorance of what we now believe to be the causal machinery of the world. Indeed Freud suggests that magical practices very closely resemble the acts of contemporary obsessional neurotics:

> ...the primary obsessive acts of these neurotics are of an entirely magical character. If they are not charms, they are at least counter-charms, designed to ward off the expectations of disaster with which the neurosis usually starts. Whenever I have penetrated the mystery, I have found that the expected disaster is death (11).

An example may make the magical viewpoint somewhat clearer. Warfare and hunting have always been cases in which tension and desire are particularly acute. Even today, in the scientific cultures of the west, members of, for example, bomber crews are apt to perform in private certain rituals in a desperate attempt to ensure a safe return. Malinowski writes (12): 'The function of magic is to ritualise man's optimism, to enhance his faith in the victory of hope over fear'. Thus we, each of us, possess some psychological insight into the practices which Frazer describes in many places in his volumes. We find, for example, a description of the behaviour of the women of a tribe of Sea-Dyaks left behind while their menfolk hunt over the East Indian seas. The women must meticulously carry out certain activities so that

(9) S. Freud, *Totem and Taboo,* in *The Complete Psychological Works of Sigmund Freud,* vol.13, Hogarth Press, London (1955).
(10) W.R.Dawson, *Magician and Leech,* Methuen, London (1929), p.56.
(11) S. Freud., *ibid.*
(12) B. Malinowski, *Magic, Science and Religion,* The Free Press, Glencoe, Illinois (1925), p.70.

the hunters may return in safety and triumph. They must, for example, wake early in the morning and open their windows as soon as it is light, or else their absent husbands will oversleep. They may not oil their hair, for if they did the warriors might slip. Sleeping or dozing during the day is also forbidden, for the men's attention must not falter. All boxes and other movables must be pushed out of the way near the walls so that no woman accidentally stumbles over them: for if this were to happen their distant husbands might also stumble and hence be at the mercy of their foes. The list of dos and don'ts goes on: it is very extensive.

In this example we can to some extent 'feel' the obsessional force which energises much magical thought and practice. We can also notice once again the belief in a mysterious 'sympathy' or system of 'correspondences' linking things and events together. The magician may not be quite certain what these correspondences might be, or how they operate, but that they exist he is certain. To make sure that he does not omit the vital correspondence he enforces as many rituals as possible. In this way magic may be regarded as both a parody and a precursor of science, for both magic and science attempt to gain predictive power over nature; both insist that beneath the ever-shifting play of appearances there is a deep set of 'correspondences', that chains of cause and effect do indeed link the phenomena of the experiential world together. It is easy to agree with Lynn Thorndike when he observes that, far from being contradictory endeavours, magic and science stand to each other as ancestor and offspring (13). Further still it is well to remember that the action at a distance which we smile at in the philosophy of a Sea-Dyak we treat in deadly earnest when we examine the philosophy of Sir Isaac Newton or attempt to plot the trajectories of astronauts.

The publication of *The Golden Bough* created a great impact in the world of classical scholarship. The customs and practices reported, the language and imagery used, threw a new light on to a field which had previously been wedded to textual criticism. Casson (14) writes as follows:

> *The Golden Bough* ... forced the scholars of the literary tradition to enlarge their vision.... Here was the inner mind of the two greatest ancient civilisations being revealed to scholars who had hitherto only examined the surface. Dark and mysterious rites and survivals, magic and superstition which would be normal in Polynesia or Australia, were seen to be working in the background of the most civilised periods of the ancient world.

It is not difficult to see that our imaginary time-traveller might experience very considerable difficulties in acclimatising to this deeply alien cultural outlook. In addition to the astronomical and astrological profundities of which de Santillana and von Dechend write, he would find himself immersed

(13) L. Thorndike, *ibid.,* vol. 1, p.2.
(14) S. Casson, *The Discovery of Man,* Hamish Hamilton, London (1939), p.242.

in a world in which the familiar 'I – it' relationship of the scientific consciousness was very largely missing. Instead the ancients 'saw' a dynamic living Nature to which the only possible relation was 'I – thou'. Yet it is from the depths of this unfamiliar culture that the modern world emerges. Thorndike maintains that it is only the vastly successful propaganda exercise mounted by sixteenth- and seventeenth-century Italians which prevents us from seeing how profoundly 'magical' were large areas of thought in Graeco-Roman antiquity (15). Admiring the lucidity and rationality of the greatest minds, we overlook this more primitive groundswell. Yet it is always there and affects the thought of even the greatest.

In this short chapter an attempt has been made to sketch in the intellectual outlook which seems to have pervaded much of the pre-classical world. For the purposes of this book two important and closely related aspects have been emphasised. First, the distinction which we are nowadays accustomed to draw between animate and inanimate creation was not clear to these early intellects. This does not necessarily mean that human attributes were just simply transposed to explain the operation of natural phenomena. It does mean, however, that the world was not envisaged to be composed of inert, passive matter, as we nowadays suppose. It was dynamic and teleological in its activity, just as a living organism is dynamic and teleological. Second, this insight was elaborated to give the notion of sympathetic correspondences between man (the microcosm) and the world (the macrocosm). Man and the world were not set over against each other as we have since come to believe; instead man was conceived to be immersed in the natural world, connected to it by multitudinous invisible bonds. These connexities were understood as aspects of the organismic character of the world. Events did not interact like billiard balls colliding with each other but by sympathy. The universe, including man, was charged with meaning and purpose; its processes were integrated by the natural sympathies of its parts.

(15) L. Thorndike, *ibid.*, vol.1,pp. 20–21.

3 THE PALAEONTOLOGY OF SOME KEY WORDS

The origins of language

That man is nowadays believed to have originated from a population of Miocene Dryopithecines was briefly mentioned in chapter 2. The root cause of this momentous event may have been a climatic change leading to extensive deforestation. Arboreal primates would in consequence have been forced out of the trees and on to the ground. There, in a world for which they were never made, they had the alternatives of living by their wits or of perishing. What terrors and emergencies they survived we shall now never know. But, clearly, survive they did. And the stratagems they used in this desperate struggle are also clear. With little bodily armament they were forced to invent and use weapons. Without the athleticism of the great carnivores and herbivores they were forced to operate in groups and to coordinate their activities. Indeed some authorities trace the primitive magico-religious notion that like attracts like and repels unlike to a psychological root in the feelings of kinship between individuals induced by this intense early socialisation.

After the passing of countless generations, after the lapse of perhaps a million years, the emergent hominidae began to recognise that natural objects—stones, bones, and so on—which they had from the first used as weapons could be improved. Thus we see emerging throughout the old world the first stone technologies. The line between a weapon and a tool cannot be clearly drawn. A palaeolithic axe could (and can) be used both for destructive and constructive purposes. Similarly we may imagine that several million years of cooperative hunting and cooperative defence forced the development of linguistic communication. In fact the development of weapons, tools and language are inextricably interfused. Hand, eye and brain acted and reacted upon each other, forcing the development of human intelligence (1). We shall see as this book progresses that this close interaction of technology and language, and thus of our perception of the world, is maintained right up to and including modern times.

It is nowadays suggested by Chomsky and his followers that all human language shares a common 'deep structure'. Linguists have, however, been

(1) See, for example, the interesting account given by C. D. Darlington in *The Evolution of Man and Society*, Allen and Unwin, London (1969).

Figure 3.1. Mesolithic hunters: Spanish cave art.

unable to analyse the babel of modern languages into fewer than about a hundred principal groups. It looks as if language may have arisen several times in several different areas and have developed partly by parallel evolution and partly by intermixture. Fortunately it is not necessary for the purposes of this book to enter into this abstruse and difficult subject. The object of the present chapter is merely to point to the early evolution of some of the key terms in biological science.

Words and organisms

Key terms in the biological sciences differ from those in the exact sciences of mathematics or physics in that they frequently carry a penumbra of rather ill-defined meaning. This often reflects the somewhat 'amphibious' nature of the subject, balanced as it is between the opposing pulls of physics and psychology. It is also the case that the connotation of many of these terms has considerably altered during the course of history. In order, thus, to gain some insight into the thought of our predecessors it is important to understand that the words they used, the tools of their thought, were often subtly different in meaning from the same words used today.

This chapter is entitled 'the palaeontology of some key words' because, although words may with justice be said to function as tools, they are in other respects rather like organisms: they evolve, they have a history. Since Darwin anatomists have recognised that, to a discerning eye, living organisms are open books revealing the vicissitudes of their ancestors. Consider, for example, the human ear. The Eustachian tube which connects the cavity of the middle ear to the pharynx can be shown to be derived from the ancestral fish's first gill cleft. The three auditory ossicles—malleus, incus and stapes—have evolved from bones of the reptilian jaw articulation. In the human scalp are sets of muscles running to the pinna of the external ear: vestiges of the mechanisms which allow many mammals to detect the place of origin of a sound. Many other such relationships, connecting the present-day structure of the human ear to its phylogenetic history, could be listed. And the ear is, of course, far from unique. All parts of the human body, from the shape of the fingernails to the direction of hair growth on the forearms, are illuminated by this approach. As it is with anatomy, so it is with physiology and behaviour. And so it is, on a vastly shorter time scale, with language.

If we compare the anatomy of an organism to a language, then the parts of which an organism is built may be compared to words and the concepts words denote. Thus in our example words may be compared to the ear's ossicles and Eustachian tube. As the latter are traced back through the vertebrate series their function in the life of the organism slowly, but in the end radically, changes. So it is with words. The word 'cause', for example, carries today only a feeble remnant of the richness of meaning it held for Aristotle.

This indeed brings out another aspect of the biological nature of words and language. So far we have been drawing an analogy between the development of language and phylogeny; one can also discern an analogy between the development of language and ontogeny. For just as during the process of individual development highly specialised organs and tissues differentiate from a comparatively homogeneous primordium, so the connotations of many of our present-day words seem to have differentiated from a far richer and more diffuse original.

Some key words

As a first example consider the use of the terms 'root', 'square' and 'cube' in present-day mathematics. Few of us nowadays recognise that they derive through Plato from Pythagorean numerology. These words, to use our biological analogy, are rather like coelacanths: surviving relics of a once complex ecology.

Consider next the word 'temper'. Here we are rather closer to the central concerns of this book. When a person behaves in an unusually violent or agitated manner we say that 'he has lost his temper'. But if we stop to think for a moment: what do we mean by 'temper'? Is there any connection with 'well-tempered steel', for example; is there any relation with, say, the 'temperance' movement? A moment's reflection assures us that there is such a connection, that it is another instance of the living fossils we have been considering. We can trace the notion back, once again, to the Pythagoreans (chapter 5); for with this sect originated the extremely influential doctrine that the personality is the outcome of a balanced interplay between opposing forces. When the balance is upset, one force comes to tyrannise over the others. The nicely balanced personality disappears: the well-judged 'temper' of the individual is lost to be replaced by extreme behaviour. Nowadays we no longer accept Pythagorean medico-physiological theory, but the term remains—remains perhaps to mislead us.

Closely associated with the notion of the personality as a well-tempered mixture of opposing forces is the somewhat more explicit physiological doctrine of bodily fluids: the humours. We shall find in chapter 5 that the humoural doctrine also first appears among sixth-century BC Pythagoreans. It remained throughout millennia as the orthodox physiological account of the personality. Indeed we still refer to phlegmatic, sanguine, melancholic and bilious individuals. We still equate ill-humour with bad temper.

Next consider the concept of 'cause'. We have already mentioned that Aristotle's concept of a cause differed considerably from that for which we reserve the term today. Here in fact we seem to have not so much the fossilised remnants of a once pervasive theory but the sole and highly vigorous survivor of a once far more extensive fauna. For Aristotle the term 'cause' had four distinguishable connotations (2). First he defines what he calls a 'material cause': 'that from which ... a thing comes into being, e.g. the bronze is the cause of the statue, etc.'; second, he alludes to the 'formal cause': the form or pattern of a thing; third, he describes the 'efficient cause': 'that from which the change or the resting from change first begins'; lastly he refers to the 'final cause': 'that for the sake of which a thing is'.

Since the seventeenth-century revolution in science, the richness of Aristotle's concept of cause has been drastically reduced. Scientists nowadays habitually only make use of the third of Aristotle's causes: the efficient cause.

(2) *Metaphysica,* 1013a25.

They believe themselves able to account for all the varied phenomena of the world in terms of a causation which connotes only an agency which produces 'change or resting from change'. Following David Hume we might call this the 'billiard ball' paradigm (3). It quite denies that final or formal causes have any reality. It admits the existence of material causes only in so far as it admits that a bronze statue could hardly exist if bronze itself did not exist. But even here it would be inclined to suggest that the existence of bronze is itself due to the operation of previous efficient causes, this time in the recesses of ancient stars.

We shall see when we come to consider the Aristotelian system in chapters 8, 9 and 10 that the Stagirite's approach is deeply influenced by the perception of nature as craftsman or artist. Thus it is not at all surprising to find that the Aristotelian concept of causation is still widely accepted in the worlds of art and craftsmanship. Consider, for example, the painter of a picture. How might we account for the work he has created? We could give a Freudian-type explanation indicating that the subject matter and its treatment are to be accounted for by some traumatic experience in the painter's infancy. Or we might give an explanation in terms of the painter's 'intentions' or 'reasons', indicating what he believed he was trying to do, or trying to express. It is likely to be the second type of explanation that the painter would employ himself. These two types of explanation might, using the Aristotelian terminology, be called examples of efficient and final causation. In this case they seem to be complementary. One treats the painter 'externally', as an object; the other treats him 'internally', analogously to ourselves.

In twentieth-century science, however, everything is treated as object. Efficient causes alone are needed. From the rich Aristotelian complex only one fragment has survived.

Another example of a word which had, and in this case still has, several different meanings is the word 'nature' itself. Lovejoy, in a famous essay (4), has dissected out over a hundred distinguishably different meanings. The stress placed upon this word during the early nineteenth century is well known. In the beginning, however, two principal meanings seem to have been important. The Ionian philosopher – scientists used the word (*physis*) to denote the phenomenal world they were trying to understand. But from the earliest times it also referred to particular things and to particular organisms. In this use it referred to the distinguishing characteristics of a thing or organism. Both these senses of the word are still very much alive today. We

(3) *An Abstract of a Book lately Published; Entitled, a Treatise of Human Nature, Etc. Wherein the Chief Argument of that Book is farther Illustrated and Explained* (1740), Cambridge University Press (1938), p.11. The movement of a billiard ball when struck by another 'is as perfect an instance of cause and effect as any which we know, either by sensation or reflection'.

(4) A.O.Lovejoy, Nature as aesthetic norm, in *Essays in the History of Ideas,* Johns Hopkins Press, Baltimore (1952).

are still accustomed to refer to physical scientists as 'natural' scientists and to the subject matter of Balzac's novels as human 'nature'. We do not find it odd to read in Hippocrates (*De humoribus*) that different climates suit different natures. These two principal meanings of the word produce some ambiguity, as we shall see, in the work of Aristotle. This thinker indeed dissected out some six different meanings of the term (5) but, in the main, concluded that it was best used to refer to those things which 'have in themselves, as such, a source of movement'.

The question as to what exactly Aristotle had in mind when he used the word movement is one which will concern us in later chapters of this book. For us today, movement is most usually applied to physical phenomena. The volcanic lava moves down the slope, furniture is moved from one house to another, a billiard ball moves across the table. But this is only part of the meaning the word had for Aristotle. Here the development through time has been rather similar to that which we noted for the term 'cause'. From a once-rich complex of meanings, natural science has selected a single restricted but precise connotation. The more ancient complex survives, however, in our non-scientific languages. We are *moved*, for example, to write a letter of condolence, we may give our support to the Labour *movement*, during a debate we may speak against a *motion* which is nevertheless *carried*. What we intend by these uses of the word is clearly very different from that intended by the uses in our first example. There is clearly no movement of spatial position, no translocation involved. In spite of this the enormously powerful influence of Galilean science almost makes us believe, when we use the terms, that such a translocation is occurring. Aristotle, however, lived two millennia before the birth of Galileo and in his mind the terms 'motion' and 'movement' seem to have carried both connotations, the physical and the psychological, unanalysed.

Again, consider the closely allied notion of *action*. In the famous Newtonian phrase 'action and reaction are equal and opposite'. Work is done when a force *acts* upon a movable object. We can all think of many examples where action has a strictly physical connotation. But we might also consider the signification of the common legal procedure of bringing an *action*, of something being *actionable*, or the common insult '*reactionary*' flung by would-be revolutionaries at those with whose opinions they disagree. The same semantic duality is apparent.

There are many such psychophysical words. We might also have considered *drive, force, energy*. One might speak of 'front-wheel drive'; on the other hand one might refer to the 'drive' of an ambitious young entrepreneur. Similar remarks might be made about *force*. Since the seventeenth century it has been a basic concept in physical science. Yet we also speak of *force* of personality, of *forceful* speech, and so on. Since the

(5) *Metaphysica*, 1015a15.

nineteenth century energy has also been a central and precisely defined scientific concept. It is also, of course, widely used to describe psychological and personality traits.

The historian of physics, Max Jammer, suggests indeed that these basic scientific concepts—force, energy, cause, and so on—were originally taken from our psychological experience. He writes:

> Force, strength, effort, power, work were synonymous as they are today in ordinary unsophisticated language. It is this consciousness of effort spent in voluntary activity which first formed the concept of force in the ancient world (6).

We might, however, in this context recall A.E.Crawley's often quoted words:

> Primitive man has only one mode of thought, one mode of expression, one part of speech, the personal. Conversely ... he is not fully conscious of personality, even his own . . . (7).

Thus it is probably not so much that the psychological is always prior to the physical, as Max Jammer implies, but that the two aspects were in the earliest times inextricably interfused. This conflation is still very apparent, as we have already noticed, and as we shall see in more detail later, in the work of the greatest of the classical philosophers—Aristotle. Only as the history of thought unfolds does the separation of the physical from the psychological become clear-cut. And, since the work of David Hume in the eighteenth century, moderns have been forced to accept that the phenomenal world, as treated by post-Galilean science, is mysterious and ultimately unintelligible. For it has become clear that the anthropomorphisms clinging to the concepts of force, energy, cause, and so on, have in fact no justification. We cannot thus enter into the processes of the world, as the ancients could, as Aristotle could, and feel them on our own pulses. Paradoxically, as physical science unravels more and more of the world's mechanism, man becomes more and more alienated. The ultimate absurdity develops when the apparatus of Galilean science is turned on man himself and, as we shall see in the later chapters of this book, treats him in consequence as another object among objects, a Cartesian automaton. One may feel that the achievement of a complete scientific understanding of man will be accompanied by the paradox of making him completely unintelligible, of transforming him into something unrecognisable.

Returning, however, to our consideration of certain key words we may examine, as a last example, the notion connoted by the word 'soul'. In any treatment of the history of biological science this notion must clearly be quite central. But, of course, the idea overflows the boundaries of biology and energises wide areas of humanistic thought. Indeed, as A E Crawley points

(6) M. Jammer, *Concepts of Force: a study in the foundations of dynamics,* Harvard University Press, Cambridge, Mass. (1957), p. 7.
(7) A. E. Crawley, *The Idea of the Soul,* Black, London (1909), p. 43.

out:

> Few conceptions can show the universality and permanence, the creative power and morphological influence which have characterised throughout history the Idea of the Soul (8).

In this book, of course, it is only the vagaries of the physiological soul that we shall follow. But once again the origin of the notion is broad and rich, containing many elements which later thought has analysed into separate and distinct strands.

Going no further back than the Homeric poems of the first millennium BC we find the concept already familiar and much used.

In the *Iliad* there is much talk of the soul. Hector dies and 'the soul went out from his limbs and flew toward the House of Hades, wailing her doom, leaving youth and manhood behind' (9).

In this and other passages the soul seems to be taken as an animating principle. Once she has departed the embattled hero lies still, is transformed into a corpse. Yet there seems to be an implication, perhaps more than an implication, that the cadaver left behind is in some sense the hero 'himself'.

The dying Hector beseeches his conqueror, 'by thy life and by thy knees and by thy parents suffer me not to be devoured of dogs by the ships of the Achaeans...'

But the iron-hearted Achilles replies that no one 'shall save thee from the dogs... The dogs and the birds shall devour every part of thee... '

It is as if Hector's soul is in some way expendable; it is as if she departs leaving the man behind to be eaten by dogs and birds.

Achilles, earlier in the *Iliad*, confirms this duality. His soul, he insists, is worth far more than all the treasures of Ilium to him:

> For by harrying may cattle be had and goodly sheep, and tripods by the winning and chestnut horses withal; but that the spirit of man should come again once it has passed the barrier of his teeth, neither harrying availeth, nor winning (10).

The equation of soul with personality does not seem to have been strongly made by the Homeric writers. Indeed, as the quoted passages hint, etymologists believe that *psyche* is derived from words meaning 'breath' or 'exhalation'. Do we not still colloquially give a blessing when an acquaintance sneezes? The conceit is, or was, that during a sneeze the soul leapt out of the nose like a lizard's tongue. And do we not still use the same word, 'expire', for death and the exhalation of breath? Once again the fossils remain, perhaps still faintly stirring, in our present-day language. It is not difficult to accept that the phenomena of fainting and dying suggested to the ancients that an animating principle departed with the breath, was indeed

(8) *Ibid.*, p.1.
(9) *Iliad*, XXII, 361 (trans. A.T.Murray).
(10) *Iliad*, IX, 410 (trans. A.T.Murray).

nothing but the breath.

Indeed, according to Adkins (11), neither in Homer nor in any other early writer is the concept of the 'spiritual' as opposed to the 'material' to be found. The psyche is conceived to be composed of a very thin, tenuous matter which at death flies out of the body and descends to Hades. This agrees well with what was said above about the conflation of the 'physical' and the 'psychological' in early philosophy. It was only when firm definitions of inorganic matter were developed that it became important to hypostatise a 'spiritual' substance.

From what has been said it is easy to see that the psyche, the Homeric breath or life soul, came to be associated with the mouth and the head. The Homeridae, however, also recognised another species of soul—a thinking and feeling soul which was located in the torso, in the region of the diaphragm. This second and more 'soulful' soul was connoted by the word *thymos*, a word having etymological connections with the Latin *fumus* and the Sanskrit *dhumas*, both of which may be translated to mean smoke. The thymos was conceived to have a controlling influence on bodily activity: it was at the promptings of the thymos that the hero sprang into activity. According to Adkins the thymos was regarded by the ancients as responsible for 'the swirling, surging—sometimes choking—sensations of anger and other violent impulses' (12). Once again one is made aware of the characteristic conflation of the physical and the psychological. This conflation is still very evident, as we shall see, in Aristotle's account of psychology and animal movement.

As we approach the beginnings of our history in the Magna Graecia of six hundred years BC, the Homeric duality slowly disappears. By the sixth century BC the psyche is well on the way towards absorbing the functions of the thymos and tends more and more to denote the psychic totality of Man. It is, however, worth noticing that the ancient duality of psychic principles has never entirely disappeared. Even today we are inclined, unthinkingly, to separate the mind from the soul. We are inclined to regard the former as the seat of reason and the latter as the seat of the emotions. This is, of course, not quite the contrast envisaged by the Homeridae and it is interesting, moreover, to note that in contrast to the Homeric scheme it is the soul which is commonly regarded as surviving death and the mind which is mortal.

However, this is to leap two and a half millennia of thought. We shall see in the following pages of this book how the concept of the soul is elaborated by Plato and Aristotle and physiologised by their Alexandrian and mediaeval successors. In the seventeenth century it is transformed into a 'subtle wind' comparable to the chemist's gas and then transmogrified first into 'nerve juice' and lastly into bioelectric potentials.

(11) A.W.H. Adkins, *From the Many to the One*, Constable, London (1970), p.15.
(12) *Ibid.*, p.17.

In this chapter an attempt has been made to show that the verbal symbols used by the ancients in their discourse about life frequently carried subtly different meanings from the same words used today. This difference reflects the conceptual change whose watershed was the instauration (to use Bacon's term) of science in the seventeenth century. In antiquity the connotations of many of the words we have considered were both wider and richer. They tended to refer to both the physical and the psychical. The word 'life' is one of the few which still seem to retain this duality. Molecular biologists and biochemists cannot, for instance, be anything but puzzled to overhear people of high intelligence seriously discussing the notion of 'life after death': apparently a straightforward contradiction in terms! Vice versa, to the religious mind the proposition that all life ends in death seems profoundly mistaken. It is this sharp ambiguity which is perhaps one of the main reasons why Shelley's question still retains its power to disturb.

4 MILETUS

The origins of biological thought are lost in prehistory. To be fully comprehensive we might have started with the vast knowledge of plants and animals which must have been accumulated by the time of the neolithic revolution; or we might have begun with an account of Imhotep, the first physician of whom we have certain knowledge, who some five thousand years ago attended Zoser, the founder of the third Egyptian dynasty. In this book, however, we shall penetrate no further back than the sixth century BC: no further back, in other words, than half the span of time separating Imhotep and ourselves.

The sixth century BC is a convenient starting line, as during this century a group of remarkable thinkers began to appear around the shores of the Eastern Mediterranean. These thinkers, although they did not live in what is now mainland Greece, were Greek. It is among this group of thinkers that we can discern the first statements of themes which were destined to resonate throughout biological thought for centuries.

In this chapter we shall look at the thought of some of the earliest of these Greek thinkers. It is important to bear in mind that these men were not scientists in the present-day sense of that term. We have noticed in the previous three chapters that the world two and half thousand years ago seemed a very different place. The approach of these early investigators was thus only partly what we should classify as scientific; it was also partly poetic. And this is why an attempt was made in chapter 1 to show the common spring of creative work in science and poetry. In both cases the effort to interpret the phenomena of the world requires an act of the imagination.

Hybrid vigour

What made the inhabitants of the Eastern Mediterranean littoral so intellectually 'live' during the sixth and fifth centuries BC? George Sarton (1) suggests a species of intellectual hybrid vigour. There is evidence that the early Ionians were to a large extent settlers from Crete. Sarton believes that this one fact goes far towards explaining the 'miracle', as he calls it, of Greek science and philosophy. It is plausible to suppose that colonists on the Ionian shore, like colonists elsewhere, would be enterprising, resourceful and free

(1) G. Sarton, *A History of Science*, vol. 1., Harvard University Press, Cambridge, Mass. (1952).

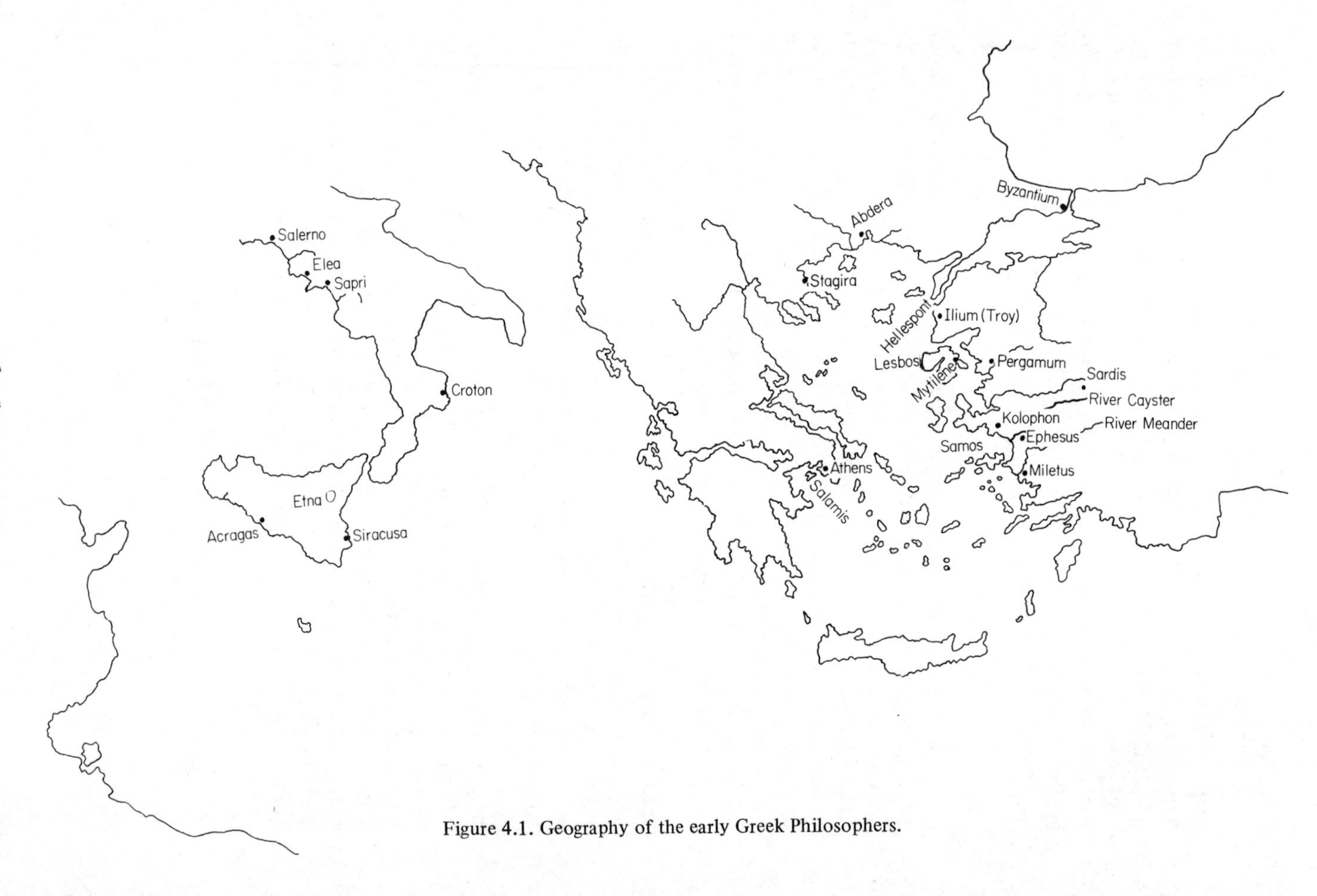

Figure 4.1. Geography of the early Greek Philosophers.

from petty bureaucratic restrictions. Moreover the Aegean coast of Asia Minor formed at that time an interface between many alien cultures. The Ionian harbours were frequented by Greek, Phoenician and Egyptian vessels. The caravan trains swaying over the Anatolian hinterland kept the coast in touch with the far recesses of Asia. From the resulting ferment of ideas, beliefs and tongues sprang a way of thought destined to transform the world.

A dramatic insight into the loosening of presuppositions, of prejudices, of dogma is gained by a perusal of the famous lines of Xenophanes of Kolophon (*fl.* 530 BC):

> Yes, and if oxen or lions had hands and could paint with their hands and produce works of art, as men do, horses would paint the forms of the gods like horses and oxen like oxen. Each would represent them with bodies according to the bodies of each.
>
> So the Ethiopians make their gods black and snub-nosed; the Thracians give theirs red hair and blue eyes (2).

The technological origins of science

Of the Ionian cities flourishing at the beginning of the sixth century BC (see map, fig. 4.1), Miletus was pre-eminent. Nothing now remains save the ruins of a great theatre and some rose pink walls around what used to be the harbour. The Meander river, true to its name, has silted up the wide gulf on to which the thriving port once faced, and the ruins now look out over a broad swampy plain. But two and a half millennia ago things were very different. Instead of cicadas and lizards there was then the bustle of a thriving commercial city.

Among the merchants and metallurgists, the navigators, engineers, architects and farmers were to be found thinkers: men whose ambition was to understand 'the nature of things'. It should not be imagined, moreover, that there was any very sharp distinction between the men of affairs and the men of thought. It was probably because they participated in, as well as lived among, the welter of practical trades that the explanations and theories they evolved were radically different from those proposed by the contemplatives of the more traditional civilisations further to the east.

Instead of the theistic mythologies of Sumeria, Mesopotamia and Egypt, the Milesian thinkers displayed what we nowadays regard as a far more 'down-to-earth' attitude. Because their lives were spent in cities quite different from the hierarchical societies prevalent in most of the civilised world, their explanatory 'paradigms' also tended to be different. Natural events were no longer explained by comparisons with the unaccountable whims of a tyrant's mood but instead by analogy with the happenings in the more familiar worlds of the artisan and engineer. The explanatory paradigm shifted, in other words, from that of personality to that of technology. The

(2) Translated in J. Burnet, *Early Greek Philosophy,* Adam and Charles Black, London and Edinburgh (1892), p. 115.

Milesian thinkers sought to account for the phenomena of the world not by invoking the will of a god or gods, but by seeking hidden causes of a type essentially similar to those with which they were familiar in their everyday lives. And because these lives were for the most part of a practical nature these early Greek thinkers have for long been regarded as the instigators of scientific thought.

The quest for the 'primary substance'

The central concern of many of these early thinkers was to establish the nature of what might be translated as the 'primary substance'. In Aristotle's words they were concerned to discover 'that from which they originally came into existence—the substance remaining unchanging underneath though subjected to changes in form' (3). It is as if the craftsman familiar with carving chairs, tables, beds, and so on, out of wood wondered whether the creator might not have done something rather similar when he created the world. The choice of 'primary substance' or *arche* differed greatly, as we shall see, from one philosopher to another. But in all cases the ambition of these early thinkers was to demonstrate that beneath the transitoriness of things, behind the shifting appearances of the world, there lay an unchanged, immutable, 'reality' or 'substance'. It is a quest we still pursue.

It is customary to accord to Thales the honour of initiating Greek and thus Western philosophy and science. His date may be established by the fact that he is credited with the prediction of the solar eclipse of 585 BC. In addition to being a philosopher–scientist, Thales was also an engineer of distinction. He is famous for altering the course of the river Halys for King Croesus. Perhaps partly because of his interest in hydraulic engineering, Thales conceived the 'primary substance' to be water. A full consideration of the reasons for this choice need not, however, detain us, for considerably more germane to the purposes of the present book are the only other alleged sayings of his which have escaped the depradations of time: 'all things are full of gods'; and 'the lodestone has life, or soul, as it is able to move iron'.

The meaning of both of these sayings has been endlessly debated. However, it seems reasonable to accept that the first saying indicates that Thales agreed with the pervasive panzoism or hylozoism of the time, and that the second implies that living things are characterised by the ability to *initiate* movement.

What is meant by panzoism or hylozoism? The first three chapters of this book have sought to outline an answer. As Mary Hesse puts it, 'nature' for these early thinkers 'is alive, reproductive and self-moving' (4). Today we strive to explain the animate in terms of the inanimate: 2500 years ago the

(3) *Metaphysica*, 983b10.
(4) M. Hesse, *Forces and Fields: the concept of action at a distance in the history of physics*, Nelson, London (1961), p.31.

problem simply did not exist. As Diogenes Laertius, the third-century chronicler of the lives and opinions of the early philosophers, puts it, 'the world' for Thales 'was animate and full of divinities' (5).

The second of Thales' reported sayings, 'the lodestone has life, or soul, as it is able to move iron', is also of considerable interest in the context of the present book. The saying suggests that Thales regarded the ability to initiate movement as one of the defining characteristics of life. We shall see the crucial importance of this notion when we come to discuss the system of the greatest of all biologists, Aristotle; and indeed the idea that living things contain within themselves, by definition, the power to initiate movement persists well into modern times. Linnaeus in his *Clavis Medicinae* of 1766 writes, 'Motion comes from Nature. No body can move on its own. The daughter of God, the soul, is the prime mover....'

The four elements

Following Thales two other important thinkers, Anaximander (*fl.* 560 BC) and Anaximenes (*fl.* 545 BC), were also natives of Miletus. These later Milesians also engaged in the search for the 'primary substance'. The detail of Anaximander's cosmology need not concern us: it was very elaborate. It is important, however, in the context of the present book to note that he was the first of the Presocratics to incorporate the four ancient elements, or 'roots', into this theory. These elements crop up throughout the history of biomedical science, and in the work of Anaximander's junior contemporary Anaximenes they function as a precursor of the atomic theory. It is thus perhaps appropriate to devote a few paragraphs to them here, when they first enter the history of science.

Aristotle (6) rationalises the tradition in the following way. He lists various sorts of tangible qualities: hot/cold, dry/moist, heavy/light, hard/soft, viscous/brittle, rough/smooth, coarse/fine. These sets of binary opposites account, he believes, for all possible sensations of touch which, for Aristotle, was the fundamental sense (7). Now the last four sets of contrarieties, says Aristotle, are all aspects of the dry/moist opposition. They are all, in other words, aspects of the distinction between the solid and liquid states of matter. The third contrariety, heavy/light, is spurious because heavy is not a genuine opposite of light: they do not annihilate each other when brought together. Thus, concludes Aristotle, the possible sets of tangible qualities are two only: hot/cold, dry/moist.

Next Aristotle goes on to show that from these two sets of contrarieties four elementary qualities may be derived:

(5) Diogenes Laertius, *Lives of Eminent Philosophers* (trans. R.D. Hicks), The Loeb Classical Library, Heinemann, London (1925), p. 27.
(6) *De generatione et corruptione*, 329b20.
(7) *De Anima*, 414b3, 435b4. The belief that touch was the fundamental and primary sense was almost axiomatic in the ancient world. The founders of the atomic theory also, as as we shall see, made this assumption.

> The elementary qualities are four, and any four terms can be combined in six couples. Contraries, however, refuse to be coupled: for it is impossible for the same thing to be hot and cold, or moist and dry. Hence it is evident that the 'couplings' of the elementary qualities will be four: hot with dry and moist with hot, and again cold with dry and cold with moist (8).

Aristotle goes on to point out that these four elementary qualities may, very satisfactorily, be attached to the four traditional elements:

cold/wet = water
cold/dry = earth
hot/wet = air
hot/dry = fire

In Aristotle and in many earlier thinkers the four elements are, in fact, idealisations—not actual, tangible, earth, air, fire and water. They are better regarded as asymptotes to which the tangible materials tend in the limit of purification.

As we have already indicated, and as we shall see later, the four classical elements play an important role in the history of biomedical science; for they were believed to have their equivalences in the make-up of the microcosm—man. The four physiological 'humours' correspond to them in the following way:

water: phlegm (secreted by the lungs and the brain)
earth: atrabile or black bile (secreted by the spleen)
air: yellow bile (secreted by the liver)
fire: blood

Ancient and mediaeval physiological psychologists accounted for human temperament in terms of balances of these humours. Thus phlegmatic personalities were caused by an excess of phlegm, melancholic by an excess of atrabile, choleric by an excess of yellow bile and sanguine by an excess of the fiery moist blood.

Microcosm and macrocosm

Anaximenes was the third and last of the eminent Milesian scientist–philosophers. He too formed an opinion about the continuing preoccupation of these early thinkers: the nature of the primary substance. Anaximenes chose air, or perhaps more accurately, mist. The manifold phenomena of the observable world derived from this *arche* by a process of rarefaction and condensation. Air, when rarefied, appeared as fire; when condensed appeared as wind, cloud, water, earth, rock. One can perhaps see in the *arche*

(8) *De generatione et corruptione*, 330a30.

chosen by both Thales and Anaximenes the half-conscious influence of the silt-laden Meander river. At another level Anaximenes' theory, wherein the variegated phenomena of the world are understood as the outcome of the spatial reorganisation of an underlying unchanging substance, has been seen as the primordium of an atomic theory.

An important corollary to Anaximenes' identification of the primary substance as 'air' follows from one of the few phrases of his work which has survived. Anaximenes writes as follows: 'As our soul being air holds us together and controls us so does wind and air enclose the whole world' (9). This is a very pregnant saying. It holds the seed of two extraordinarily influential themes: firstly the microcosm/macrocosm correlation, and secondly pneumatism.

The microcosm/macrocosm duality held, as we have already noticed, that the body of man (the microcosm) was analogous in important ways to the universe (the macrocosm). The correspondence is well expressed by Aristotle in the *Physics* (252b25): 'If this can occur in an animal, why should not the same be true of the universe as a whole? If it can occur in a small world it can also occur in a great one . . .' (10).

We have seen in previous chapters that the analogy pervaded the ancient world. Seneca, for example, in his *Naturales Quaestiones* (64 AD) drew many detailed parallels between the body of man and that of the universe. Much the same parallels are brought out some fifteen hundred years later by Raleigh in his *History of the World* (1614). Thus we read how man's blood

> which disperseth itself by the branches of the veins through all the body, may be resembled to those waters which are carried by brooks and rivers over all the earth, his breath to the air, his natural heat to the enclosed warmth which the earth has in itself... the hairs of man's body, which adorns or over shadows it, to the grass which covereth the upper face and skin of the earth... (11).

These similarities, which we now recognise as fanciful and which were indeed beginning to be recognised as fanciful even in Raleigh's day, were for long taken quite seriously. We shall see how the misanalogy bedevilled the sciences of matter until well into the European Renaissance, strongly retarding the growth of a scientific chemistry. We have all heard of the astrological nexus whereby it was maintained that the behaviour of the microcosm was already prefigured in the wheeling firmament: prediction of the fate of the microcosm could thus, in principle, be achieved by astronomical observation, by, to use another linguistic fossil, *consideration*. Finally the microcosm/macrocosm correlation played its part, possibly in

(9) G.S. Kirk and J.E. Raven, *The Presocratic Philosophers,* Cambridge University Press (1971), p. 158.

(10) An excellent account of the microcosm/macrocosm duality is to be found in J. Needham, *Science and Civilisation in China,* vol. 2, Cambridge University Press (1964), p. 295.

(11) Quoted by E.M.W.Tillyard, in *The Elizabethan World Picture,* Chatto and Windus, London (1948), p. 85.

this case a constructive part, in the birth of modern physiology. William Harvey was steeped in Aristotle, as were his physiological peers and immediate forerunners. Circular movement held, as we shall see, a peculiarly important position in the Aristotelian philosophy. The cosmos was conceived to be 'held together' by the circular movements of the heavenly bodies. The circular movement of the blood in the microcosm played, in Harvey's estimation, and it must be admitted still to a certain extent in our contemporary estimation, the same role.

The second long-persisting misunderstanding which may be traced back to the times of Anaximenes and beyond is the doctrine of pneumatism. In fact this doctrine seems to be but a special case of the microcosm/macrocosm duality referred to above. Anaximenes states in the passage quoted that the vital principle animating living bodies is identical with that which pervades the whole world. Anaximenes uses the word *'pneuma'* and this, in addition to denoting air, also denotes breath. Anaximenes is far from alone in making this correlation. Ideas about the 'breath of life' were common to all the civilisations of the old world. We noticed something of the prevalence of this group of ideas in chapter 3 when discussing the origin of the concept of the soul. Furthermore, according to Joseph Needham (12), the *pneuma* of the Greeks is equivalent to the *prana* of the Indians, the *ch'i* of the Chinese and, later on, the *ruh* of the Arabs. With Anaximenes, however, the *arche*, which underlies the macrocosm, also forms the essential principle of the living body. We shall find, as we go on, that the concept of *pneuma*, the breath of life, reappears in Latin as *spiritus* and remains to confuse chemists and physiologists right up to the time of Harvey and beyond.

Break-up of the Ionian colonies

In the middle of the sixth century BC the Ionian King, Croesus, with the help of his engineer, Thales, attacked the Persian emperor, Cyrus, at Cappadocia. He was defeated and pursued back to his capital, Sardis. This proved unable to withstand the Persian siege and quickly fell into Cyrus's hands. Thenceforward Sardis became a provincial capital of the Persian Empire. From this centre Cyrus's generals soon gained control of all Ionia. The Greek cities fell one by one and the inhabitants were subjugated. Miletus was no exception. The Persian force, however, was a land force and had at this time no naval strength. Many Ionians were thus able to escape the Persian hegemony by crossing the Aegean either to mainland Greece or to one of the many colonies established during the preceding century in Sicily or southern Italy. Consequently it is to the thinkers of this western diaspora that we turn in the next chapter.

(12) In *The Chemistry of Life* (ed. J. Needham), Cambridge University Press (1970).

5 THE ITALIAN SHORE

The western diaspora

Between about 750 BC and 550 BC Greek colonies were established not only along the Aegean coast of Asia Minor but also further west along the coasts of southern Italy and Sicily. Some of the most famous names in the history of European philosophy and science worked in these colonies. Pythagoras, an emigrant from the Ionian island of Samos, settled at Croton situated on the 'instep' of Italy. Today a single standing column is all that remains of the Greek city. Empedocles was a prominent citizen of Acragas at the western end of Sicily. Xenophanes, whom we have already mentioned, was driven from his Ionian birthplace of Kolophon, and spent the rest of a very long life wandering in Sicily and southern Italy. Parmenides is eternally associated with Elea, the remains of which are now being excavated between Paestum and Sapri.

It can be argued that the philosophical temper of these 'western' Greeks was significantly different from that of their Ionian forerunners and contemporaries. Whereas the latter were concerned to achieve an intellectual grasp of the nature of things, 'of the things that exist', the westerners seem far more concerned with the saving of their own souls. Their mental bias seems less open, less, perhaps, technological and more religious, more introspective. Whether it is possible to advance a sociological reason for this change of mood, whether the break-up of the once-confident Ionian colonies before the Persian force is causal, is an interesting speculation, but a speculation which we have certainly no space to pursue in this chapter. For our purposes it must suffice to note that these westerners contributed a number of important ideas to the ferment from which the classical, mediaeval and indeed modern syntheses have emerged.

Pythagoras

The earliest and probably the greatest name among this group of 'western' philosophers is that of Pythagoras. He is believed to have been in his prime about 530 BC. His is a shadowy, half-legendary existence. Indeed this is a direct consequence of his method, for his essential teaching was restricted to a small group of intimates and these disciples were sworn to secrecy. There

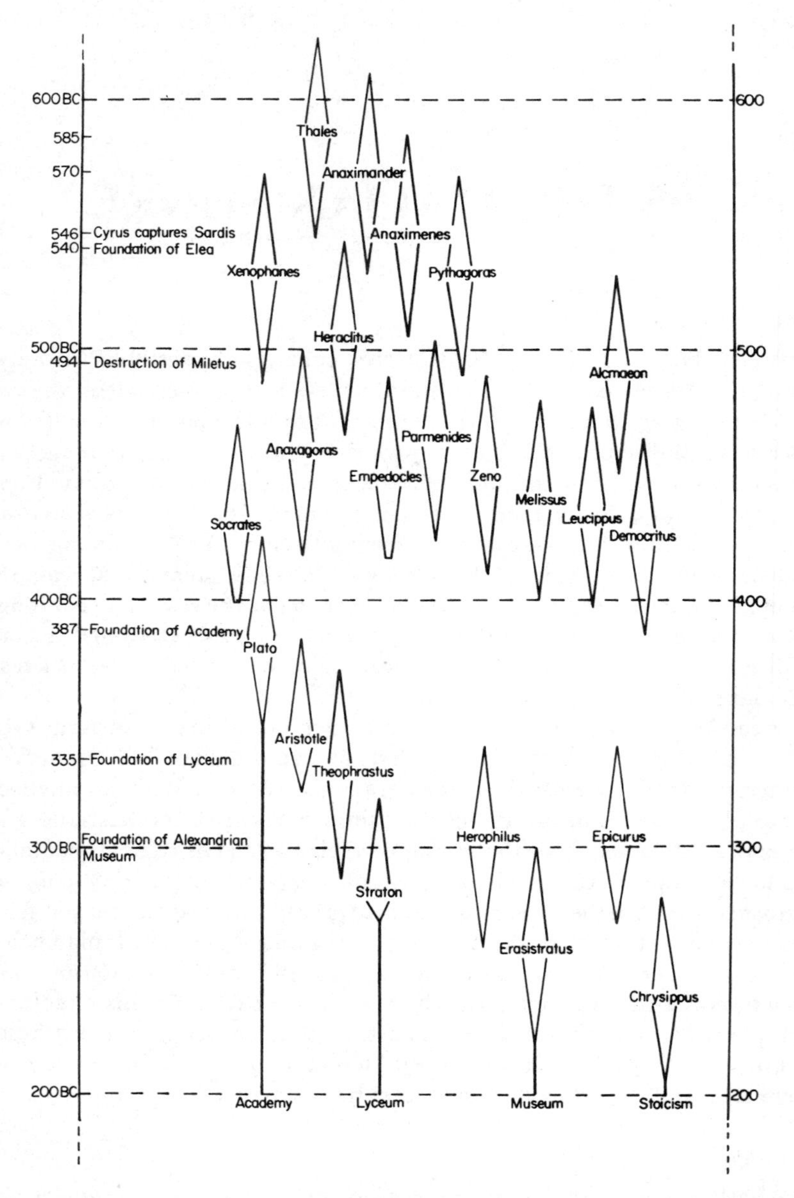

Figure 5.1. Chronology of the early Greek philosophers. The precise dates of many of the early Presocratic philosophers are difficult to establish. It is traditional to suppose that they reached their maximum force of mind at the age of forty. In consequence the diagram has been drawn to show each career reaching its greatest amplitude at the age of forty and thereafter dwindling away to terminate at eighty. Where exact dates of birth and death are known these have of course been used.

were severe penalties for those who divulged the esoteric knowledge to outsiders.

In spite of this it is well known that the core of the Pythagorean philosophy consisted of mathematics. Indeed one of the most often quoted tenets of the sect was that 'number is the essence of all things'. According to Aristotle (1) one of the most striking pieces of evidence brought forward in defence of this thesis was the discovery that 'the attributes and ratios of musical scales were expressible in numbers'. The tradition suggests that the discovery was made by measuring off lengths of string on a monochord (a single-stringed musical instrument) and observing how the resultant note varied. It is not difficult to appreciate how significant the discovery of an exact relationship between a measured length of string and a sensation must have seemed to a mathematically-orientated sect. It must have come with the force of a revelation to find that the basic intervals of contemporary music were generated by string ratios of 1:2, 3:2 and 4:3. It must have seemed that the numerical ratio was the 'essence' and the sensation merely derivative, secondary. And it seems likely that we can trace the origin of the distinction between primary and secondary qualities, a distinction which was to prove so portentous in the hands of Galileo and his successors, to this Pythagorean revelation.

In addition to their mathematical and numerological interests the early Pythagoreans also concerned themselves with medicine. The principal doctrine of the Pythagorean medical school seems to have been that health depends on an equilibrium or harmony *(isonomia)* of forces within the body (2). When the harmony is upset one of the microcosmic forces begins to tyrannise over the others and the equilibrium is destroyed *(monarchia)*. The results are seen in the symptoms of disease. It is clear that the Pythagoreans are using a rather different microcosm/macrocosm analogy from the one discussed in chapter 4. Instead of an analogy being drawn between man and the physical world, man is compared to the political state. This analogy has, of course, been quite as influential as the former. Writers from Plato to Hobbes have made full use of it.

One of the earliest physicians to hold the theory outlined above was Alcmaeon (*fl.* c. 500 BC). Alcmaeon was a native of Croton and was thus almost certainly influenced by the Pythagoreans.

Alcmaeon

Alcmaeon is of considerable interest in the context of this book as he has strong claims to be the founder of neurophysiology. There is evidence that he dissected the brain, optic nerves and eye. The eye, he says, 'sees by water and the fire in it'. The water, he believes, comes from the brain, along the optic

(1) *Metaphysica*, 985b23.

(2) This has been regarded by some authorities as the earliest recorded hint of the modern concept of physiological homeostasis.

nerves, and is returned to the brain along the same nerves, this time carrying the fire or light which is before the eyes (3).

For Alcmaeon the 'seat of sensations is the brain'. This organ contains the 'governing faculty' and here all sensations are 'somehow fitted together'. Alcmaeon's physiology is thus profoundly cerebrocentric, unlike that of Aristotle and others who, as we shall see later, supported a cardiocentric system. The brain's task, according to Alcmaeon, was to synthesise the sensations transmitted to it from the organs of special sense: the ears, eyes and nose. Putting these sensations together to form patterns was, for Alcmaeon, the nature of thought; the storage of sensations was the function of the memory.

It is clear that Alcmaeon, although sharing Croton with the Pythagoreans, did not share the latter's mystical predispositions. His outlook appears more down-to-earth, more matter-of-fact. It is an outlook, indeed, which shows strong similarities to that of the Hippocratic physicians who practised an empirical medicine on the Ionian littoral somewhat later in the fifth century BC.

However, one metaphysical and profoundly significant saying is attributed to Alcmaeon. In a resonant phrase he identifies the cause of man's mortality as an inability to 'join the beginning to the end'. What does he mean by this?

The context seems to be that of an ancient complex of ideas about motion. It was widely believed that the perfect motion was circular; all other types of motion were in some way flawed. It may be that this notion had arisen from a primitive fascination with the motion to be observed in the heavens (4): a motion which moderns blinded by the glare of metropolitan living have largely forgotten. But for the ancients the silent perpetual wheelings of the heavenly bodies were profoundly impressive. This *perpetuum mobile* was also to be observed in the cycle of the seasons: summer follows spring, winter follows summer, spring follows winter and so on and on. Perhaps, too, the technological significance of the wheel also contributed towards this ancient belief that circular motion was somehow impeccable, more perfect than any other type (5). Unlike any other type of motion it has no beginning and no end: it seemed to be eternal. Motion in a straight line, however, has both beginning and end. It seems likely, therefore, that it is this deep-rooted complex of ideas which lies behind Alcmaeon's otherwise rather cryptic saying. If the end and the beginning of a human life could only somehow be joined we should have motion in a circle: personal immortality—'in my end is my beginning'.

We shall follow some of the ramifications and ensuing confusions attendant on this deeply ingrained belief in the 'perfection' of circular motion

(3) K. Freeman, *The Presocratic Philosophers,* Blackwell, Oxford (1946), p. 137.
(4) Aristotle, *De Anima,* 405a30.
(5) See C. Singer, E.J. Holmyard, A.R. Hall and T.J. Williams, *A History of Technology,* vol. 2, Clarendon Press, Oxford (1955), p. 590.

in subsequent pages of this book. We shall notice it strongly at work in the Aristotelian system and even in the achievement of the great instaurator of modern physiology—William Harvey.

The Empedoclean physiology

If William Harvey is generally regarded as the fountainhead of the modern era in physiology we might, perhaps, assign a similar position in the world of classical antiquity to Empedocles. The fragmentary remains of his writings seem to outline one of the earliest self-consistent systems of physiology.

Empedocles is believed to have flourished about 450 BC at Acragas (now Agrigento) in Sicily. He, like all these early thinkers, was very far from being a narrow specialist or even simply a physician. W K C Guthrie apostrophises him as 'one of the most complex and colourful figures of antiquity ... he appears at once as philosopher, mystic, poet, political reformer and physician' (6). Legend consigns him to a fiery death in Etna: a fate immortalised by the doggerel:

> Great Empedocles, that ardent soul,
> Leapt into Etna and was roasted whole.

However, it is his physiological rather than his theological beliefs (which were responsible for his fiery end) that are of interest to us here.

An entry into the fragments of his physiology which have come down to us is perhaps best obtained by considering one of his best known and most striking passages. It is a passage in which, according to Burnet (7), Empedocles anticipates not only Harvey and Torricelli but also the experimental method of modern science. In it he draws a sustained analogy between the processes of respiration and the operation of a contemporary device for lifting water from one vessel and transferring it to another. This device was called a 'water carrier' or 'clepshydra'. It seems to have been a narrow-necked vessel pierced by a number of holes (fig. 5.2).

To transfer water or wine from one vessel to another the clepshydra is immersed in the fluid and the top of the narrow neck covered with the palm of the hand. The carrier is then withdrawn and when the neck is unstopped the contained liquid rushes out. This procedure seemed to Empedocles to illuminate one of the central processes in animal physiology:

> So do all things inhale and exhale: there are bloodless channels in the flesh of them all, stretched over their bodies' surface, and at the mouths of these channels the outermost surface of skin is pierced right through with many a pore, so that the blood is kept in but an easy path is cut for the air to pass through. Then, when the fluid blood rushes away thence, the bubbling air rushes in with violent surge: and when

(6) W. K. C. Guthrie, *A History of Greek Philosophy*, vol. 2, Cambridge University Press (1965), p. 123.
(7) J. Burnet, *Early Greek Philosophy*, Black, London (1892), p. 264.

the blood leaps up, the air is breathed out again, just as when a girl plays with a klepshydra of gleaming brass. When she puts the mouth of the pipe against her shapely hand and dips it into the fluid mass of shining water, no liquid enters the vessel, but the bulk of air within, pressing upon the frequent perforations, holds it back until she uncovers the dense stream; but then, as air yields, an equal bulk of water enters. In just the same way, when water occupies the depths of the brazen vessel and the passage of its mouth is blocked by human hand, the air outside, striving inwards, holds the water back, holding its surface firm at the gates of the ill-sounding neck until she lets go with her hand; and then again (the reverse of what happened before), as the breath rushes in, an equal bulk of water runs out before it. And in just the same way, when the fluid blood surging through the limbs rushes backwards and inwards, straightway a stream of air comes in with swift surge; but when the blood leaps up again, an equal quantity of air is again breathed back(8).

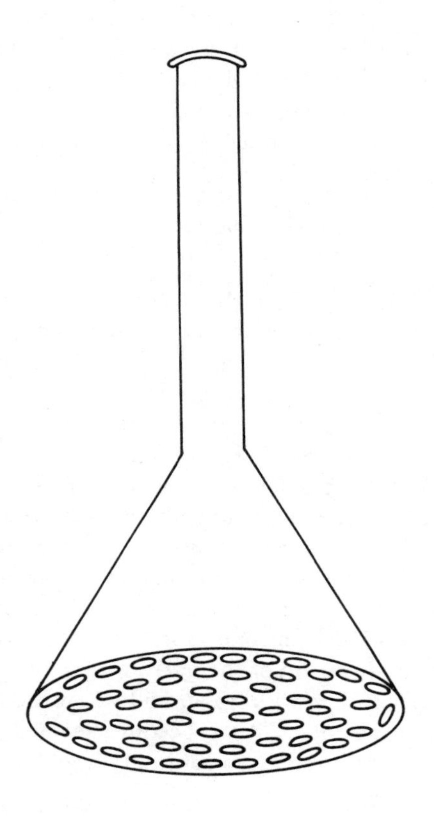

Figure 5.2. Clepshydra

(8) Reported by Aristotle (*de Respiratione,* 473b9), translated in G.S.Kirk and J.E.Raven, *The Presocratic Philosophers,* Cambridge University Press (1971), p.341.

The theory expounded in this passage is crystal-clear. The rhythm of breathing—so obvious a feature of man and his domestic animals—is a consequence of the ebb and flow of the blood. As blood surges forward, the air in the multitudinous tubes is driven out (expiration); when the blood ebbs back, air is drawn in (inspiration). Classical scholars differ on the question of exactly where the air is drawn in and expelled. The translation above indicates that the tubes open all over the body surface. Some authorities, for example W K C Guthrie, suggest that Empedocles locates the tubes at the back of the nostrils. This translation would certainly make more sense to a modern reader. However, this is perhaps a detail, for whether the pores open into the nasal or pharyngeal cavity or whether, like the sudorific glands, they open all over the body's surface makes no difference to the central conception of the Empedoclean physiology: breathing is a consequence of the ebb and flow of the blood.

We are now in a better position to understand the significance of Burnet's opinion concerning the 'modern' character of Empedocles' thought. For, as we have seen, Empedocles, as perhaps befits one of the original atomists, compares the organism, not to society, not to a psychological process, but to an aspect of contemporary 'technology'. This is a great theme. We have already seen it emerge in chapter 4 and we shall see it reappear at many places later in this book. We shall see, for example, Descartes compare the organism to seventeenth-century mechanisms, and in the twentieth century we all know how pervasive is the analogy drawn between brains and computers. It has, indeed, been suggested that the Harveyan revolution in physiology was itself partly occasioned by the contemporaneous interest in hydraulic engineering (chapter 14). It seems as if we only achieve a full understanding when we have constructed and used a device ourselves. This insight can then be carried across to account for the operation of natural entities. If this analysis is correct it follows that the development of our understanding of natural processes is closely bound up with the evolution of our technology.

Returning, however, to the fifth century BC and the Empedoclean physiology we can see that it, like its Harveyan successor, takes the cardiovascular system to be the most important of the body's systems. Indeed Empedocles' conception of the importance of the cardiovascular system goes much further than that of Harvey. For Empedocles also conceives the system to form the physical substratum of the mind. He is reported to have written that 'men think mainly with the blood...'.

This position cannot but sound strange to modern ears. It was, however, by no means uncommon in antiquity. We shall see that Aristotle himself subscribed to a version, albeit a rather more sophisticated version, of the same theory. It is thus worthwhile examining the rationale of the Empedoclean psychophysiology.

'Men think mainly with the blood', wrote Empedocles, 'because there the elements are most thoroughly mixed.' This, too, sounds unfamiliar to the modern reader. What, then, was Empedocles' theory?

Empedocles lived a little later than Parmenides, whose seminal thought we shall discuss in the next chapter. Empedocles' philosophy had thus, like all post-Parmenidean philosophies, to withstand the radical Eleatic critique. Empedocles sought to escape the Parmenidean *reductio ad absurdum* by means of a primitive atomic theory. His atoms were the four ancient elements (see above). The phenomena of this world—movement, coming-to-be, passing-away, and so on—were saved by postulating that what in fact was happening was a continuous collocation and dislocation of the underlying elements. This aggregation and disaggregation of the underlying elements was caused by two forces—love and strife.

Coupled with this primitive atomism Empedocles had developed an interesting theory of sense perception. Like later atomists he believed that all objects of perception continuously gave off effluvia or images of themselves. These tenuous but quite material effluvia fit, if they are the right shape and size, into appropriate pores in the sense organs. Once within the body these effluvia are able to interact with the body's substance. Aristotle reports his theory as follows:

> With earth we see earth, with water water, with air divine air, with fire consuming fire; with love do we see love and with strife destructive strife(9).

In short, like is perceived by like.

If we wish to gain some insight into the background for this distinctly unmodern concept, we probably need look no further than the ideas of sympathetic magic and the microcosm/macrocosm correlation discussed in previous chapters. Indeed, in the context of ethics we are still sometimes told that good perceives good in others, evil evil: birds of a feather flock together.

From whatever half-conscious background the Empedoclean theory emerges, its relation to his cardiovascular psychophysiology is clear; for in order to perceive the happenings in the macrocosm the microcosm has to be built of similar elements, and, according to Empedocles, of all the body's tissues it is the blood which consists of the most thorough mixture of the elements. Moreover the nature of the mixture controls the sensitivity and intelligence of the individual. Thus, according to Theophrastus, he writes:

> Men think mainly with the blood, for there the elements of the body are most thoroughly mixed. Those, then, in whom the mixture is equal, or nearly so, with the elements neither too far apart nor too small nor too large, are wisest and keenest in perception, and so in proportion, are those who come nearest to them; whereas those in the opposite state are witless (10).

Thus we see why it is that Empedocles conceives that 'men think mainly with the blood'. Indeed he goes further still and asserts that 'the blood

(9) Aristotle, *Meteorology,* 1000b6.
(10) Kirk and Raven, *ibid.,* p.344.

around men's hearts is their thought'.

We saw above that Empedocles' theory of perception involved the passage of material 'images' from the outside world through the sense organs into the body. His theory is in fact in some ways a remarkable prefiguration of Helmholtz's nineteenth-century doctrine of 'specific nerve energies'. For he asserts that

> ...one sense cannot judge the objects of another, since the passages of some are too narrow, for the object perceived, so that some things pass straight through without making contact while others cannot enter at all (11).

In other words, each sensory pathway was designed in such a way that only the proper 'percept' would fit.

But more important than these metaphorical foreshadowings of contemporary neurophysiology is the psychophysiological theory these fragments imply. It seems that for Empedocles, as for his Abderan successors (chapter 6), the body – mind conundrum, which has generated so much heat in modern philosophy, was not seen as a problem. Indeed it seems misleading even to label his theory as 'psychophysiology', for he seems to make no distinction between the physical and the mental. We have seen that he identifies 'thought' with 'the blood around men's hearts'. We shall return to this monistic view in the next chapter when we examine the more fully documented theory of Democritus.

Before leaving Empedocles, however, mention must be made of his evolution theory. Once again one is struck by its 'modern' character—a character, indeed, which does not reappear until the nineteenth century AD. Unlike many of his contemporaries and successors who conceived that a creator had designed the world and its inhabitants with some end in mind, Empedocles seems to have taken a considerably bleaker view. However crude and fantastic it now seems, Empedocles does appear to have proposed a type of evolution by natural selection for the origin of living forms. He suggested that the various parts and organs of the body originated independently of each other and only later came together to form the organisms we now recognise. Aristotle, who as we shall see was very critical of atomism and of evolution theory, and tried to poke fun, reports as follows: 'Here sprang up many faces without necks, arms wandered without shoulders, unattached, and eyes strayed alone, in need of foreheads' (12). Over the course of time, according to Empedocles (still reported by Aristotle), these bits and pieces came together in various combinations. In many cases monsters resulted: 'man-faced ox progeny' and so on. But sometimes more appropriate combinations occurred and these survived, 'being accidentally compounded in a suitable way; but where this did not happen, the creatures perished and are perishing still...' (13).

(11) *Ibid.*, p. 343.
(12) *De Caelo*, 300b30.
(13) Aristotle, *Physics*, 198b29.

It is fascinating and important for the argument of this book to notice this hint that even before Leucippus and Democritus had fully developed the atomic theory a linkage between atomism and evolution by natural selection had been perceived by the philosopher of Acragas. Could it be that Empedocles had pondered the contradiction which the exquisitely organised bodies of animals and plants seemed to pose to a theory which saw the visible world as emerging from the random coming together and falling apart of invisible particles 'without the aid of the gods'? This was certainly a point which the great philosopher – biologist of the Lyceum pressed, and pressed hard, against the atomists; and it is a point which was not fully answered until, as we shall see, the twentieth century AD.

It would be wrong to end this brief outline of the Empedoclean physiology, psychology and biology without emphasising that the system which we glimpse through the few fragments which have come down to us appears to be remarkably coherent and plausible. The phenomena of respiration, the blood, the heart, the senses, thought itself, are welded together to provide a satisfying picture of how the body 'lived'. The life sciences are, moreover, tied into a cosmology, a physics and an anthropology so that Empedocles forms a good example of Guthrie's remark that each of the Presocratic thinkers 'presents us with a system marvellously coherent down to its smallest details' (14).

For, as must by now be obvious, the Presocratic thinkers were, on the whole, very different from present-day scientists and philosophers. The central effort, which they never lost sight of, was to achieve an imaginative vision of the world in which they found themselves. The broader and more self-consistent, the better. This is indeed one of the great dissimilarities between ancient and modern thought. Nowadays the scientist is content if he can solve a small, often to the outsider a trivial, problem; the Presocratics would have been contemptuous of such seeming pusillanimity: they strove for nothing less than a complete synthesis.

(14) Guthrie, *ibid*, vol.2, Preface,xii.

6 ELEA AND AFTER

Being and not-being

Did the universe have a beginning and will it have an end? Or has it always been and will it always be more or less the same, perhaps changing in certain shifting localities but remaining, overall, unchanged? In other words, do we support the 'big bang' theory or the theory of continuous creation?

These are questions in modern cosmology and as yet admit of no certain answer. If anything the consensus seems to be moving towards an acceptance of the concept of an initial universe nucleus and an initial devastating explosion. Quite recently, too, the notion of a reciprocating universe has been gaining favour. In this theory phases of universe expansion and contraction are regarded as alternating. We live during a phase of universe expansion. Aeons ahead the expansion will come to an end and contraction to the unimaginable nucleus commence.

But consider this seminal world nucleus. Or is it, as we said, unimaginable, inconceivable? *Where* could it be? What will be or were its *surroundings*? What would it mean to say that it moved? These, we may think, are schoolboy questions, philosophically naive, innocent of Einsteinian theory. But they point towards the Eleatic critique. For somewhat similar problems of being and nothingness puzzled the wits of the philosophers of Elea.

The paradoxes of Parmenides

Living in the first half of the fifth century BC, Parmenides is often regarded as a pivot in the history of human thought. Before the advent of Parmenides the Ionian thinkers had, on the whole, been content to account for the 'sensible in terms of the sensible'. This was indeed, as we saw in an earlier chapter, their great advance on the mythopoeic thought of yet earlier civilisations. Parmenides showed that this was only half what was necessary. When he had completed his work, the philosophers of nature had ever after to show that their theories were also intelligible; and these two bars at which a theory must answer are still the basis of scientific thought today. A theory, to be a scientific theory, must satisfy both the facts and the intellect.

What did he mean by 'intelligible'? It is probably easier for us to understand what he considered to be unintelligible. The contrary would thus be what he regarded as intelligible. And what Parmenides considered to be unintelligible he defined time and time again. At the very beginning of his poem he writes as follows:

> Come now, and I will tell thee – and do thou hearken and carry my word away – the only ways of enquiry that can be thought of: the one way that it *is* and cannot not-be, is the path of Persuasion, for it attends upon Truth; the other that it is *not* and needs must not-be, that I tell thee is a path altogether unthinkable (1).

In short, Parmenides is insisting that it is utterly impossible to conceive that which is not. Nothing is not a thing and to treat it as such leads to absurdities. Readers of Lewis Carroll will remember the play made with the concept of 'nobody'. 'Nobody' passed Alice on the road and consequently 'nobody' should have arrived first, and so on. Parmenides is making the same point. He drives home again and again the nub of his argument—if something is unthinkable then it cannot exist: 'For you could not know that which does not exist (because it is impossible), nor could you express it; for the same thing can be thought and can exist'.

But what if we apply this intellectual discovery to the events of the natural world? If the application is made rigorously, the world the senses show us can only be dismissed as an illusion. For the most obvious feature of the world is that it is a never-ending series of happenings, a flux of events, a scene of growth, change and decay, of night following day, of season succeeding season, of mountain building and of the flow of water eroding mountains into the sea. But if what is *is*, and what is-not *is-not*, how can this be? How, in other words, is change possible? This is the Parmenidean challenge to all succeeding thinkers.

Consider change of place. This clearly involves an object moving from one position to another. But how is this possible? Between the body's initial position and its final position lies a space, and this space cannot *ex hypothesi* be empty. It must be full for, as we have seen, emptiness, vacuity, nothingness, just simply cannot exist. How, then, can an object force its way through this totally compact, interstice-less, body?

Consider change of quality. This involves, say, a hot object becoming cold or a bright object becoming tarnished. How can this be? How can an object turn into what formerly it was not? Where has the tarnish come from? Where has the heat gone to? 'Brightness', a more recent poet has said, 'falls from the air.' For Parmenides this would have been unthinkable and hence impossible. A bright object exists, it *is*, it cannot on the instant cease to exist—this is unthinkable. *Ex nihilo nihil fit*, nothing can come of nothing; *mutatis mutandis*, something cannot end as nothing.

Consider change of size. Growth, ripening, diminution and decay are probably more anxiously watched in primitive societies than in our own urban technocracies. Yet in order to account for these phenomena we have to suppose that the body, in growing, combines with something other than

(1) G.S. Kirk and J.E. Raven, *The Presocratic Philosophers* (1971), p. 269.

itself. But the only thing other than corporeal body is vacuity, 'non-being', and this, as we have seen, is unthinkable and hence non-existent. Similarly we have to suppose that decrease in size is brought about by the lapse of what *is* into what *is not*: again, clearly, an impossibility. Thus we are driven to conclude that birth, growth, diminution and decay are, like change of quality and change of place, illusory, waking dreams which torture the senses. What is real can only be

> A well rounded sphere, from the middle everywhere of equal strength, for it need not be somewhat more here and somewhat less there, for neither is there non-being to prevent it reaching its like, nor is there being so that it could be more than being here and less than being there ... unregenerated, and imperishable, whole, unique, immovable and complete (2).

Do we not recognise the aboriginal world-nucleus? The only difference is that this was, apparently, unstable. Hence the present scene. But why it should have been unstable theoreticians are not yet prepared to say.

Retreat to atomism

That the world-nucleus did suffer an inaugural catastrophe cannot be gainsaid. For some reason the centre could not hold and we are today perched precariously on the flying fragments. And these fragments are, of course, atoms, or more precisely the particles of which atoms are composed. This, too, is prefigured, metaphorically, in the subsequent history of the Parmenidean theory, for the intellectual outcome of the Eleatic argument was the atomic theory of matter. The eternal, changeless, dense sphere of being which Parmenides preached was shattered, first rather tentatively by Anaxagoras and Empedocles, and then more definitely by Leucippus and Democritus, into a myriad invisible 'dense, changeless, spheres', a concept which reappears hardly changed as Isaac Newton's 'solid, massy, hard, impenetrable, movable particles'.

However, this saving of appearances, although immensely successful, even in the ancient world, did not genuinely meet the Parmenidean insight. For, as we have just seen, Parmenides based his argument that the changeable world of appearances must be illusory precisely on the fact that the existence of a void was inconceivable. Nothing was not a thing, it could not be thought without self-contradiction, it could not exist. Yet the Democritean atomic theory posited, as we shall see, just this: that there were atoms *and* a void in which they moved. This proved a stumbling block for many and, in particular, as we shall see, for Aristotle. Indeed it is possible to see in the Aristotelian philosophy a thorough-going rebuttal of the Democratean theory.

Long before Aristotle, the immediate followers of Parmenides—Melissus and Zeno—fought hard against those who would dilute the master's doctrine.

(2) Fragment 8 (trans. J. Burnet), *ibid.*, p. 176.

The paradoxes of Zeno are well known. They were intended to turn the tables on those who made mock of the Eleatic 'one'. The arrow flies through the air, or so it would appear. But at any instant it must surely be somewhere, it must surely be in some place. How can it get, therefore, from one place to the next? Greek thinkers were not prepared to be brainwashed into accepting unintelligible happenings such as quantum jumps! The phenomenon was not open to rational understanding: *ergo* it must be illusory. Achilles chases the tortoise. But when he reaches the place from which the animal started, the tortoise is a little further on, and when he reaches this place the tortoise is a little further on still, and so *ad infinitum*. Swift as Achilles may be he will, according to Zeno, never sum the infinite series and catch the tortoise. Consider the noise made by dropping a grain of millet upon the ground. It would be inaudible. Yet if we dropped a sack of millet we should hear it well enough. But how can this be? The multiplication of zero can never yield anything but zero. By these means Zeno pushed home the Eleatic critique of the senses. The dilemma is clear. If we accept the information provided by the senses as true, then the intellect is false; if we accept the intellect as true, then sensory information is false. Consider finally what is involved in the belief that there is a plurality of things in the world. On the one hand we must accept that the world contains a *definite* number of things. We may never be able to determine this number but we are bound to believe that it exists. But on the other hand, says Zeno, between each thing another thing is always to be found. Between each point on a line there is always another and so, once again, *ad infinitum*. But if this is the case then the number of things in the world cannot be definite. It must be infinite. Once more we have a contradiction. Once more we must conclude that the world we are accustomed to, the world we live in, is illusory. What is real can only be the changeless Parmenidean sphere of being.

A pandaemonium of particles

We have already seen that some form of atomism provides one means of escape from the Parmenidean analysis. It involved, as we noted above, fudging the issue by accepting the possibility that what did not exist nevertheless did exist: by accepting, in other words, that not all existents were corporeal. Thus by introducing the revolutionary idea that a void, nothing, could exist, was *something*, first Leucippus (*fl.* 430 BC) and then his pupil Democritus (*fl.* 420 BC) were able to account for the phenomena of change, of coming to be and passing away.

Although Leucippus was probably born in Miletus and is said to have been instructed by Zeno at Elea, he certainly died in Abdera. It was at Abdera that he founded the school of which Democritus was the most famous pupil. Abdera was situated on the coast of Thrace slightly east of the island of Thasos. Aristotle, as we shall see later, was also a native of Thrace, and it is consequently not surprising to find that he paid particular attention to the atomists. Thus one of the best and probably most trustworthy accounts of the origin and significance of atomism is to be found in *De Generatione et Corruptione* (3):

> For some of the early philosophers thought that which is must of necessity be one and immovable; for the void is not-being; motion would be impossible without a void apart from matter; nor could there be a plurality of things without something to separate them .... But Leucippus thought he had a theory which was in harmony with sense and did not do away with coming into being and passing away, nor motion, nor the multiplicity of things. He conceded this to experience, while he conceded on the other hand to those who invented the *One* that motion was impossible without the void, that the void is not-being and that no part of being is not-being. 'For', said he, 'that which is strictly speaking being is an absolute plenum; but the plenum is not one. On the contrary there is an infinite number of them and they are invisible owing to the smallness of their bulk. They move in the void and by their coming together they effect coming into being, and by their separation passing away.'

It is clear that Aristotle sees the atomic theory as a direct response to the pressure of the Parmenidean argument. By allowing, *contra* the Eleatics, that a void could and did exist, the atomists believed themselves able to account for the phenomenal world. The variegated happenings in this world could be explained as the outcome of the fortuitous coming together, entanglement, loosening and falling apart of the atoms.

Fortuitous: this aspect of the theory was ever after to be challenged. How could the ordered structures of the world, and most especially the bodies of living organisms, be formed by the mere fortuitous coming together and falling apart of atoms? As we shall see, Aristotle, the biologist and embryologist, was particularly scathing: 'There are some', he wrote (4), 'who make chance the cause of both these heavens and of all worlds.' To the Stagirite this seemed absurd and untenable. To the close student of organic form and the development of organic form it was quite obvious that exact design, exact planning were involved, that shaping forces were at work. Other philosophers attacked the Democritean position from the opposite direction. If every happening, they argued, is the outcome of a chain of 'billiard ball' collisions and rebounds then, in principle, every happening is strictly predictable. Then what becomes of the freedom of the will? Later Epicurus, the great populariser of atomism in the ancient world, introduced the somewhat inelegant notion that the motion of the atoms was not entirely knowable. Like today's subatomic particles, they were liable to swerve from their trajectories at unpredictable times and in unpredictable ways. In this way he believed, and his Roman successor Lucretius also believed, that the 'bonds of fate were snapped' and the conviction that one is free to choose one path rather than another explained.

Returning, however, from the Rome of about 50 BC to Abdera some three

(3) 325a2, 325a23.
(4) *Physica,* 196a24.

hundred and fifty years earlier, it is important to notice some other aspects of the original atomic theory. First, how was it that the atoms came together to form and reform the transient structures of the world? Democritus considered that they swirled together in whirls and vortices. Perhaps he was thinking of the swirls and eddies of dust particles in a shaft of sunlight. The Democritean atoms, besides being infinite in number, were also of a multitudinous variety of shapes. Thus when they swirled together those of complementary shape did not rebound but clung together. By a process of accretion visible bodies formed.

There are, of course, considerable intellectual difficulties in this theory. Where does the original motion bringing the atoms together come from? Aristotle, in particular, was interested to know the answer to this question. Democritus conceived the atoms to have moved in the void from eternity. He posited no first cause. The Democritean universe was thus on this count also, as again Aristotle perceived, thoroughly non-teleological.

But there is a further difficulty about this Democritean idea. For what can conceivably cause the swirling together of the atoms to form visible bodies? The motes in a sunbeam are, after all, brought together by atmospheric eddies. Can there be an eddy in the void? Later on, atomists like Epicurus and Lucretius introduced, as we noticed above, the intellectually unsatisfactory notion of an unpredictable sideways swerve. This, in addition to (allegedly) meeting the free-will problem, also accounts for the collisions necessary for the formation of macroscopic objects. For otherwise, as Lucretius points out,

> Everything would fall downwards like raindrops through the abyss of space. No collision would take place and no impact of atom on atom would be created. Thus nature would never have created anything (5).

It will be clear from the above short discussion that an essential element of modern atomic physics was missing from the Abderan theory. The Democritean atoms could only interact by contact. Like billiard balls they cannoned into and off each other, and if of appropriate shapes became entangled with each other. Nowadays we recognise what would have seemed quite unintelligible to the ancient atomists: that bodies can act where they are not. What we could mean by saying that a body can influence the behaviour of another body across a void would have seemed to them quite unclear. We have just seen how strange it strikes us, and also many of the ancients, not only to admit that non-being nevertheless *was* but also that that which was-not exhibited eddy currents like the atmosphere. Does it not strike us as strange, as it surely would have done the ancients, to hear that the void, non-being, nevertheless possesses the properties of an electric or magnetic or gravitational field—that, indeed, it can show the properties of waves (6)?

(5) Lucretius, *The Nature of the Universe* (trans. R. Latham), Penguin, Harmondsworth (1951), p. 66.
(6) For example, the electron waves of Schroedinger's theory.

Spiritual atoms

The Abderan philosophers were not solely concerned with macrocosmic phenomena; they also showed considerable interest in the nature of the microcosm. Indeed, Democritus is famously depicted surrounded by the bodies of animals which he is in the process of dissecting.

The fragments of Democritean physiology which have survived suggest that the Abderans applied their materialistic principles to the small world quite as much as to the great. The human body was conceived to consist of atoms just as was the world at large. Interspersed among the body's atoms were, however, spiritual atoms. These atoms resembled those of fire in being spherical and very small. The resemblance ensured that, like the constituents of a flame, they could move with utmost rapidity: 'swift as thought'. Aristotle reports Democritus' opinion thus: '... the round ones (atoms) are soul, because shapes of this kind are best able to slip through anything and to move other things by their own movements' (7). Concentrations of these 'soul atoms' formed the mind. Classical scholars differ about where this concentration was located. Some authorities believe the Abderans placed it in the heart, others believe they located it in the brain.

The phenomenon of respiration was related to this physiological atomism, for the Abderans believed that the mercurial soul atoms tended to escape from the body and had to be both restrained, as far as possible, and replaced. This was the function of respiration. The current of air both held back those atoms about to escape and replaced those atoms which had escaped. The relationship to the pneumatic theory discussed in chapter 4 is clear.

The Democritean theory of perception has also been remarkably influential. The Abderans, as their atomic principles demanded, attempted to treat all sensations as special cases of the fundamental sense of touch. They supposed that not only tangible objects but also the objects detected by the nose, ears and eyes exerted their effects by impact. Today we are prepared to accept much of Democritus' account so far as scents and sounds are concerned. Odoriferous objects are believed to emanate particles which have to impinge upon an olfactory epithelium in order to exert their effects. Noisy objects, we now believe, exert their effects via an intervening medium: the air. But the objects of sight are different. Indeed light has always been a central theme in the history of science. Democritus' treatment of vision is markedly Empedoclean. He believes that visible objects perpetually emanate 'husks' or 'idols' of themselves which, shooting through the air with unimaginable rapidity, continuously impinge upon the observer. His theory is somewhat complicated, however, in that he conceives that the observer is also active and that the air between the eye and the object 'is contracted and stamped by the object seen and by the seer'. It is this 'stamped' air which affects the observer. Once the image has impinged upon the seer, it is able to

(7) *De Anima,* 403b31.

mould the quicksilver soul atoms into its own shape. This, according to the atomists, is the *terminus ad quem* of perception. Movement of the slippery soul atoms is regarded as the physical basis of thought.

These fragments of Abderan 'psychophysiology' confirm the impression we obtained from Empedocles. The early atomists do not seem to be troubled by our concern about the relation of 'mind' to 'matter'. 'Reality' seems for these thinkers to have been all on one level: minds, percepts, sensations, were as material as the stones and the stars.

However, if the full mind – matter dichotomy was still unrecognised, the closely related distinction between primary and secondary qualities was well understood. We have already seen that the Pythagoreans regarded number as the essence of all things and sensation as derivative. Parmenides, too, had divided his work into two parts: the *Way of Truth*, to which belonged the intelligible *One*, and the *Way of Seeming*, an account of the sensible world. Democritus was quite explicit and quite modern in his statement of the situation: 'by convention are sweet and bitter, hot and cold, by convention is colour; in truth are atoms and the void...' (8). The dichotomy is crystal-clear. Sounds, scents, colours are merely how the world seems to us: the truth is different. It is instructive to notice how similar this is to the Galilean position outlined in chapter 14. 'I hold', wrote the great Italian, 'that there exists nothing in external bodies for exciting in us tastes, odours, sounds except sizes, shapes, numbers and slow or swift motion' (9). Democritus, like Galileo, recognised that sensations like hot and cold, sweet and bitter, bright and dull, depended very much on the condition of the beholder: atoms and the void alone remained constant. This firmly-grasped distinction marks an important step towards the perception of the dichotomy between the mental and the physical—a perception which, as we shall see in chapter 7, was considerably sharpened by the work of Socrates. And with this very modern distinction between primary and secondary qualities, between truth and opinion, we may leave the two philosophers of Abdera.

Epicureanism

We shall see, in the next chapter, that towards the middle of the fifth century BC, 'Men gave up enquiring into the works of nature and philosophers diverted their attention to political science and the virtues which benefit mankind' (10). In consequence the atomic theory was not developed far beyond the point at which Democritus left it. Indeed it is a little difficult to see how it could ever have been much more than a speculative theory in antiquity. The technological environment had not developed to a level which suited much further evolution. The theory was, however, taken up by one of

(8) Democritus, Fr. 9, quoted in Kirk and Raven, *ibid.*, p.422.
(9) Galileo Galilei, *Il Saggiatore*, Q. 48.
(10) Aristotle, *De Partibus Animalium*, 642a25.

the most influential philosophers of antiquity: Epicurus (341–271 BC).

Epicurus probably had an ulterior motive in publicising the atomic theory, for, like many other fourth-century philosophers, he was far more concerned with the nature of the microcosm than with the nature of the macrocosm. He was far more concerned, in the unsettled times at the end of the fourth century, to discover the conditions for a contented life than to prosecute research into the world about him. Thus, although he was a convinced follower of the Democritean theory, he did not develop atomism much beyond the level at which the great Abderans had left it. Indeed the additions he did make, for example the idea that atoms swerve from their trajectories (mentioned above), were regarded, even in antiquity, as detractions rather than improvements.

The ulterior motive which Epicurus had for popularising the atomic theory lay, according to Farrington (11), in a passionate desire to counteract the tendency towards theocracy emanating from the Platonic Academy. Readers of the *Republic* and the *Laws* will remember that Plato sought to overcome the internecine strife of his times by arguing for a close, tightly organised society. In order to achieve this society he was quite prepared to place ideology above the disinterested search for truth if this were conducive to greater contentment amongst the citizens. He was quite prepared to weld the citizens together by means of a myth, a State myth, even if the latter were in fact a lie.

Epicurus believed this to be a most dangerous development. Against Plato's 'organic society' with its technique of the 'noble lie' he wished to argue for the 'open society'. He wished to reassert the old Ionian freedom and disinterested pursuit of truth. He saw that the atomic theory, with its discrete, independent units, provided the best theoretical underpinning for this counter-attack. He saw very clearly that this theory implied that there was no 'Big Brother' watching, merely atoms and the void. In the famous *Sovereign Maxims* which were widely circulated in antiquity we read, for example, that

> ... it would be impossible to banish fear on matters of the highest importance, if a man did not know the nature of the whole universe, but lived in dread of what the legends tell us. Hence without the study of nature there is no enjoyment of unmixed pleasures (12).

The ethical significance of the atomic theory in the Epicurean philosophy is also well brought out by its Roman propagandist Lucretius Carus (c. 100–55BC). His great poem *The Nature of the Universe* seems to have been inspired by a burning desire to outface the superstitions propagated by priests and mystagogues. In the introduction to the first book of his poem we find the following paean to the founder of the atomic theory:

(11) B. Farrington, *The Faith of Epicurus,* Weidenfeld and Nicolson, London (1967).
(12) Diogenes Laertius, *The Lives of Eminent Philosophers,* vol. 2, The Loeb Classical Library, Heinemann, London (1925), p. 667. Diogenes Laertius is believed to have compiled his *Lives* at the beginning of the third century AD.

> When human life lay grovelling in all men's sight, crushed to the earth under the dead weight of superstition whose grim features loured menacingly upon mortals from the four quarters of the sky, a man of Greece was first to raise mortal eyes in defiance, first to stand erect and brave the challenge. Fables of the gods did not crush him, nor the lightning flash and the growling menace of the sky. Rather they quickened his manhood, so that he, first of all men, longed to smash the constraining locks of nature's doors. The vital vigour of his mind prevailed. He ventured far out beyond the flaming ramparts of the world and voyaged in mind throughout infinity. Returning victorious, he proclaimed to us what can be and what cannot: how a limit is fixed to the power of everything and an immovable frontier post. Therefore superstition in its turn lies crushed beneath his feet, and we by his triumph are lifted level with the skies (13).

Lucretius goes on to describe some of the excesses and terrors to which superstition leads. He reminds his readers of the frightful sacrifice of Iphigenia by her father Agamemnon to ensure a propitious outcome for the Trojan war. He outlines the fearful scenes the priests have painted of eternal torture in the fires of hell for the souls of the damned. All this, he says, vanishes like the morning mist before the strong beams of the theory he is about to set out. Nothing can ever be created by divine power out of nothing'; 'Nothing exists that is distinct from body and from vacuity'; 'Nature resolves everything into its component atoms...' There are no malevolent gods, no father-figures; nor is there an after-life, judgment, heaven and hell. The undiscovered country from whose bourne no traveller returns need puzzle the will no longer: it is undiscovered because it does not exist.

If it is true that Epicurus developed his philosophy in conscious opposition to Plato's attack on the 'open society', his advocacy of the 'noble lie', then it is only just that when a theocratic system was ultimately established his doctrine should be banned. This indeed proved to be the case: the spread of Christianity during the dark and middle ages outlawed Epicureanism as an atheistic creed. Only one manuscript of Lucretius' great poem is known to have survived the ages of faith. Although atoms never entirely vanished from the European consciousness, they were conventionally regarded as theologically suspect. When Pierre Gassendi finally resurrected the atomic theory in the seventeenth century, he did so by publishing a translation of the tenth chapter of Diogenes Laertius' *Lives of Eminent Philosophers* (see above). This chapter is, and always was, one of the best and most comprehensive sources of Epicureanism. Gassendi, a Roman Catholic priest, sought to soften the ecclesiastical backlash by expunging, as far as possible, Epicurus' materialistic metaphysics. Thus after a dormancy of two millennia the theory which was to play so important a part in the rise of modern science had hardly changed from that promulgated by the two philosophers of Abdera.

(13) Lucretius, *ibid.*,p.29

7 THE *TIMAEUS*

The just and the unjust

In a well-known passage A N Whitehead apostrophises Plato as the seminal philosopher of the western world, to whom all later philosophers stand as a series of footnotes.

By far the greatest influence on the life and thought of Plato was the teaching and example of Socrates. But Socrates, as we noticed at the end of the last chapter, must be held largely responsible for diverting the attention of philosophers from the phenomena of the macrocosm towards an analysis of the microcosm—man. Socrates recounts (1) how, as a young man, he had had a 'prodigious desire' to understand physical science. Later, however, he came to despair of ever arriving at the truth in this field and consequently heeded a version of the Delphic oracle's advice to 'first know thyself'. Xenophon puts it this way:

> Socrates did not even discuss that topic favoured by the talkers, 'The Nature of the Universe', and avoided speculation on the so-called 'cosmos' of the professors, how it works and the laws which govern the phenomena of the heavens: indeed he would argue that to trouble one's mind with such problems is sheer folly (2).

Indeed he came to regard such activities as akin to madness. For, says Xenophon, he was astounded that these men could not see that the problems and riddles they wrestled with were insoluble:

> Some hold that *what is* is one, others that it is infinite in number: some that all things are in perpetual motion, others that nothing can ever be moved at any time: some that all life is birth and decay, others that nothing can ever be born or die.

'Nor', says Xenophon, 'were these the only questions he asked about such theorists.' To Socrates, living in the turmoil of the fifth century BC, the enquiries of the nature philosophers seemed peculiarly useless: a luxury when the far more pressing concerns of government and social organisation demanded investigation. Xenophon reports Socrates as saying:

(1) *Phaedo,* 96–7. Here and in subsequent quotations from Plato I have used the translation by B. Jowett, Clarendon Press, Oxford (1892).
(2) *Memorabilia,* 1 (1), 10–15 (trans. E. C. Marchant), The Loeb Classical Library, Heinemann, London (1923).

> Students of human nature think they will apply their knowledge in due course for the good of themselves and any others they choose. Do those who pry into heavenly phenomena imagine that, once they have discovered the laws by which these are produced, they will create at their will winds, waters, seasons and such things to their need? Or have they no such expectation, and are they satisfied with knowing the causes of these phenomena? (3).

Socrates was born about 470 BC and was executed by the Athenians in 399 BC. In 480 BC, ten years before Socrates' birth, Themistocles had rallied the Athenian navy at Salamis and defeated the Persian force in that reef-strewn gulf. Xerxes had hurried from the throne he had had set up on Mount Aegialeus, from which he had expected to watch his own admiral gain the victory, had hurried back through Greece to the bridge of rafts across the Hellespont and had left the Athenians to govern themselves. It was the art of governing the consequent democracy which pressed upon the mind of Socrates. The recently developed schools of the Sophists taught how to win an argument: Socrates sought to achieve a much more difficult ambition—by argument to ascertain the truth. He sought with passion to ascertain the essence of justice, of virtue, of happiness, or, to quote Xenophon once more:

> His own conversation was ever of human things. The problems he discussed were, What is godly, what is ungodly; what is beautiful, what is ugly; what is just, what is unjust; what is prudence, what is madness; what is courage, what is cowardice; what is a state, what is a statesman; what is government, and what is a governor ... (4).

Knowledge and opinion

Yet, although the subject matter had altered from the macrocosm to the microcosm, the major presuppositions of the Ionian philosophy remained unchanged. The Presocratics, as Galileo sarcastically observed, prized above all else those things which were 'impassable, immutable, inalterable'. Thus we feel ourselves on familiar territory when we read the following rhetorical exchange near the beginning to the *Timaeus*:

> What is that which always is and has no becoming; and what is that which is always becoming and never is? That which is apprehended by intelligence and reason is always in the same state; but that which is conceived by opinion with the help of sensation and without reason, is always in process of becoming and perishing and never really is (5).

The *Timaeus* is, in fact, believed to be one of Plato's later works, composed some time after the death of Socrates. The insistence on the priority of being over becoming is, however, to be found in all the dialogues which Plato reported and extends, as we shall see, into the work of his own pupil,

(3) *Memorabilia,* 1 (1), 16.
(4) *Ibid.,* 1 (1), 16.
(5) *Timaeus,* 28.

Aristotle. The passage quoted above occurs during a discussion of the work of the world's creator. A little further on in the same book we read that

> If the world be indeed fair and the artificer good, it is manifest that he must have looked to that which is eternal; but if what cannot be said without blasphemy be true, then to created pattern. Everyone will see that he must have looked to the eternal; for the world is the fairest of creations .... And having been created in this way the world has been framed in the likeness of that which has been apprehended by reason and mind and is unchangeable ...

It is clear from these passages that, for Plato, only invariant entities—essences, forms, ideas, archetypes, and so on—are fully intelligible; all else belongs to the world of perishable things about which we can have nothing better than a mere opinion. Plato thus shares the Pythagorean, Parmenidean, epistemology which accords the title of knowledge only to that of which we have the manifest right to be sure; and, among those things of which we have the manifest right to be sure are, preeminently, the tautologous truths of mathematics. Hence, as is well known, Plato caused to have inscribed above the entry to the Academy the familiar rubric stressing the prior necessity of geometrical knowledge in those who would become his pupils.

The doctrine of anamnesis

This epistemology is, in fact, central to the whole Socrato-Platonic endeavour. Only by supposing that 'forms' of virtue, justice, statesmanship, and so on, existed and were analogous to mathematical forms (and thus immutable) could the ethical relativities generated by the civil strife following the Periclean age be overcome (6). Thus the Pythagorean theory of knowledge is not only implicit in all the Socratic dialogues which Plato chronicled: it is essential to their very existence.

Its centrality is evident in the so-called Socratic method; for Socrates invariably assumes that his interlocutor really knows the answers to his own questions and perplexities. Socrates arrogates to himself only the position of midwife (7) whose function is to terminate pregnancy by bringing forth the infant truth. Socrates' method consists of a process of subtle and lengthy cross-questioning by means of which the inconsistencies and fallacies in his victim's mind are cleared away until, finally, the truth is laid bare. In other words, the method, like a legal trial, is based on the fundamental tenet that self-contradictory assertions are acknowledged to be inadmissible. The method is clearly appropriate and excellent for clearing up confusions in thought, for defining what is meant by 'good' or the 'just man', or even, as in

(6) See, for example, *Republic,* Book 1: Thrasymachus' view that might is right, that 'justice is nothing more than that which is advantageous to the stronger faction', is summarily dismissed.
(7) *Theaetatus*, 149.

the *Meno*, for showing that an unlettered slave possesses, latent 'within himself', the axioms of geometry. But it is equally clear that it is quite useless for advancing those areas of knowledge which depend on observation and the use of instruments like the telescope or microscope.

For Socrates and Plato it is only ignorance which leads to strife and evil doing. Education would show men that their true interest lies in order and harmony; and education, as implied in the previous paragraph, is essentially a process of 'drawing out' and making clear to the conscious mind truths which had formerly lain latent and unrealised. Determining the truth about some matter is thus basically a process of reminiscence. But where and when was this latent knowledge implanted? In the *Phaedo*, where Socrates is movingly shown calmly arguing away the last few hours before his execution, the answer is outlined.

Socrates' ambition, as we have seen, was steadily directed towards illuminating the truths of the moral world. His entry to the problem posed by the doctrine of reminiscence is thus through the ethical calculus of good and bad. As he sits waiting for the gaoler to bring him his last drink he points out to his friends that the genuine reason for, and cause of, his predicament is the fact that the Athenians have thought it good to sentence him to death and he in his turn has thought it good to accept their judgment. He relates how in his youth he had devoured the works of Anaxagoras because the latter had taught that the mind governs all things. But, he says, he merely found that Anaxagoras, like other Presocratics, treated the mind as a very tenuous but still material stuff. This disillusioning experience had, as we have already noticed, turned him off physical science. For, he says, left to themselves his bones and muscles, and any other material of which he was composed, would long since have taken themselves off. But he still awaits the hemlock: therefore the moral universe is something other than the material, is in fact something which lies behind, or posterior to, the material.

Into the mouth of one of the circle of friends with whom Socrates is spending his last hours is put the Pythagorean doctrine that the soul is the 'harmony' of the body. If this analogy is true then the soul cannot survive the body any more than the harmony can survive the lyre. Socrates, however, is able to counter this theory by pointing out that the lyre is the cause of the harmony but the body is not the cause of the soul. The moral is posterior to the physical; the soul is posterior to the body.

The behaviour of the body is thus caused by the soul. It is this which prevents Socrates breaking from prison. Moreover the latent knowledge, like the Kantian categories, cannot have been obtained through the senses. For the type of knowledge possessed, for example mathematical knowledge, is used to judge the phenomena of the sensible world and hence, he says, cannot be derived from it. Indeed, as the post-Renaissance development of physical science shows, the entire method consists in the construction of ideal models—point masses, frictionless contacts, reversible processes—and of 'explaining' the observed phenomena as more or less close approximations to these paradigms. Thus the recollected knowledge cannot, according to Socrates (8), have been gathered since birth: for the very use of our sensory faculties depends on a preexistent apparatus for judging how far our sensations approach or fall short of certain standards. And, says Socrates, were we not all born with the full use of our faculties? It follows, he concludes, that 'our souls must have existed without bodies before they were in the form of man and must have had intelligence'. The argument finishes with the following passage:

> ...if, as we are always repeating, there is an absolute beauty and goodness, and an absolute essence of all things; and if to this, which is now discovered to have existed in our former state, we refer all our sensations, and with this compare them, finding these ideas to be preexistent and our inborn possession – then our souls must have had prior existence ... There is the same proof that these ideas must have existed before we were born as that our souls existed before we were born; and if not the ideas then not the souls (9).

The argument as Socrates is careful to point out depends on establishing that the standards—beauty, truth, equality, and so on—do have 'a real and absolute existence'. It was, of course, the ambition of Socrates, throughout his career which, even as he argued, was drawing to its violent end, to dissect out these paradigmatic standards from the adulterated and confused form in which they were exhibited by the world.

An aristocrat's interpretation

We must now leave the compelling figure of Socrates, of whom Plato said that of all the men he had known he was 'the wisest, justest and best', and return to the less sympathetic person of Plato himself. With Plato philosophy tends to retreat from the joyful colloquy with everyday problems and affairs which was so characteristic of Socrates. We begin, in place of the immediacy of the market place, the battle of wits, the cut and thrust of debate, to sense the tedium of long chains of argument—argument composed in the study and not subject to the sudden critical attack or the sceptical comment. The long shadow of the scholastic disputation, pilloried by Francis Bacon in a famous passage as laborious spiders' webs of learning, seems to fall across the clear Athenian scene.

Unlike Socrates, whose father and mother had been sculptor and midwife respectively, Plato came from the aristocracy on both sides of his family. Some authorities (10) detect the influence of this social background in his philosophy. The ideal state, the Republic, is constructed on rigid class lines. The rulers know what is best for the ruled and, as we have already noticed in chapter 6, are quite prepared to propagate what they know to be false in

(8) *Phaedo,* 75–6.
(9) *Phaedo,* 77.
(10) See, for example, A.W. Gouldner, *Enter Plato,* Basic Books, New York (1965).

order to coerce the rebellious lower classes (11). Nowhere is this sociological bias better displayed than in the *Timaeus*; for the fructifying analogy throughout this work is, as we shall see, that between society and an organism. The tripartite division of Plato's ideal state makes its appearance, in only slender disguise, in the tripartite organisation of the body.

The *Timaeus* is a strange and, to our minds, in parts a fantastic book. It is unfortunate that it was the only work of Plato known to the middle ages. It was translated into Latin by Chalcides about AD 350, principally because the neo-Platonists believed it to be a religious tract. In consequence it has had a quite disproportionate influence on western thought.

How seriously are we supposed to take it? Plato and Socrates before him were, as we have seen, mainly interested in logical truths. The realm of nature was merely contingent. Nothing in it was certain, nothing that could be said about it was more than mere opinion, mere conjecture. Again we catch an echo from Parmenides. The founder of the Eleatic school had also, after he had expounded his theory of the *One*, seen fit to develop a cosmology—*The Way of Seeming*—which he admitted to be irrational but was, he said, as good as any other. Thus, when Timaeus sets out to tell his 'likely story' and the tale unfolds replete with numerology, perfect geometrical figures, allusions to gods and demigods, we should not perhaps be too surprised. Nevertheless the contrast between the theories of Timaeus and those of the great Milesians is striking. In place of their passionate attempt to explain the sensible in terms of the sensible, to 'see' nature through the eyes of artisan and technician, we have a natural philosophy couched in terms more fitting to an aristrocrat and slave-owner.

Timaeus' likely tale

What, then, were Timaeus' conjectures about the life sciences? We find that he makes his approach by considering the genesis of things, by rehearsing how they have come to be, how they have, in other words, been created. Here, we find, a connection is made with some of the ethical and epistemological concerns discussed in the preceding section. The moral is posterior to the physical. The end governs the means. What end had the creator in mind when he created the world? To diffuse the 'good'—that is, order, harmoniousness—as widely as possible. Right away we see a deep contrast with the Democritean universe. In place of atoms blindly running, colliding, entangling, in the void we have a shaping intelligence. The argument of the *Phaedo* is here generalised to all creation.

Next the creator noticed that no unintelligent creature was fairer than an intelligent creature and, furthermore, that 'nothing devoid of soul can be intelligent' (12). Consequently, says Timaeus, the creator placed intelligence

(11) *Republic,* 459.
(12) *Timaeus,* 30.

in the soul and then placed the soul in a body. Thus 'the soul in origin and excellence (is) prior to and older than the body and (is) its ruler and mistress, of whom the body was made to be the subject' (13). The identity of view between Timaeus and the Socrates of the *Phaedo* is clear.

The description of the body which follows seems in contrast markedly un-Socratic. The Pythagorean beliefs of Timaeus are made obvious. Emerging from some rather turgid passages of numerology comes a description of a body made of a mixture of the four ancient elements—earth, air, fire and water—and spherical in shape. We may recall the discussion in chapter 5 in which it was seen that for many ancient thinkers the circle was regarded as the 'perfect' geometrical form. Similarly Timaeus takes the sphere to be the perfect configuration for a solid body. The spherical body which Timaeus describes represents the whole universe. The soul was united with this corporeal universe by complete interpenetration, analogously to the aether of nineteenth-century science. There is no need to stress again the Parmenidean overtones of this account.

Having designed and created the world, the creator's next task was to people it with living creatures. He envisaged four major groups corresponding to the four ancient roots: fire, air, earth and water. To the realm of fire correspond the gods, to the air correspond the birds, to the earth terrestrial and pedestrian animals and to the water, fish. However, the Demiurge conceived his task to be complete when he had fashioned 'the heavenly race of gods'. He expressly left the task of fashioning the rest of animate creation to these newly created lesser immortals, reserving for himself only the privilege of sowing the divine seed into the minds of mortal men. The lesser gods, with no great originality, proceeded to fashion mankind on strictly analogous principles to those used by the creator himself in modelling the universe.

First of all they bethought themselves as to how best to construct a receptacle for the divine spark bestowed directly by the creator. The Greek pantheon may have been quarrelsome, but it contained at this time no revolutionaries comparable to those imagined by Milton, and consequently it was clear that no improvement could be made on the pattern laid down by the original creator. Thus it was decided to make the divine receptacle as nearly as possible spherical in shape. Thus was the head devised (14). Furthermore, in order to prevent this globous head from tumbling helplessly about among the 'high and deep places of the earth', the rest of the body—the arms and legs and torso—were constructed.

After a rather lengthy digression in which Timaeus discusses the origin and nature of the elements and how they are apprehended, he returns to describe the fabrication of the rest of the body. He is careful to point out that his opinion is only probable and, furthermore, in one of the few exhortations

(13) *Ibid.*, 34.
(14) *Ibid.*, 44.

to scientific research to be found in Plato, 'will be rendered more probable by investigation' (15).

The divine spark having been incorporated in the quasi-spherical vessel of the head, the lesser gods fixed in the rest of the body a soul of a very different nature—a mortal soul 'subject to terrible and irresistible affections'. Why they should have thought this an intelligent or amusing operation is nowhere explained. However, having fashioned this lesser soul subject to the temptations of pain and pleasure, rashness, fear, anger and irrational hope, they had the sense to isolate it from the divine, aristocratic intelligence already housed in the head. Having inserted this soul into the torso they constructed an isthmus—the neck—which, they calculated, would prevent it reaching the head and subjugating the mind. Furthermore, this lower soul was itself binary, consisting of a better and a worse part. The apprentice gods therefore set a diaphragm across the torso confining the 'more noble' part of the mortal soul—that part which is 'endowed with courage and passion and loves contention'—in the thorax so that it has the possibility of entering into some communication with the calm ruler in the head. The other and less noble part of the soul, that which hungers and desires meat and drink, they isolated below the diaphragm in the abdominal cavity.

Timaeus does not give a reason for this tripartite division of the soul. Readers of the *Republic* will, however, find it familiar. It will be recalled that Plato's ideal state was divided into three classes: the guardians, the soldiers and the hoi polloi. The guardians were distinguised by the possession of wisdom and cool reason. The warriors displayed the warlike virtues of passion and courage. The proletariat spent their time 'getting and spending', exhibiting in this the low-grade characteristics of avarice and cupidity. The function of the first two classes was to protect and guide the vulgar citizenry. In so doing it is clear that the two upper classes needed to cooperate and communicate with each other. This is reflected, as we saw above, in the psychophysiology of the *Timaeus.* In the *Republic*, indeed, there is some uncertainty as to where exactly the distinction between passion and reason can be drawn. Plato would have recognised Mr Valiant-for-Truth; he esteemed moral courage and would have commended the aristocratic virtue of defiance in defeat. Passion and reason seem in these cases difficult to disentangle.

In the *Republic* Plafo draws explicit analogies backwards and forwards between his ideal state and the psychological organisation of man. The *Timaeus*, we should remember, is an explicit sequel to the *Republic*. It seems clear, therefore, that Timaeus' seeming omission is merely due to the fact that the tripartite nature of man had been established in the earlier dialogue. It is of considerable interest in the present context, however, to note how this extremely influential psychophysiological theory arose from a comparison of man and the political state. The amphibious nature of biology

(15) *Ibid.*, 72.

was referred to in an earlier chapter. Nowadays it is much more usual to draw analogies with physical or computer science. But we shall find that Plato is far from being the only thinker to use a sociological paradigm. Biologists of the eminence of Charles Darwin and Rudolph Virchow have also made use of this analogy.

Returning, however, to the *Timaeus* we find that the immortals went further than merely confining the two lower parts of the soul in different parts of the torso. They arranged matters so that the heart and the great blood vessels were placed in the thorax at the command of the 'warrior' soul. Hence when passion or courage was aroused, and not least when the cephalic soul commanded, the message was easily transmitted around the whole body. Moreover, knowing that apprehension of wrong or evil from without or of desire from within would make the heart swell and beat in a fury, the gods surrounded it with the 'soft and bloodless' tissue of the lungs. These, by receiving 'breath and drink' cool the wildly beating heart and also cushion it 'like a soft spring'.

A similar ingenuity was displayed in the arrangements made for the lower, concupiscent, soul. This lesser luminary was, according to Timaeus, 'bound down like a wild animal' in the region between the diaphragm and the navel. Here it was allowed to feed at 'a sort of manger' and in this activity, far from the council chambers and committee rooms of the intellect, remain in swinish content. However, the gods foreknew that this soul was both unruly and uncouth and, moreover, would by no means listen to reason. Thus, in order that the higher parts of the soul might exercise control, the gods designed the liver and placed it in this lower chamber of the body. The cerebral guardian was able to control the appearance of the liver, on occasion making it wrinkled and rough, twisting its lobes about and 'closing and shutting up the vessels and gates (causing) pain and loathing', thus frightening the bewildered mooncalf of the lower soul into acquiescence. On other occasions the converse happens. The intellect makes the liver 'smooth and bright and free' and infuses a natural sweetness in place of the former bitterness and bile. Poor Caliban quickly forgets the terrible 'threatening and invading', the 'twisting and contorting', the frightening colouring of the liver when angry, and trustingly settles down 'happy and joyful', as Timaeus says, 'to pass the night in peace and to practise divination in sleep' (16). Thus, by a judicious combination of big stick and carrot, the proletarian lower soul is kept in his allotted place, quietly feeding at his manger.

After this pioneering account of the threefold nature of man, Timaeus goes on to explain exactly where, in his opinion, the souls and the body are united. We have already noticed that this does not seem to have been an acute problem in antiquity, as the concept of an immaterial substance was not yet clear. Nevertheless Timaeus does propose a locality for the psychophysical

(16) Ibid., 69–72

bond: the marrow. 'The bonds of life', he asserts, 'which unite soul and body are made fast there' (17). The part of the marrow which 'like a field' was to receive the immortal soul was made spherical and called the brain. The rest of the marrow, destined to receive the mortal soul, was constructed in the form of cylinders. Around these cylindrical configurations the creator fashioned first the bones, so that the vital marrow became enclosed in a protective armour, and then the rest of the body.

Timaeus describes the production of bone—apparently achieved by a mixture of culinary and metallurgical arts—and then describes how a globe of bone was built to house the brain. An opening was left in this globe so that the marrow of the neck and back—the spinal cord—might communicate with the marrow of the skull—the brain. Around the spinal marrow, protecting it, were fashioned vertebrae. Between the several vertebrae and between the other bones of the body, joints were interposed. The joints, according to Timaeus, were endowed with movement. It is a little difficult to be sure whether this movement is conceived to proceed from the joints or whether, as we now understand, the joints are regarded as mere passive agents of the musculature. Some authorities (18) believe that Plato did not perceive the significance of the muscles in producing movement, believing that the function of the 'flesh', as he calls it, was to protect the 'brittle and inflexible' bone. Certainly Timaeus asserts that the creator contrived the flesh because he believed it would

> serve as a protection against the summer heat and the winter cold, and also against falls, softly and easily yielding to external bodies like articles made of felt; and containing in itself a warm moisture which in summer exudes and makes the surface damp, would impart a natural coolness to the whole body; and again in winter by the help of this internal warmth would form a very tolerable defence against the frost which surrounds it and attacks it from without (19).

The action of muscles in producing bodily movement is nowhere stated. Indeed, it is easier to interpret the physiology of the *Timaeus* as attributing the power of initiating movement to the tendons rather than to the muscles. 'The members', we read (20), are bound together 'by the sinews, which admitted of being stretched and relaxed about the vertebrae...make the body capable of flexion and extension, while the flesh would serve as protection... ' The notion that it is the sinews which cause animal movement has, as we shall see, a long and distinguished history. It remained for Steno in AD 1660 finally to refute the idea that the tendons played some active part in animal movement. The overall impression given by this section of the *Timaeus*,

(17) *Ibid.*, 73.
(18) E. Bastholm, *The History of Muscle Physiology*, Munksgaard, Copenhagen (1950). Bastholm points out that the conviction that the muscles were merely packing or supporting tissue while the tendons were active in producing skeletal movements is also commonplace in the Hippocratic writings.
(19) *Timaeus*, 74.
(20) *Ibid.*, 74.

however, does not suggest that it is either the tendons or the muscles which are the prime movers but the joints themselves. And this ties in well with Timaeus' belief that the soul is located in the bone marrow. At the joints the prime mover, the marrow-soul, is well placed to work the levers of the body's frame (21).

It is clear that Plato's neuromuscular physiology is still at a very primitive level. He shows little or no interest, moreover, in actually ascertaining the facts of anatomy still less of prosecuting experiments into physiology. His sensory physiology is similarly poorly developed: he seems to regard sense organs such as the ear and the eye as gateways through which the world 'outside' enters into direct contact with the mind. Finally his cardiovascular and respiratory physiologies are couched in metaphors which make them difficult for moderns to understand. We read (22) of a basketwork or 'creel' of fire which is spun around two lesser webworks—one in the thoracic and one in the abdominal cavity. He connects up this imaginative idea of an incandescent vascular system with a theory of respiration and a theory of digestion. Expiration and inspiration, he believes, take place not only through the mouth and nostrils but also through the whole body surface. Now Timaeus follows Parmenides in believing that a vacuum is an impossibility. Thus when air is expelled during expiration, it is expelled into a plenum, and it is thus bound to push other air back into the space it has just vacated. But what expels the air in the first place? Timaeus maintains that the 'internal fountain' or 'creel' of fire generates heat which strives to escape to its natural place outside the body. This forces air out of one or other of the two possible openings to the exterior. Once the air begins to escape other air, as explained above, immediately starts to enter the other opening. Thus 'a circular motion swaying to and fro is produced by the double process which we call inspiration and expiration'. This tidal movement of air in and out of the body fans the internal fires, and in particular it fans the abdominal ebullience so that

> ever and anon moving to and fro, [it] enters through the belly and reaches the meat and drink, it dissolves them . . . and pumps them as from a fountain into the channels of the veins, and makes the stream of the veins flow through the body as through a conduit.

The reader may be disposed to agree at this point with Xenophon's report of Socrates' strictures on the 'talkers'. Nonetheless the vague outlines of a system of vegetative physiology can be made out in these cloudy passages. It is clear, as has been apparent in earlier chapters, that a connection between respiratory and cardiovascular physiology had been discerned, however dimly, from the earliest times. Our Pythagorean philosopher attempts to tie the processes of digestion into this system.

(21) It will be recalled that it is the starting point of the argument that it is the soul which moves the body and not *vice versa.*
(22) *Timaeus,* 78–80

But in spite of the dreamlike and metaphorical nature of the account one thing is very clear: the overriding type of explanation employed is an explanation in terms of final causes. The whole system is deeply embedded in a teleological metaphysics. Everything happens with an end in view. In this respect it has a markedly less 'modern' character than the earlier Empedoclean account (chapter 5). But partly because of the vast influence of Athenian philosophy, it remained for centuries an influential interpretation. This teleological character of Plato's life science is perhaps nowhere better epitomised than in his account of the origin of species. Instead of an Empedoclean groping towards a type of Darwinian theory, the intellectual atmosphere is charged with the notion of living beings struggling to mount towards perfection. Those that fail are suitably punished. The whole scene is supervised by the apprentice gods.

The young gods, in their wisdom, recognise that men will need food. In consequence they create the plant kingdom. Plants only need and hence are only endowed with the lowest of the three Platonic souls. The origin of the various animals is explained in the context of metempsychosis, a doctrine which was strongly held by the Pythagoreans. Thus we read (23) that 'of the men who came into the world, those who were cowards or led unrighteous lives, may with reason be supposed to have changed into women in the second generation...', and further on that 'the race of birds was created out of innocent light-minded men, who, although their minds were directed towards heaven, imagined in their simplicity, that the clearest demonstration of things above was to be obtained by sight'. The tetrapods originated from men who 'had no philosophy in any of their thoughts', and the 'polypods' and apoda originated from the least philosophical of all. But the worst doom, reserved for those whose ignorance and senselessness was total, was to return as fish, oysters and other aquatic animals.

(23) *Ibid.*, 91-2.

8 ARISTOTLE: METAPHYSICS

Biography of a philosopher

By Dante called 'the master of them that know', Aristotle by common consent bestrides the ancient and late mediaeval worlds (1). Born at Stagira in Thrace of a medical family, he entered the Platonic Academy in 367 BC when seventeen years of age. For twenty years he worked in close contact with the elder philosopher by whom he is reputed to have been called 'the intellect of the school'. On Plato's death in 347 BC he composed a eulogy in which he commended the philosopher as 'the man whom it is not lawful for bad men even to praise, who alone, or first of mortals, clearly revealed by his own life and by the methods of his words, how to be happy is to be good' (2).

In spite of Aristotle's long association with the Academy and in spite of his preeminent intellectual gifts, the succession as head of the Academy passed to another. Aristotle left Athens and spent the next eleven years in a variety of occupations, including two years studying the marine biology of the Ionian shore (3) and four years at the Macedonian court tutoring the boy who was to become Alexander the Great. Returning to Athens in 336 BC, Aristotle established his own school or, better, research institute, which he named the Lyceum. Here in the final thirteen years of his life Aristotle lectured and put together the treatises which have immortalised his name (4). Aristotle spent the last year of his life on the island of Chalcis, whither he had removed himself for political reasons, saying that he would not offer the Athenians a

(1) Aristotle's place in mediaeval thought is chequered. During some periods the Church banned his works. Only after Moerbeke's direct translations from the Greek originals were used by Thomas Aquinas in the middle of the thirteenth century did his influence finally become paramount. Thomist philosophy still forms the basis of Roman Catholic theology.
(2) Encyclopedia Britannica Inc., *Great Books of the Western World* (ed. R.M. Hutchins), Chicago, London, Toronto (1952): *Aristotle,* biographical note. I have used the translations collected in the *Great Books* series for all subsequent quotations from Aristotle. Further details may be found in the bibliography.
(3) Aristotle is believed to have spent the major part of these years at Mytilene on the island of Lesbos.
(4) There is some controversy about the precise way in which the Aristotelian treatises were composed. Were they written by Aristotle himself, or his assistants, as material for his lectures? Or were they notes taken by his auditors and subsequently used for disputation in the school? It is probable that both suppositions are correct: one explanation applying to one treatise, the other to another.

second chance of sinning against philosophy. He died at the age of 62 in 322 BC.

The analysis of sensation

Aristotle's outlook seems to have differed from that of either of his two great predecessors. We may speculate on the reason for this difference. We have noticed that he came from a medical family. It is also believed that this family originated in Ionia. It is likely, therefore, that his family were heirs to the great tradition of Hippocratic medicine which had flourished on the shores and isles of Ionia for a century. This tradition was strongly empirical in its orientation and, in the main, scorned the metaphysical theorising of the philosophers. We find elements of this attitude in the thought of Aristotle. His approach is far more directed towards the things of this world than was that of Plato; it is far more concerned with natural science than was that of Socrates. Indeed, it has been said that in Aristotle the original impulse of Ionian natural philosophy reaches its ultimate flowering.

An introductory indication of this difference of outlook can be found in the Stagirite's approach to that most central of Socratic and Platonic concerns: ethics. In the *Nichomachean Ethics* he chides his elders for expending all their time and effort in trying to determine what is the essential nature of virtue, goodness, and so on, when what is really needed is a description of good behaviour:

> Since, then, the present enquiry does not aim at theoretical knowledge like the others (for we are enquiring not in order to know what virtue is, but in order to become good, since otherwise our enquiry would have been of no use), we must examine the nature of actions . . . (5).

Aristotle's passionate concern, even in the most abstract reaches of his thought, was always with the phenomena of this world.

This cast of mind is epitomised at the outset of his most profound work: the *Metaphysics*. At the very beginning of this tract we find the following passage:

> All men desire to know. An indication of this is the delight we take in our senses; for even apart from their usefulness they are loved for themselves . . . (6).

Evidently, although Aristotle is composing the book which is commonly believed to have given the term metaphysics to European languages, his mental attitude is far different from that denoted by the present-day pejorative use of the term. The quoted statement would be quite acceptable to 'hard-headed' contemporary scientists.

But Aristotle's intellect ranged far more widely than that of any present-day scientist. In each field of his colossal intellectual undertaking we find, however, the same meticulous approach. T S Eliot puts it well when he

(5) *Nichomachean Ethics,* 1103b30.
(6) *Metaphysica,* 980a.

writes:

> Aristotle, in whatever sphere of interest, looked solely and steadfastly at the object . . . he provides an eternal example not of laws, or even of method, for there is no method except to be very intelligent, but of intelligence itself operating on the analysis of sensation to the point of principle and definition (7).

Eliot is here discussing the Aristotelian criticism of poetry, but his words apply to the whole Aristotelian corpus. The attitude which Eliot describes is that of a man with no ulterior motive, with no axe to grind, of a man who strives only to let the facts speak for themselves.

Nevertheless, although we may feel this passionate attempt pervading Aristotle's works, we may doubt its total success. Aristotle's work, as we shall see, constitutes a complete and closely articulated world system. Like all his predecessors and contemporaries he was under pressure to develop a total and comprehensive philosophy. Specialisation, as we have already several times remarked, is a comparatively modern phenomenon. It was, however, impossible to create such a system in Aristotle's time. We may doubt, indeed, whether it will be ever possible. But the attempt to construct such a system, then as now, often leads to recalcitrant facts or opinions being remoulded closer to the heart's desire. Aristotle, like any other such system-builder, cannot be judged entirely innocent on this count.

Many subsequent commentators have, for example, accused Aristotle of forcing his own interpretations on to the work of his philosophical predecessors. Aristotle's meticulous habit is to consider the opinions of his forerunners before setting forth his own. In this preliminary criticism of his predecessors he sometimes seems to warp their ideas so that they are made to seem infant Aristotles groping for the truth which he alone, and in the fullness of time, has grasped. Furthermore, twenty years spent in the Academy could scarcely have left intellect and imagination unmarked. Indeed, many have seen in the Aristotelian system a development and enrichment of the Platonic theory of forms. For Plato the latter were static, mathematical, eternal; for Aristotle, as we shall see, the forms were, very importantly, principles of activity and vitality. Thus we may feel that the Stagirite's attempt to 'see the world steadily and see it whole' was made, to some extent, through the lenses of Platonic spectacles. The facts, which he attempted to let speak for themselves, spoke, perhaps, with a Platonic accent.

The classification of knowledge

The works which nowadays constitute the Aristotelian corpus are believed to have formed the basis for lecture courses put on at the Lyceum (see footnote (4), p.67). Thus we may suppose that Aristotle was subjected to the lecturer's familiar discipline of subdividing and analysing his material in order to achieve a clear and simple exposition. It will be obvious that there is a world

(7) T.S. Eliot, *The Sacred Wood*, Faber and Faber, London (1920).

of difference between this and the Socratic dialogue. And perhaps because of this difference in method, Aristotle glimpsed a profound truth, a truth which had not been emphasised by his predecessors. He saw that the very possibility of attaining knowledge depended on the mind's ability to separate out groups of interrelated phenomena from the Heracleitean flux of events. These phenomena fell naturally together because they shared common principles and were describable in common terms: their treatment, in other words, constituted a single science. Thus although Aristotle may with justice be regarded as the supreme generalist of all time, he was also, paradoxically, responsible for initiating that compartmentalisation of knowledge which has become so marked a feature of the modern world. He is quite clear, for example, that explanatory concepts developed in one branch of knowledge do not necessarily apply in another: 'Nothing can be demonstrated except from its "appropriate" basic truths' (8).

On the other hand this initial necessity to subdivide knowledge into manageable, reasonably homogeneous units did not prevent Aristotle from incessantly pointing out the connections between the specialisms. Aristotle's opus is full of cross-references, full of allusions backwards and forwards. Reading his work one begins to grasp the subtle and profound manner in which the gigantic range of his thought forms a coherent pattern. We follow in his treatises the unfolding of the great lecture course at the Lyceum in the final third of the fourth century BC. We move from formal logic through the special sciences to politics, rhetoric, ethics and poetics. We begin to appreciate how each topic relates to and depends on the others and how each fits into the general scheme—a scheme which some still see as the only genuine alternative to the modern technologico-scientific system.

A predilection for biology

One of the core organising concepts of the Aristotelian scheme seems to have been a profound intuition about the nature of life and living things. This intuition is in some respects primitive. We have already discussed the panpsychisms and panzoisms prevalent among the earlier Presocratics. We have seen that for many of these early thinkers the ultimate explanatory paradigm was provided by the living organism, or even by consciousness, rather than by the mechanisms of inert inorganic matter. The same position, it can be argued, was, in the last analysis, held by Aristotle. But Aristotle's thought was far from primitive in all other respects. He spun from this primal intuition a web which it took the scientific revolution of the seventeenth century to break.

We may perhaps speculate that Aristotle's deep interest in life and living things arose from his early formative experiences in a medical family. We have also noticed that he spent two of the most creative years of his life, after

(8) *Analytica Posteriora*, 75b36.

leaving the Academy, studying marine biology. Certainly, if we look at the whole range of his work, we find that a large proportion is biological. These encyclopaedic works—*The History of Animals, The Parts of Animals, The Motion of Animals, The Gait of Animals*—are remarkable collections of observations, some only confirmed in the nineteenth century of our era. There was no comparable mass of detailed observational data available to Aristotle in the subjects of physics and chemistry. The techniques necessary to build up a comparable mass of information were simply not available, were not to be invented, in fact, for millennia. But Aristotle's predilection for biology goes deeper than this. It is not just that there was rather little material to ponder in the inorganic realm. We can feel some of the force of Aristotle's devotion to the life sciences in the following passage from *The Parts of Animals* where he admonishes those who regard biology as trivial if not downright distasteful:

> We must not recoil with childish aversion from the examination of the humbler animals. Every realm of nature is marvellous: and as Heracleitus, when the strangers who came to visit him found him warming himself at the furnace in the kitchen and hesitated to go in, is reported to have bidden them not to be afraid to enter, as even in the kitchen divinities were present, so we should venture on the study of every kind of animal without distaste; for each and all will reveal to us something natural and something beautiful (9).

'Nature does nothing in vain'

Notwithstanding T S Eliot's laudatory comments, it is hardly possible for a philosophic mind to be ultimately dispassionate. Not even the genius of Aristotle could stand completely outside the framework of things, could achieve the god-like status Archimedes imagined when he called for a fulcrum to move the world. For every thinker some facts must be more equal, more basic, than others. In order to begin the task of making sense of the world's phenomena the philosopher, like other men, has to select among, grade and rank, the 'raw' impressions which his senses present to him. Aristotle was, himself, fully aware of this. 'It is hard', we read in the *Posterior Analytics* (10), a work in which Aristotle discusses scientific methodology,

> to be sure whether one knows or not; for it is hard to be sure whether one's knowledge is based on the basic truths appropriate to each attribute – the differentiae of true knowledge . . . . I call the basic truths of every genus those elements in it which cannot be proved.

What, then, are the basic truths on which the Aristotelian system is based? Perhaps the layman has penetrated, as is often the case, to the centre of the Aristotelian position with the tag: 'Nature does nothing in vain'. Things do not happen at random but with an end in view. And this end is, indeed, for Aristotle one of the most important reasons for their happening; it is, in other words, the final cause. It will be recalled from chapter 3 that Aristotle's use of

(9) *De Partibus Animalium,* 645a15.
(10) *Analytica Posteriora,* 76a25.

the term 'cause' was much wider than is customary today.

Nature does nothing in vain. This short phrase expresses a concept of extraordinary richness; a concept, moreover, sharply at variance with the outlook of twentieth-century science; a concept at variance, indeed, with the nascent science of Leucippus and Democritus.

Aristotle, as we noticed in chapter 6, reacted strongly against the Democritean vision. Some authorities believe that this reaction was due to the twenty formative years he had spent at the Academy. And it is certainly true that the Aristotelian doctrine of final causes bears a strong family resemblance to the teleological metaphysics we discussed in the last chapter. But, in addition to this Platonic influence, we may perhaps infer that part of the aversion to atomism was the aversion of a biologist to some very premature biophysics.

Aristotle, the biologist, was unable to understand how the random shiftings of 'billiard-ball' or even hooked atoms could account for the phenomena he knew best: the structure and function of animals, the ordered sequential processes of embryology, the similarity of parent and offspring, the intelligent life of men in society. It just didn't make sense.

Aristotle thus rejected atomism, rejected notional particles randomly moving in an unthinkable void, and constructed instead a subtle, profound and fascinating philosophy which seemed to him to fit the facts he was most familiar with better: a philosophy, in other words, which grew from biological rather than physical first principles.

What is 'substance'?

Aristotle, then, was not in the modern sense of the term a 'reductionist'. He was not persuaded that the ultimate explanation of things is written in the mathematical abstractions of submicroscopic physics. The phenomenon which seemed to him most basic in the apparent flux of the world was the unity and persistence of the individual living being. For Aristotle agreed with Plato and the Eleatics that the object of knowledge must of necessity be fixed and immutable. If this were not so, he argued, our efforts to know, to be sure about something, would be doomed to perpetual frustration; at best we could attain only more or less likely opinions: 'The proper object of unqualified scientific knowledge is something which cannot be other than it is' (11).

The uniqueness of each individual living being is a biological fact which has in recent years become very common knowledge. Most of us are nowadays aware that the major difficulty holding up the advance of 'spare-part' surgery is the fundamental ability an organism has to recognise and reject foreign tissue. Biologists understand that this rejective power is but another example of the body's ability to recognise and reject disease-causing micro-organisms. Furthermore, this immunological defence

(11) *Ibid.*, 71b15.

is based upon an ability to recognise alien molecular *pattern*: the reaction is triggered not by size, or mass, or velocity, but by shape, conformation, pattern.

Aristotle, of course, knew nothing of the modern science of immunology, or of its adjuncts in organ transplantation, but his perception that the living organism is a unique pattern or vortex of matter is fully consistent with these insights of modern biology. Like the modern biologist he would argue that the pattern is more significant than the raw materials of which an organism is composed. Conceivably, too, Aristotle is thinking not only of the biological facts but also of our own introspective experience of ourselves. We always appear to ourselves, except in pathological cases, as single unitary individuals. Aristotle draws on these psychobiological realities to provide one of the 'basic truths' of his philosophy.

It is the *Metaphysics* which, to quote Werner Jaeger (12), 'expresses (Aristotle's) ultimate philosophical purposes', and, according to the same authority, 'any study of the details of his doctrine which does not start from this central organ must miss the main point'. Now the *Metaphysics*, according to its author, is 'about substance': 'Indeed the question which was raised of old and is raised now and always, and is always the subject of doubt, viz. what being is, is just the question, what is substance' (13).

Aristotle's answer to this question is set out in the central books of the Metaphysics: books 7, 8 and 9 (14). He starts out by saying that 'substance is thought to belong most obviously to bodies', and of bodies both the bodies of plants and animals and also those of the classical elements: fire, water, earth and air. After some analysis he concludes that it is the substratum of things which is 'in the truest sense their substance' (15). For it is this which is not predicated of anything but of which all other things are predicated.

However, the Aristotelian substratum is not quite our substratum. We, because we are all nowadays atomists, think immediately of hard, massy, impenetrable particles. Aristotle proceeds further. He asks, in effect, what is it that the atoms, or if you like, the subatomic particles, are composed of? If we take 'length, breadth and depth' away, what is it that we are left with? Yet quantities are not, as Aristotle says, themselves substances; on the contrary, substances are precisely that to which they belong, or of which they are predicable. If they are removed we seem to be left with something quite unknowable: 'Neither a particular thing, nor a particular quantity nor

(12) W. Jaeger, *Aristotle* (trans. R. Robinson), Oxford University Press (1934).
(13) *Metaphysica*, 1028b2.
(14) The *Metaphysics* has been called a 'desperately difficult work' (W. D. Ross). Modern scholars suggest that the text we have was assembled from notes taken down by students at the Lyceum and perhaps committed to memory for use in disputation and so on.
(15) *Metaphysica*, 1029a2.

otherwise positively characterised; nor yet the negation of these' (16).

Now we begin to see the biological colouring of his thought. For, according to Aristotle, one of the most important characteristics of substance is form, configuration. He returns to this notion in several other parts of his system. At the very outset of his course, in the first of the series of discussions on methodology—the *Categories*—he devotes considerable space to a consideration of 'substance'. 'Substance', we read once again, 'in the truest and primary and most definite sense of the word is that which is neither predicable of a subject nor present in a subject' (17). But in this work we are enlightened with a specific example of what the Stagirite has in mind: 'the individual man or horse'. This is what Aristotle means by a substance. Take a further example: Socrates is, in Aristotle's philosophy, irreducible; he cannot be predicated of anything else; other things, on the other hand, may be predicated of him. And throughout the rest of the discussion the reference is always the same. 'All substance appears to signify that which is individual' (18). And the most frequently quoted examples of such individuals are living individuals; a particular man, a particular horse, a particular ox.

Aristotle's concept of substance thus comes clear when we recognise that the paradigmatic case is the living individual, whether man or animal. Substance, for example, admits for Aristotle of no contrary. For, says he, what could be the contrary of an individual man or ox? Substances cannot vary in degree. Socrates can never be more or less than Socrates. 'A man', Aristotle writes, 'cannot be more or less a man either than himself at some other time or than some other man' (19). Substances are unique in that contrary qualities can be predicated of them. An individual, he writes, 'is at one time white another time black, at one time warm another time cold...' (20). The individual remains all the while unchanged, unique and unmistakable.

Returning to the *Metaphysics* we find Aristotle developing the idea that individuals cannot be defined (21). They are the elementary data into which our experience may be analysed and from which our theories may be synthesised. To define an individual is therefore impossible: they are irreducible to any more elementary concept. In the same way a modern physical scientist would not be able to 'reduce' a concept such as, for example, 'energy' into more elementary concepts.

But what is it which makes a thing a substance rather than, as Aristotle says, a 'mere heap' of parts? Why should we refer to some things as

(16) *Ibid.*, 1029a25.
(17) *Categoriae*, 2a10.
(18) *Ibid.*, 3b10.
(19) *Ibid.*, 3b35.
(20) *Ibid.*, 4a20.
(21) *Metaphysica*, VII, 15.

substances and to others as mere bits and pieces, without unity: mere potencies? Aristotle believes that the answer lies in the formal cause. It will be remembered from chapter 3 that Aristotle believed himself to have made a considerable advance over his predecessors in disentangling four meanings for the term 'cause', and that one of these meanings was that of the 'formal' cause—the form or pattern which makes a thing what it is. Aristotle explains what he means in the following passage:

> . . . the question is *why* the matter is some definite thing; e.g. why are these materials a house? . . . and why is this individual thing, or this body, having this form a man? Therefore what we seek is the cause i.e. the form by reason of which the matter is some particular thing and this is the substance of the thing (22).

Students of Plato will recognise that Aristotle has in the back of his mind the Platonic theory of forms—but with the difference that whereas for Plato the forms were static, immanent and eternal, for Aristotle the form is a causative agent. It is by virtue of this cause that a substance is what it is. The Stagirite illustrates his meaning by means of two examples. He points out that both a syllable and flesh have a unity which is not evident in their several parts. 'ba', he says, 'is not the same as "b" and "a", nor is flesh the same as fire and earth' (23). If the syllable or the flesh is analysed into its units then, although the parts, the elements, exist, the unitary whole, the individual syllable or flesh, has disappeared. The syllable or flesh thus, according to Aristotle's analysis, consists of something over and above its constituent parts. This additional 'something' is not just another part, another constituent, or else it, too, could be analysed into its units and so *ad infinitum*. We should be no nearer determining what it is which makes the flesh or syllable a unitary whole and not a mere congeries of parts. No, for Aristotle, the additional 'something' is an 'activity' which forms the otherwise inchoate matter into a specific identifiable pattern, a recognisable individual. 'It would seem', says Aristotle,

> that this 'other' is some thing, and not an element, and it is the *cause* which makes *this* thing flesh and *that* a syllable. And similarly in all other cases. And this is the *substance* of each thing (for this is the primary cause of its being) . . . (24).

Aristotle goes on to point out that substances may in this way be distinguished from mere parts or elements. For, he says, those things which are classified as substances are 'formed in accordance with a nature of their own', and it is this 'nature' which is to be identified with the Aristotelian substance. 'An *element*, on the other hand, is that into which a thing is divided and which is present in it as matter; e.g. "a" and "b" are the elements of the syllable.'

(22) *Ibid.*, 1041b5.
(23) *Ibid.*, 1041b15.
(24) *Ibid.*, 1041b25.

Substance as 'creative act'

The foregoing brief examination of Aristotle's most searching work—the *Metaphysics*—indicates how far Aristotle's deepest insight into the nature of things differed from the modern insight; indicates, indeed, how far it differed from the Abderan vision outlined in chapter 6. In the next chapters we shall see how the concept that substance is at the same time substratum and the cause of a thing's being what it in fact is, forms a foundation for the Stagirite's physics and physiology. We shall see that the insight which located the root of things in an activity, a causal act, leads to a radically different science from that with which we are nowadays familiar. Instead of a pre-Einsteinian explanation of things in terms of an inert and lumpish matter and an invisible, intangible energy, Aristotle sees a unitary 'energetic matter'—substance. In the last analysis being, for Aristotle, *is* the creative act in virtue of which a thing is what it is (25).

(25) See J. Owens, *The Doctrine of Being in the Aristotelian Metaphysics*, second edition, Medieval Studies of Toronto Inc. (1963), p. 470ff.

9 ARISTOTLE: PHYSICAL SCIENCES

The physical treatises

> We have already discussed the first causes of nature and of all natural motion, also the stars ordered in the motion of the heavens and the physical elements – enumerating and specifying them and showing how they change into one another – and becoming and perishing in general. There remains for consideration a part of this enquiry which all our predecessors called meteorology. It is concerned with events that are natural, though their order is less perfect than that of the first of the elements of bodies. They take place in the region nearest the motion of the stars. Such are the milky way and the comets and the movements of meteors. It studies also all the affections we may call common to air and water, all the kinds and parts of the earth and the affections of its parts .... When the enquiry into these matters is concluded let us consider what account we can give, in accordance with the method we have followed, of animals and plants both generally and in detail. When that has been done we may say that the whole of our original undertaking will have been carried out (1).

Thus Aristotle; carefully referring back and forth, subdividing knowledge, classifying, placing his proposed discourse in relation to what has gone before and to what has still to come, pointing out connections, putting before his readers the vast interlocking apparatus of his *episteme*. The 'original undertaking' to which he refers is nothing less than an account of the entire phenomenal world: an account, to use the words of a more recent philosopher, of 'all that is the case'.

The method of attack is to proceed from the most general ideas towards particularities (2). But these general ideas are not inherent or *a priori* in any way. They are the first ideas to present themselves, and only later are they subjected to critical analysis. Aristotle uses the example of children who at first call all men 'father' and all women 'mother' and only later distinguish a unique man as father and a unique woman as mother (3).

The physical treatises consist of four major works: *Physics, On the Heavens, On Generation and Corruption,* and *Meteorology*. In this chapter we shall concern ourselves chiefly with the *Physics*. This work itself consists of two parts. The first part (books 1 – 4) deals with 'nature' in general terms, the second part (books 5 – 8) discusses motion.

(1) *Meteorologica,* 338a20.
(2) *Physica,* 184a24.
(3) *Ibid.,* 184b10.

The subject matter of physics

For Aristotle, physics was the science of nature (4). What, however, did Aristotle mean by this term? Not, apparently, exactly what we mean today when we speak of nature (or 'Nature'). Throughout the *Physics* and in other works it seems that the connotation of this term is closer to that which we intend when we speak of 'human nature', 'a kindly nature', and so on. This more biological, more psychological connotation aligns Aristotle's physical treatises with his *Metaphysics* which, as we saw in the last chapter, has the same overtones.

Aristotle's discussion of nature occurs in the context of the Eleatic challenge. No classical philosopher could escape the Parmenidean argument. Certainly Aristotle, nurtured as he had been in the Platonic Academy, could not disregard this radical challenge to the very possibility of a *science* of nature. Indeed, the pressure of the Eleatic argument must have persistently nagged the mind of one who was at once the originator of logic and of many empirically-based sciences. Perhaps we can sense something of this pressure behind the somewhat exasperated references Aristotle makes to their theories:

> Their premises are false and their conclusions do not follow . . . accept one ridiculous proposition and the rest follows – a simple enough proceeding. We physicists, on the other hand, must take for granted that the things that exist by nature are, either all or some of them, in motion . . . (5).

This quotation in fact brings us straight to the heart of the Aristotelian physics. The essential first principle of nature, the subject matter of physics, is movement. It is the 'static' which is an abstraction: it is this which is unreal, not motion. Aristotle's endeavour is to save the phenomena from the Eleatic scepticism by reversing the argument. Instead of positing a changeless, spherical, eternal plenum as alone real, and the evidence of the senses as an illusion, Aristotle posits movement as alone real and the Eleatic *One* an abstraction. He points out that 'To prove what is obvious from what is not is the mark of a man who is unable to distinguish what is self-evident from what is not' (6).

Aristotle, the descendant of an Asclepiad family, is convinced that the Eleatics have hold of the wrong end of the stick. He agrees with Heraclitus that 'knowledge enters through the doors of the senses'. It is the Eleatic *One* which is illusory, not the reports of our eyes and ears. Accordingly he agrees with the haughty Ephesan that motion cannot just be legislated out of the world. It is of the essence. Yet, on the other hand, we have seen that for twenty years he had studied under Plato, that moreover he is the founder of the great subject of logic, and that consequently he is bound to believe that

(4) *Ibid.*, 184a15.
(5) *Ibid.*, 185a10.
(6) *Ibid.*, 190b5.

we can know only that which is unchanging. The great Abderans had found a way out of this seeming impasse. The Stagirite, as we have already seen and as we shall see in more detail later, could not accept their theory either. He tried for another solution.

Potential into actual

If motion was, for Aristotle, the most fundamental datum of experience, the datum with which physicists were concerned above all else, his concept of motion was nevertheless far broader than is usual today. It can be argued, indeed, that Aristotle treats all movement as instances of the biological paradigm of organic growth:

> The fulfilment of what exists potentially, in so far as it exists potentially, is motion – namely, of what is alterable *qua* alterable, *alteration;* of what can be increased and its opposite what can be decreased (there is no common name), *increase* and *decrease;* of what can come to be and can pass away, *coming to be* and *passing away:* of what can be carried along, *locomotion* (7).

Nowadays, of course, we tend to look at things from the other end. Since the seventeenth century the Democritean world-view has been predominant. We see organic growth and diminution, increase and decrease, birth and death as the outcome of the 'locomotion' of submicroscopic particles. These particles tend also to be thought of as Newtonian atoms blindly running, bouncing, colliding. This, it can be appreciated, is very far from the Aristotelian position.

The Stagirite 'sees' translocation as just one species in the genus of the potential becoming actual. A body in movement is first *potentially* and later *actually* at the finishing line. The teleology implicit in this view is clear. It persisted well into the sixteenth century AD. The famous algebraist Tartaglia believed that a falling body speeds to its appointed end much as a traveller accelerates at the end of his journey (8). Circular movement, however, escapes this neat analysis and Aristotle, as we shall see, reserves a very special place for it in his philosophy.

The analysis of movement is thus, as we have just seen, in terms of potentiality and actuality. These are, accordingly, very important concepts in the Aristotelian scheme of things. They receive their fullest exposition in the *Metaphysics*. In book 9, Aristotle presents a number of examples through which we can grasp his meaning. The examples are to do with house-building, with the carving of statues, with teaching, with doctoring, with embryological development: a very 'unphysical' set of examples, we may nowadays think. They all reflect Aristotle's deeply felt conviction that 'Nature does nothing in vain'. He writes as follows:

(7) *Ibid.*, 201a10.
(8) See S. Drake and I.E. Drabkin, *Mechanics in Sixteenth Century Italy,* University of Wisconsin Press, Madison, Milwaukee and London (1969), p. 20.

> Actuality is to potency as that which is building is to that which is capable of being built, and the waking to the sleeping, and that which is seeing to that which has its eyes shut but which has sight, and that which has been shaped out of matter to matter, and that which has been wrought up to the unwrought (9).

In the same way the terms actuality and potentiality may be applied to the fully carved statue and the wood or stone from which it has been carved; or to the man of science who is still potentially a man of science even when taking part in athletics simply because he remains all the time capable of setting himself to study.

Meditation on art, on biology, on the workings of the conscious mind have led Aristotle to this account of change or movement—a set of examples radically at variance with the familiar billiard-ball paradigm which lies at the back of the common philosophy today. In all Aristotle's examples there seems to be a development towards an end: 'that for the sake of which...'. The embryo chick still housed in its egg is potentially a rooster; the block of stone in the sculptor's yard is potentially a statue; the mind knows itself to be drawn on by the hoped-for end.

To moderns Aristotle's system can seem a mere playing with words. Certainly a pile of bricks may potentially be a building. So what? This seems to offer no sort of an explanation of the process which leads to the completion of the building. But Aristotle's mind worked within a totally different system of concepts from ours. The analysis of change in terms of potentiality and actuality links in with other parts of his system to form an organic and convincing account of the world, a system which many still see as the only genuine alternative to the Democritean.

The embryological paradigm

Aristotle connects the potentiality/actuality analysis to the rest of his philosophy in the following way. He proposes first of all that in an important sense the *actual* is prior to the *potential.* The chicken, in short, comes before the egg. It is impossible, says Aristotle, to build, to write, to play the harp, or to study nature if one is not first a builder, a writer, a musician or a scientist. The actuality of being a philosopher, for example, must obtain before the building of a philosophical system can commence. In this sense the end governs the means. Aristotle writes:

> The man is prior to the boy and the human being to the seed; for the one already has its form and the other has not, and because everything that comes to be moves towards a principle i.e. an end (for that for the sake of which a thing is, is its principle, and the becoming is for the sake of the end), and the actuality is the end, and it is for the sake of this that the potency is acquired (10).

The actuality, for Aristotle, is thus the fully formed adult to which the juvenile stages point. As a close student of embryology who had examined

(9) *Metaphysica,* 1048b1.
(10) *Ibid.,* 1050a5.

many stages of the chick's development within the egg, Aristotle could not but be a profound teleologist. Indeed he explicitly makes fun of those who suggest that the intricate forms of the living body arise by chance and coincidence. We have already noticed (chapter 7) how he ridicules the ideas of Empedocles who had suggested that all possible combinations of organs were tried out during evolution and that the ineffective ones, the 'man-faced ox-progeny' and the 'olive-headed vine-progeny', became extinct. Although we nowadays take a deepened version of this idea very seriously indeed, it was to Aristotle absurd (11). It seemed obvious to Aristotle, the biologist, that 'teeth and all other natural things either invariably or normally come about in a given way'. This could not be just due to random 'chance' and 'coincidence'; it was inconceivable that we should each possess incisors for cutting, canines for tearing, molars for grinding, and so on, merely by chance.

Aristotle does not, of course, overlook the occasional development of monsters. He explains the origin of such unfortunates in a very modern way. He suggests that the developmental process becomes in some way disorientated or misdirected. Today we might suspect the lack of an enzyme whose presence was essential to the orderly sequences of the embryological process. Aristotle makes use of the same general type of explanation:

> Mistakes come to pass even in the operation of art . . . the doctor pours out the wrong dose . . . . If then in art there are cases in which what is rightly produced serves a purpose and if where mistakes occur there was a purpose in what was attempted only it was not attained, so must it be also in natural products, and monstrosities will be failures in purposive effort (12).

These deeply understood embryological instances impel Aristotle to take morphology as primary and physiology as secondary—an epiphenomenon. It is the form which genuinely activates the process: it is the adult form which draws the juvenile out. It is the actuality which breaks the chrysalis of the merely potential. It causes the fulfilment of that which exists potentially: 'the fulfilment of what exists potentially, in so far as it exists potentially, is motion . . .'.

We can now begin to see how this set of concepts, this vision of nature, ties in with the central concept of the *Metaphysics*: substance. It will be recalled that we concluded chapter 8 with the idea that substance, a thing's essence, is itself an *activity*, the creative activity which ensures that an entity remains what it in fact is. We saw that the idea of form and substance were closely allied. This alliance becomes more striking when we remember that, for Aristotle, to understand something was to understand its causes (13) and that one of the causes which figured large in the Aristotelian scheme was the formal cause.

(11) See *Physica,* 2, 8.
(12) *Ibid.,* 199b1.
(13) *Metaphysica,* 982a1: 'Philosophy, wisdom, is a knowledge of causes'.

The embryological analysis of the concept of cause

In chapter 3 some account of Aristotle's notion of cause was given. It will be remembered that in place of the modern unitary concept the Stagirite recognised four different aspects of causation: the material, the efficient, the formal and the final. Now, so far as things in motion are concerned—and this, it will be recalled, is what Aristotle considered physics to be all about—the last three of these 'causes' tend to coincide: 'for the "what" and "that for the sake of which" are one, while the primary source of motion is of the same species as these (for man generates man)' (14).

This line of argument is difficult to follow if we do not allow Aristotle his biological orientation. If we try to use the Aristotelian framework to account for the expansion of an ideal gas or for the movement of a projectile, we shall be in difficulty; if, however, we use it to account for the embryological development of an organism or the moral endeavour of a man, the fit is convincing.

It has already been stressed that the embryo and juvenile are such only because they point towards the adult. If the juvenile becomes, for some reason 'fixed', as in the axolotl or in other neotenous forms, then it can, by definition, no longer be regarded as a juvenile. Thus, argues Aristotle, the form of the adult is the 'why' of the developmental stage: it is both the final cause in point of time and the formal cause in point of shape. Similarly, continuing with the embryological example, the efficient cause is to be closely identified with the formal and final causes. This cause, which unlike the other two acts externally, initiates the whole process. But it resembles very closely the adult morphology which, as we have just seen, is both formal and final cause. For the act of fertilisation must be carried out, as our biologist – philosopher was well aware, by a member of the same species—by a member, moreover, closely similar to the adult form of the juvenile. 'The father', says Aristotle, 'is the cause of the child' (15). Humans propagate humans, oxen oxen and vines vines. The form recurs generation by generation without mistake.

The conflation of formal, final and efficient causes is also to be found in moral activity. 'The what' and 'that for the sake of which' are obviously closely allied, if not identical, for the plan present in the mind is both 'formal' and 'final' in the sense of being the reason for acting. The efficient cause, that which initiates rather than guides the activity, is in this case internal rather than external. It is the optative: 'oh, that *this* might be', and hence fully identical with the formal and final causes already mentioned.

Thus we may conclude that for Aristotle the paradigm of movement, of change, is not the movement of a projectile or the expansion of an ideal gas, but the development of an organism, the 'movement' of an animal or human

(14) *Physica,* 198a25.
(15) *Ibid.,* 195a30.

towards what it considers to be its 'good'. This is very interesting for it seems that nowadays we have hold of quite the other end of the stick. The gathering intellectual movement since the time of Galileo has taken the motion of the projectile as paradigmatic and sought to explain the development of organisms and the moral behaviour of men in its terms; the Aristotelian, on the other hand, seeks to explain the movement of a projectile in terms of a paradigm taken from the biological realm. It has often been noted that the Aristotelian physics is really biology in disguise; could it be that modern psychology makes the equal and opposite mistake?

A voidless cosmos

In the last sections of this chapter, before discussing Aristotle's theoretical biology in chapter 10, the Stagirite's treatment of locomotion and the flight of a projectile is outlined. This was always the weakest link in the system and was, consequently, the link through which Aristotelianism was broken in the sixteenth and seventeenth centuries AD.

We have noticed that all post-Parmenidean philosophers had of necessity to take some position *vis-à-vis* the 'One'—for if the reports of the senses were simply illusory, it was nonsensical to pursue seriously the empirical study of nature: what was 'really' 'out there' was not in any respect at all what 'appeared' to be 'out there'. Leucippus and Democritus had saved the phenomena by accepting what seemed to be a logical impossibility—the existence of a void. They proposed, so it seemed to the Greeks, that what was not nevertheless, in some way, *was*. This also seemed intolerable to the Stagirite.

Aristotle believed that there were no 'gaps' in our existential experience, just as there are no lacunae in the animal body. There is an all-pervading process. The biological realm, moreover, is a realm of order. A mind which, as we saw on page 71, found beauty in organic forms where others felt only distaste, could hardly feel at home among the Abderan pandaemonium of atoms. For Aristotle, the world was in the original and true sense of the terms a *universe*, a *cosmos* (16). Everything had its appointed place and there was a place for everything. Everything, indeed, according to the Stagirite, had a *natural* place (17). It was thus possible not only to displace things in the natural world but also to *misplace* them. In both cases the *natural* movement of the object is back towards its 'appointed' place. *Vice versa*, if an object was observed to move from its natural place, it was necessary to look for a cause. We shall see that this movement not only did not necessitate a void, it proved the exact opposite: it proved that a void could not exist! Nature, for Aristotle,

(16) See A. Koyré, Galileo and Plato, *J. Hist. Ideas,* 4 (1943), 400–28.
(17) The notion that all things have a 'natural' place is perhaps to be associated with Aristotle's strong classificatory instincts. His analysis of motion in terms of displacement from, and return to, a 'natural' place has also, of course, strong teleological implications.

unlike the modern physicist, abhorred a vacuum.

Aristotle's account of translocation depends on his notion of 'place'. His argument is to be found in book 4 of the *Physics*. He asks first, 'does place exist'? Answering this in the affirmative, he goes on to question what it is: is it matter or form? Place seems to contain matter, so it cannot itself be matter; on the other hand it is thought to be separable from matter, so it cannot be form. He concludes that place seems to be coincident with the boundaries of a thing. He winds up by pointing out that were it not for locomotion, the movement of a body from one place to another, the concept of place would have been unnecessary. That movement occurs is, however, for Aristotle incontrovertible and so he ultimately defines 'place' in the following way: 'If a body has another body outside and containing it, it is in a place and if not, not' (18).

This is a pregnant definition. It implies that the world is, for Aristotle, dense. His universe is a plenum, crammed like a cubist painting with closely adposed interlocking bodies; for everything must be in a place, and this entails by his definition that everything is tightly surrounded by other bodies. This vision is very much unlike that of contemporary physicists. Readers of Sir Arthur Eddington will recall that the physicist's table is very largely void with a few insignificant centres of matter; mention has already been made of the astronomer's universe: a vast space with here and there comparatively minute amounts of matter.

Returning, however, to Aristotle we find that the Stagirite, having established to his own satisfaction a definition for place, feels himself in a position to tackle questions about the void. After considering the opinions of his predecessors for and against its existence he sums up as follows:

> Since we have determined the nature of place and void must, if it exists, be place deprived of body, and we have stated both in what sense place exists and in what sense it does not, it is plain that on this showing void does not exist either unseparated or separated; for the void is meant to be, not body, but rather an interval in body (19).

Aristotle thus believes himself to have neatly demonstrated the impossibility of the void. Moreover, he believes himself to have demonstrated its impossibility in such a way as to completely undermine the Abderan position. In contrast to the atomists, who insisted that the hypothesis of a void was necessary in order to save the phenomena of movement, the Stagirite now goes on to propose the exact opposite. The phenomena of movement could only be saved if a void did *not* exist. This was a brilliant tactic. The atomists had thought the unthinkable in order to save movement from the Eleatic scepticism. Now their position was attacked from the rear: the hypothesis they had erected in order to validate experience was shown to do no such thing.

Aristotle's argument consists of two parts. First he maintains that it is

(18) *Physica,* 212a32.
(19) *Ibid.,* 214a16.

simply fallacious to suppose that it is necessary to posit a void in order to account for movement. He points out that what is 'full', containing no void, can easily undergo qualitative change. Furthermore he asserts that even movement with respect to place does not entail that there be an interval between things, 'for', he says, 'bodies may simultaneously make room for one another' (20). We shall examine the implications of this latter concept more fully later.

But this is only the first half of Aristotle's argument, for he wishes to show that the Abderan hypothesis is not only unnecessary but also vicious. He wishes to show, in other words, that the existence of a void is not only unnecessary to save the phenomena but could *not* in fact do so. He wishes to show that movement in a resistive medium is the only type of movement conceivable.

He rests his argument on the observation that the void, being 'non-existent and a privation of being', cannot have differentiated parts. In other words, there cannot be different *places* in a void: for how would they differ? Hence a body put into a void, if this is even possible—and it would seem on the Aristotelian analysis that it is not—cannot move, for it cannot exchange one place for another. He goes further. He makes use of the idea that bodies have natural places, an idea which, we have already seen, he strongly supports. The natural place of fire is in the empyrean, that of earth is beneath one's feet: the natural movement of fire is thus upwards and of earth downward. But in the void if there is no place there is, *a fortiori*, no *natural* place. There is no differentiation, no up or down, no middle. 'Either', he concludes, 'nothing has a natural locomotion or else there is no void' (21). And Aristotle firmly believed in the validity of natural locomotion: indeed, the concept is fundamental to his system.

It is very interesting to note, as Sir Thomas Heath points out in his posthumous book on Aristotle's mathematics (22), that the remainder of the Stagirite's argument against the void, read in the opposite sense to its intention, closely prefigures the Galilean analysis:

> We see that bodies which have a greater impulse either of weight or of lightness, if they are alike in other respects, move faster over an equal space, and in the ratio which their magnitudes bear to each other. Therefore they will also move through the void with this ratio of speed. But this is impossible; for why should one move faster? (In moving through plena it must be so; for the greater divides them faster by its force. For a moving thing cleaves the medium either by its shape, or by the impulse which the body that is carried along or is projected possesses.) Therefore all will possess equal velocity. But this is impossible (23).

Quite so: why indeed should one move faster than the other? This is the point of the perhaps apocryphal tale of the weights dropped before passing Aristotelians in seventeenth-century Pisa.

(20) *Ibid.,* 214a29.
(21) *Ibid.,* 215a13.
(22) T. Heath, *Mathematics in Aristotle,* Oxford University Press (1949).
(23) *Physica,* 216a12.

Again, in another passage, we read:

> . . . no one could say why a thing once set in motion (in a vacuum) should stop anywhere; for why should it stop *here* rather than here? So that a thing will either be at rest or must be moved *ad infinitum* (24).

Exactly. What for Aristotle seemed an ultimate *reductio ad absurdum*, by which the Abderans were exposed as the propagators of a false theory, was taken by Galileo and later even more explicitly by Isaac Newton as one of the core concepts of the new physics. 'Everything proceeds at rest or in rectilinear motion unless acted upon by a force.' Newton's first law does away with the necessity, the Aristotelian necessity, of positing *external* movers, replacing them with an 'internal' force: *inertia*.

It is clear, therefore, that the Stagirite's deepest intuition about the world was that it was a well-ordered and closely structured whole. There were no intervals, no vacuities, no 'fallings from us, vanishings...'. The universe was a concrete and pressing reality. And yet, of course, the most fundamental datum of experience for the Stagirite was, precisely, movement. He is utterly scornful of those Eleatics who, like Melissus, argued that because a void could not exist, movement was therefore impossible and the reports of our senses consequently fallacious.

Aristotle had thus to explain how movement could occur in a resistive medium. We have already seen that he points out that no difficulty arises in the case of qualitative change, and that so far as locomotion is concerned, 'bodies may simultaneously make room for one another'. We can perhaps catch Aristotle's meaning if we think of, say, a submarine slipping through the ocean. The water gives way ahead and closes again around the stern. The motive power, of course, is generated by the submarine's engines. Aristotle recognises the force of this point and writes as follows:

> Everything that is in motion must be moved by something. For if it has not the source of movement in itself it is evident that it is moved by something other than itself, for there must be something else that moves it (25).

The projectile

Aristotle's explanation of locomotion does, however, lead him into one very difficult corner, for it makes it almost impossible to account for the movement of a projectile. The Stagirite's explanation of this commonplace phenomenon bedevilled dynamics for millennia. He proposes that the air in front of a projectile is displaced, swirls round and impels the projectile from behind. In this way he thinks to get around the obvious fact that projectiles, by definition, do not possess a source of movement within themselves. This proposition, known as antiperistasis, was so obviously absurd that it was satirised throughout the ages until it finally dropped out of science altogether

(24) *Ibid.*, 214a12.
(25) *Ibid.*, 242b25.

in the sixteenth century AD.

Aristotle's explanation of movement in a resistive medium fits into yet other parts of his system. In the background is the ancient idea, strongly held in the Lyceum, that circular motion is more perfect than any other. Only this type of motion, returning as it does to its starting point, can image the eternal (26). And, ultimately, in a compact well-ordered universe, such as Aristotle envisaged, circular movement is the only type possible: a type of movement where, as he writes, an object moves from its allotted place in a circle back to that place. Movement in a straight line is unnatural and violent. The projectile is thus, on this count also, something of a cuckoo in the carefully wrought Aristotelian nest.

Nevertheless, for many everyday commonsense instances of locomotion, the Aristotelian analysis, requiring that 'everything that is in motion must be moved by something', is very attractive. The Stagirite's position is admirably illustrated by Duhem in his great work on the history of science, ***Système du Monde***:

> At the Piraeus Aristotle observes a group of haulers; their bodies bent forward, they pull with all their might on a rope attached to the prow of a ship. Slowly the boat approaches the shore with what appears to be a constant speed. Other haulers arrive and take hold of the rope in addition to the first group; the vessel now cuts the water more rapidly than it had before. But suddenly it stops. The keel has struck the sand. The men who were sufficient in number and strength to counteract the resistance of the water cannot overcome the friction of the keel on the sand. To pull the boat up on the shore they need reinforcements (27).

As Duhem observes, the Aristotelian system seems to account for this and similar everyday occurrences very nicely. It is, perhaps, small wonder that the Aristotelian dynamics gained widespread acceptance. Indeed, as Alexandre Koyré points out, 'Aristotelian physics... forms an admirable and perfectly coherent theory which, to tell the truth, has only one flaw (besides that of being false): that of being contradicted by everyday practice, by the practice of throwing' (28).

(26) See chapter 5, p. 38; Aristotle, *De Caelo,* 269a20ff: 'The circle is a perfect thing ... circular motion is perfect and divine.' For Plato circular motion was 'the moving image of eternity.'

(27) Quoted by Sir David Ross in the commentary to his translation of the *Physics,* Oxford University Press, pp. 31–2.

(28) Koyré, *ibid.,* p.411.

10 ARISTOTLE: LIFE SCIENCES

Biological research at the Lyceum

We have stressed throughout these last two chapters that the Aristotelian system is permeated with biological insight and founded on what seems to us biological intuition. This is very far from being a novel observation. Charles Darwin, in a well-known passage, apostrophises Aristotle thus: 'Linnaeus and Cuvier have been my two gods, though in very different ways, but they were mere schoolboys compared with old Aristotle' (1). Indeed this aspect of Aristotle, that he was, to paraphrase Marjorie Grene (2), the only great philosopher to philosophise out of a deep background of biology, is the reason why three chapters of this book are devoted to his thought.

There is no need to rehearse at length the encyclopaedic range of Aristotle's biological endeavour. The vast mass of observational detail upon which the Stagirite's theoretical biology is based is collected in the *History of Animals*. Indeed this undertaking seems too great to have been the work of only one mind or one pair of hands. It seems more probable that this gigantic compilation, like the collection of 158 constitutions and the complete list of winners at the Pythian games, was the output of a research school. In addition to the encyclopaedic *History of Animals* and Theophrastus' companion volumes on botany, there is evidence that research into medicine, physiology and anthropology was carried out at the Lyceum. Menon, one of Aristotle's associates, was commissioned to write a history of medicine, and the Stagirite's work is full of reference to the Hippocratic writings which were being put together at that time on the other side of the Aegean. More than fifty different animals are believed to have been dissected at the Lyceum, and an anatomical work, perhaps used as a dissection guide, is thought to have existed (3).

This brief résumé of the biological interests of the Lyceum shows that the Peripatetics had abundant hard fact on which to base a theoretical biology. Theirs was no armchair speculation: the question of old, as of today, cried out from dissection table and physiological lab—'What is life?'

(1) *Darwin's Life and Letters* (ed.F.Darwin), vol. 3, John Murray, London (1888), p.252.
(2) M. Grene, *A Portrait of Aristotle,* Faber, London (1963).
(3) See W. Jaeger, *Aristotle* (trans. R. Robinson), Oxford University Press (1934).

The concept of nature

We have already noticed that Aristotle's system is itself rather like an organism. The parts are all interdependent, and interlock to form a functional whole. Thus the exegete is in somewhat the same position as the analytical biologist. It is difficult to know quite where to start, which part to isolate first from the cooperative system. And, as with analytical biology, the study of isolated parts can be criticised on the grounds that they have been placed in unnatural conditions, cut off from their normal milieu. One way of overcoming this difficulty is to examine them from as many different angles, in as many different lights, as possible. Thus one way to gain an entry into Aristotle's metabiology is to pick up and examine from a slightly different angle one of the concepts introduced in the last chapter: the concept of nature.

It will be remembered from chapter 9 that the Peripatetic concept of nature was somewhat different from our own. It was, we saw, closely connected to the Stagirite's wide concept of movement or activity. Thus we may read:

> For nature is in the same genus as potency; for it is a principle of movement – not, however, in something else but in the thing itself, *qua* itself (4).

This passage expresses a concept which must by now be very familiar, a concept which seems to run as a *leitmotiv* throughout the Aristotelian system. 'Natural things' possess within themselves, as a defining characteristic, a principle capable of initiating movement.

Now we saw in chapters 8 and 9 that the concept of movement had, for Aristotle, a very wide denotation. In the *Categories*, for example, we read that 'there are six sorts of movement: generation and destruction; increase and diminution; alteration; change of place' (5). These sorts of change may be grouped as substantial change; quantitative change; qualitative change; and change of place or locomotion. Aristotle regards all things which possess the innate ability to initiate one or other of these changes as belonging to his category of 'natural things'. We find this concept clearly stated at the outset of book 2 of the *Physics*:

> . . . of things that exist, some exist by nature, some from other causes. 'By nature' the animals and their parts exist, and the plants and simple bodies (earth, air, fire and water) – for we say that these and the like exist by nature (6).

All the things mentioned possess a feature in which they differ from things which are not constituted by nature. Each of them, according to the Stagirite, has *within itself* a principle of motion and stationariness (in respect of substantial, quantitative, qualitative or positional change). On the other hand, a bed, a coat or anything else of that sort—in so far as it is a product of art—has no innate impulse to change. 'But in so far as they happen to be

(4) *Metaphysica,* 1049b6.
(5) *Categoriae,* 15a10.
(6) *Physica,* 193a15.

composed of stone or of earth or of a mixture of the two, they *do* have such an impulse and just to that extent.'

This is clear enough and harmonises well with the views expressed in other parts of the Stagirite's work. The philosopher next, after a short digression in which he asserts that the existence of 'nature' is an obvious, irreducible and axiomatic fact of experience, asks the obvious question: what, in a natural body, do we identify with its 'nature'? Some of his predecessors, he observes, took the material substratum of a natural body to be its nature. A bed, they would have been inclined to say, if allowed to rot away might later send up shoots. But these shoots would develop into wood and not into another bed! The true nature of a bed is thus wood; the bed is merely the configuration or pattern forced upon the wood by the carpenter and hence merely accidental (7).

Aristotle, as would be expected, does not favour this view. He finds it very odd to equate the nature of a bed with the material of which it is made. He is, of course, inclined to take the form of an object to be its 'nature'. This, after all, is what distinguishes a *bed* from a *table* both of which may be wooden. He writes as follows:

> The word 'nature' is applied to what is according to nature and the natural in the same way as 'art' is applied to what is artistic in a work of art. We should not say in the latter case that there is anything artistic about a thing if it is a bed only potentially, not yet having the form of a bed; nor should we call it a work of art. The same is true of natural compounds. What is potentially flesh or bone has not yet its own 'nature' and does not exist 'by nature', until it receives the form specified in the definition, which we name in defining what flesh or bone is (8).

In this passage we meet one of Aristotle's commonest procedures: the comparison of the world of nature with the world of art and craftsmanship. The paradigm he has in mind is the activity of the artist, or carpenter, or potter, in impressing form on the formless. The analogy is made quite explicit in *The Parts of Animals* where he writes without qualification that 'just as human creations are the products of art, so living objects are the products of an analogous cause' (9). Nowadays, since Descartes, we use a different paradigm. We still compare the bodies of living organisms to artefacts, but to engineering artefacts: to organs, to watches, to computers. More importantly, we no longer bear in mind that a watchmaker or an electronic engineer was responsible for building the paradigmatic artefact. In a sense the Platonic vision has triumphed over the Aristotelian: the Stagirite's active formative cause has vanished from the intellectual scene.

Returning, however, to the Lyceum we see that it is the form of a certain class of objects which Aristotle takes to be identical with their 'nature'. We have already seen how deeply the Stagirite responded to the beauty and fitness for purpose of living things:

(7) *Ibid.*, 193a15.
(8) *Ibid.*, 193a31.
(9) *De Partibus Animalium*, 641b15.

> We should venture to study every kind of animal without distaste . . . . Absence of haphazard and conduciveness of everything to an end are to be found in nature's works in the highest degree and the resultant end of her generations and combinations is a form of the beautiful (10).

Thus we should probably not be very far out if we assume that it is the forms of living animals that he has in the forefront of his mind when he speaks of 'nature'. It will be recalled from the last chapter that the Peripatetics believed that the form of an adult animal was an important cause of its being what it in fact was. We saw that 'form' or 'morphology' was regarded as active—a 'source or cause of movement'. And this, of course, is what Aristotle originally defined 'nature' to be.

It is clear that the Stagirite's concept of 'nature' is closely allied to his concept of 'substance' which we watched him define in the *Metaphysics*. Whether we enter the Aristotelian system through the analysis of substance in the *Metaphysics* or of nature in the *Physics*, the result is the same. We see that the guiding principles of the system consist in an interlocking series of notions which appear to us latter day mechanists to be derived from a study of embryology, craftsmanship and introspective psychology. The form is the cause of an entity being what it in fact is: it exists before the entity itself exists (compare the Socratic account of the soul, chapter 7) and draws it forth from inchoate potentiality into full actuality. As we noticed at the outset Aristotle's whole system is extremely closely articulated. It resembles, indeed, his own vision of the natural world, a vision of a world in which everything has its appointed place and appointed function—a vision, in fact, of a world-organism where everything happens for a purpose, for the good of the whole. In an important passage at the end of chapter 8 of the second book of the *Physics*, Aristotle sums up his position: 'The best illustration is of a doctor doctoring himself: Nature is like that' (11).

The doctor possessing the art of medicine is able to use it with the purpose of curing his own indisposition. Nature is like that. Natural things have within themselves a principle which works on the material of which they are composed with an end in view. This self-reflexive aim, this purpose which moulds the material substratum, is the *good* of the natural object.

Shorn of its reference to a bygone metaphysics this conclusion is, in fact, strikingly modern. Nowadays we talk of homeostasis and negative feedback and think of the self-governing mechanisms which are basic to our technology. But what is this but 'a doctor doctoring himself'? Moving outside the field of physiological and biochemical regulations, we can meet the same concept in ecology. Consider the following sentence written by an eminent contemporary ecologist: '...this self-repairing, constructive process of nature represents a type of equilibrium that approximates an open steady state' (12). Again, what is this but 'a doctor doctoring himself'? Aristotle's vision of the

(10) *Ibid.*, 640a23.
(11) *Physica*, 199b30.
(12) P.B. Sears, Utopia and the living landscape, *Daedalus*, 94 (1965), 474–86.

living world is remarkably similar to ours: it is only that it appears through the screen of a metaphysics radically different from our own (13).

Aristotle on the soul

Very closely allied to Aristotle's concept of nature is his concept of soul. Indeed, it does not seem that any thorough-going distinction can be drawn except that the soul is rather more explicitly a characteristic of living things. Nature, it will be recalled, was also an attribute of the four classical elements and other aspects of the world which are not commonly regarded as being alive. But the soul, we read in the first paragraph of the work which Aristotle devoted to it, is 'in some sense the principle of life' (14).

If this is so, the soul is the central concept of biology and its critical analysis must form the core of any metabiology. Indeed, the history of the accelerating erosion of this concept is, in one of its aspects, the history of the transformation of biology into a complicated physics and chemistry. This, as we shall see, forms another of the central themes of the present book.

Aristotle fully recognises the importance of a study of the soul (15), but he is under no illusions about the difficulty of the undertaking: 'To attain any assured knowledge about the soul is one of the most difficult things in the world' (16). Some authorities have indeed detected an evolution of the Aristotelian concept of the soul from an early through a middle to a mature doctrine. It is suggested (17) that the earliest view expressed in the *Eudemus* is an elaboration of the two-substance view favoured by the Platonic Academy of the Stagirite's youth (see chapter 7). This view, in fact, still crops up in *On the Soul*, a work which is generally considered to have been composed late in Aristotle's career. Succeeding this early view is the notion of the soul as the 'source of movement'. This view permeates a great deal of Aristotle's biological work and, in the guise of 'nature' or 'substance', is to be found, as we have already seen, throughout his system. Finally, it is said, this concept shades into a mature and subtle theory in which the soul is regarded as the 'actuality' of the body. A close study of the Stagirite's works suggests, however, that the latter two positions are indistinguishable. We have already seen how in the *Metaphysics* and in the *Physics* 'form' is equated with 'actuality', and this in turn equated with motive 'cause'.

In *On the Soul*, Aristotle starts his account in his customary manner by reviewing the ideas of his predecessors. Thus we read of the atomists who likened the spherical atoms of the soul to the motes of dust dancing in a sunbeam: for, they said, just as the latter are always in motion, so must be the

(13) The rediscovery of the notion that 'nature' is a 'self-repairing constructive process' is, it should be noted, quite recent. It is unlikely that the biological orthodoxy of the late nineteenth century would have seen it in this way (see chapter 19).
(14) *DeAnima,* 402a5.
(15) *Ibid.*, 402a1.
(16) *Ibid.*, 402a10.
(17) See M. Nuyens, *L'Evolution de la psychologie d'Aristote,* Louvain (1948).

soul-atoms, for it is the soul which initiates movement in the animal body. On the other hand, Aristotle points out, a second group of early psychologists were more impressed by the observation that what has soul in it perceives than by the observation that it moves. Many of these early thinkers, as we have noticed in earlier chapters, held the belief that like is perceived by like. In consequence, like Empedocles, they fashioned the soul from mixtures of the elements. An extension of this view, says Aristotle, is held by those who believe that the soul is a kind of harmony. Aristotle discusses and criticises all these earlier theories and sums up a little tartly: 'Let the foregoing suffice as our account of the views concerning the soul held by our predecessors; let us now dismiss them and make as it were a fresh start, endeavouring to give a precise answer to the question: what is soul'?(18).

Aristotle begins immediately. His answer is succinct and closely reasoned. It is based on the interlocking system of ideas developed in his logical, physical and metaphysical works. He believes that it synthesises in a new and more powerful scheme the gropings of the Presocratics which he has collated in his preliminaries.

His initial move stems from the *Metaphysics*. He brings out his carefully worked concept of substance. He reminds us that substance is the essential 'this-ness' of a body or, to use Marjorie Grene's transliteration, the 'being-what-it-is' of a body. In this sense, as we have already seen, substance is closely allied to 'form'; or, to use the Aristotelian terminology, substance is the 'actuality' of a body whereas matter is merely a body's 'potentiality'. Aristotle weaves these concepts into his account. The argument is so close and compact that it must be quoted in full:

> Of natural bodies some have life in them and others not; by life we mean self-nutrition and growth (with its correlative, decay) . . . But since it is also a *body* of such and such a kind, vis. having life, the *body* cannot be soul; the *body* is the subject or matter, not what is attributed to it. Hence the soul must be a substance in the sense of the form of a natural body having life potentially within it. But substance is actuality and thus the soul is the actuality of a body as above characterised (19).

Aristotle's meaning is quite clear. It is summed up in Edmund Spenser's famous line: 'For soul is form and doth the body make'.

Aristotle, however, wants to go further still and make his concept quite unambiguous. He points out that all the bodies which we are accustomed to call alive are complicated and highly organised entities. Thus to obtain a completely unambiguous definition of the soul Aristotle asserts that it is 'the actuality of a natural organised body'. In this way he distinguishes soul from nature which, it will be remembered, is the actuality not only of living bodies but also of the four classical elements.

It is clear that according to this formula soul and body are inseparable. Aristotle has escaped the intolerable paradoxes of the two-substance theory:

(18) *De Anima,* 412a1.
(19) *Ibid.,* 412a17.

where and how can soul act on body, and *vice versa*? (20) The soul is the 'essential whatness' of an organised body. In a famous passage, Aristotle writes:

> Suppose an axe were a natural body, its 'essential whatness' would have been its essence and so its soul; if this disappeared from it, it would have ceased to be an axe. . . . As it is, it is just an axe; it wants the character which is required to make its whatness or formulable essence a soul; for it would have had to be a natural body of a particular kind, viz. one having *within itself* the power of setting itself in motion or arresting itself (21).

This is a very illuminating passage. It summarises succinctly some of the points we have been labouring. Soul is form, but it is also causal—it has the power to initiate movement. The self-consistency of the intricate system of concepts outlined in previous pages is clear.

It is also clear that Aristotle's mature theory of the soul is very materialistic. It is very far from the orthodox dualism implied by the thanks offered by many ancients for not having been born, for example, a woman, a barbarian or an animal (Socrates); that view is exemplified in Plutarch's report of Empedocles' words: 'Since the soul has come hither from elsewhere he (Empedocles) euphemistically calls birth a sojourn abroad—the most comforting of all names; but in truth the soul is a fugitive and a wanderer banished by the decrees and laws of the gods' (22). For the mature Aristotle this idea is an absurdity. Soul and body come into existence and cease to exist together. It must be borne in mind, however, that when we use the term materialistic of Aristotle's position, his materialism is very different from ours. The common philosophy today sees a wide gulf fixed between the nature of the microcosm and the nature of the macrocosm. That this was not the case for the Peripatetics surely needs by now no further labouring. Viewed in relation to the rest of the Aristotelian system, the theory put forward in *On the Soul* can be seen as a precise and consistent answer to the question, 'What is life'?

Subdivisions of the soul

For the alumni of the Lyceum, as we have seen, 'what has soul in it differs from what has not in that the former displays life' (23). But the Peripatetics, as we emphasised at the beginning of this chapter, were far too well versed in biology to forget that animals do not exhaust the category of living things. It would have been intolerable to have had to assert that plants were not also living beings. It follows from Aristotle's definition that they, too, must possess souls.

(20) It has already been pointed out that the Stagirite's views on the soul seem to have altered during his life. Thus in *De Anima,* 413a5, we find a two-substance view expressed: the soul is to the body as the sailor is to his ship. And if this view is correct, Aristotle admits, the question of how soul and body interact is extremely puzzling.
(21) *De Anima,* 412a13.
(22) Plutarch, *De Exilio,* 17, 607D; quoted in Kirk and Raven, *ibid.*, p. 359
(23) *De Anima,* 413a22.

The biologists at the Lyceum saw the living world to be stratified into three main levels: plants, animals and men. These three levels were defined by the following criteria.

The plants showed that they possessed a self-moving mover, as the Aristotelian theory required, by exhibiting the phenomenon of growth. This, as we saw on p. 89, was one of the aspects of the Stagirite's wide concept of motion. Growth, according to Aristotle, is due to the faculty of self-nutrition. The possession of this faculty thus defines the lowest order of living beings.

The members of the animal kingdom have added to this basic faculty the faculty of sensation. Aristotle was too good a biologist to fall into the elementary trap of supposing that locomotion was the characteristic which distinguished animals from plants. He was well acquainted with zoophytes and sponges, and recognised that although sedentary these creatures were nonetheless animals (24). They were animals because they were sensitive.

Finally, he points out, 'certain living beings—a small minority—possess calculation and thought' (25). This order of living beings, says Aristotle, includes man and possibly another order like him or superior to him.

It follows from this analysis of the living world that the 'highest' form, man, possesses all three of these faculties of the soul. This is the basis of the Aristotelian physiological psychology. The human soul, the cause of all the body's activity, is divided into three parts: vegetative, animal and rational. It will be recalled from chapter 7 that Plato also propounded a trinitarian concept of the soul in the *Timaeus*.

The cause of movement

The soul, as we have been stressing, was for Aristotle, as for Plato, a 'self-moving mover', a 'source of motion or rest'. This motive character was, of course, firmly associated with each of its three parts. The next question, therefore, which Aristotle asks is: what initiates the movement, what triggers the motion? Aristotle, as we have seen, was very scornful of those who, like the atomists, considered that random, chaotic movement lay behind all things. Nature, we remember Aristotle insisted, does nothing in vain. Whatever else of modern science the Stagirite anticipated, he never anticipated the second law of thermodynamics. Thus we recognise the consistency of his thought when we read in *On the Motion of Animals*: 'All living things are moved with some object so that this is the term of all their movement' (26). Consequently, says Aristotle, if we are to understand the initiation of movement in living organisms, we must penetrate behind the 'thinking, perceptive and nutritive powers' to the objects of thought,

(24) This view persisted into the writings of the early Greek theologians. Nemesius of Emesia (*fl.* AD 390) points out that the 'sea-nettle' is an animal because it possesses an animal's sense of touch. See Wallace-Hendril, W., *The Greek Patristic View of Nature*, Manchester University Press (1968).
(25) *De Anima*, 413a20.
(26) *De Motu Animalium*, 700b15.

perception and nutrition. For, as he says in another passage, 'The mover must exist before the moved, the begetter before the begotten and nothing is prior to itself' (27). This line of thought will by now be familiar.

Aristotle considers the faculties of the soul *seriatim*. He starts with a discussion of nutrition. With this faculty he couples that of reproduction. He asserts that an animal or plant reproduces itself 'in order that, so far as its nature allows, it may partake in the eternal or divine. That is the goal to which all things strive...'. This objective is also, for Aristotle, the objective of the vegetative soul in nutrition—for, in this process, material which is at the outset very unlike, very foreign to, the organism is transformed into the body of the organism:

> It's a very odd thing –
> As odd as can be –
> That whatever Miss T eats
> Turns into Miss T . . . (28).

This appears to the Stagirite to be just another aspect of the body's yearning to partake in 'the eternal or divine', to perpetuate its pattern or form. This is the most basic feature of a living being. Food, the object of this psychic power, is essential if the form is to persist; in its absence the organised pattern of matter inevitably ceases to be.

Having discussed nutrition and the vegetative soul, Aristotle passes next to a consideration of the faculty characteristic of animals—sensation. His method does not alter. He considers in turn the objects of the senses of sight, hearing, taste, smell and touch. He believes that sensations are caused by the interaction of the sense organs with something external to the body—the objects of sensation (29).

Finally Aristotle arrives at the ratiocinative soul. His argument is complicated, but in essence the distinction between the ratiocinative and the sensitive soul lies in the fact that, whereas the latter is moved by particulars, the former is moved by universals. The universals, moreover, are not 'out there' in the external world but somehow within the mind. This is understandable enough. We all accept that the average man nowhere exists, only quirky individuals; we are all aware that the 'average family' with its 2.37 children is, necessarily, present only in the statisticians' imagination.

All three divisions of the soul are thus moved by appropriate objects. In order to exert their effects the objects must appear worth moving for: that is they must appear 'good'. Clearly, however, as Plato often stressed, appearances are sometimes deceptive. Hence Aristotle concludes that the objects which trigger the soul's motion are the appropriate real or apparent 'goods' of the several faculties.

We see in this analysis once again the profoundly teleological view of

(27) *Physica*, 202a14.
(28) Walter de la Mare, 'Miss T', in *Collected Poems* (ed. Richard de la Mare), Faber, London (1969), p. 146.
(29) *De Anima*, 418a5.

nature which is so characteristic of Aristotle. The real or apparent 'good' of the various faculties of the soul remains stationary. In Aristotle's phrase, it 'moves but is unmoved'; it is an example of an unmoved mover. The appropriate faculty of the soul, however, yearns towards it: it is moved and causes movement. It causes the organism to move in such a way that the end, the real or apparent good, is achieved.

Physiological psychology

But how does the moving soul cause the body to move? Aristotle's answer is to be found in *On the Motion of Animals*. In this short and closely argued book we find that the Stagirite's first concern was to find within the body a stationary 'platform' from which movement could be initiated. He illustrates the point at issue by alluding to the story of Boreas sitting in his boat and, by blowing on the sail, moving the boat forward. Aristotle perfectly appreciates Newton's second law: action and reaction are equal and opposite. Only by standing on an immovable shore could Boreas have impelled his boat across the ocean.

But where is there such an immovable platform within, for example, the human organism? Aristotle starts his analysis by considering the use of a stick or rod to hit or move objects. A stick, he points out, is an inanimate object and therefore cannot, by definition, possess a soul. It is thus impossible for the soul, the initiator of motion, to exist anywhere in the stick even at its proximal end, as Plato seems to have imagined his marrow soul to lie in the joints. But, Aristotle goes on, the stick lies in the same relation to the hand as the ulna does to the elbow joint; and the same argument, he insists, can be applied to all the body's joints. They are all merely physical mechanisms. They must be pushed or pulled by some more central agent.

Aristotle continues his analysis. Left and right sides, he says, are symmetrical and, moreover, can be moved simultaneously. It is thus absurd to suppose that the left side activates the right side, or *vice versa*. The motorium must be in the centre. Similarly with the upper and lower halves of the body: the motorium, consequently, must not only be in the midline but at the midpoint of the body. This, Aristotle points out quite candidly, is convenient, for in this position is also to be found the body's sensorium. At this point it becomes clear that Aristotle sides with Empedocles against Alcmaeon and Plato in taking the heart to be the psychological centre of the body.

The arguments by which Aristotle concludes that the body's centre of sensation is the heart are partly psychological, partly physiological and partly embryological. In *On the Parts of Animals* we read the following passage:

> For the heart is the first of all the parts to be formed, and no sooner is it formed than it contains blood. Moreover, the motions of pain and pleasure, and generally all sensations, *plainly* have their source in the heart . . . . Again, as neither the blood itself, nor yet any part which is bloodless, is endowed with sensation, it is *plain* that

> that part which first has blood and which holds it as it were in a receptacle, must be the primary source of sensation (30).

Plainly. Similar, though rather more physiological, arguments may be read in *On Youth and Old Age.* In this short work it is pointed out that if the heart, or, in non-sanguinous animals, a part corresponding to the heart, is where the 'life' of an animal is concentrated, then it is evident that 'the principle of sensation must be located there too, for it is *qua* animal that an animal is said to be a living thing, and it is called animal because it is endowed with sensation' (31).

Thus, returning to our exploration of Peripatetic notions on the origin of animal motion, we need only remember that for Aristotle the soul is a 'moving mover' to foresee the next step in the argument. For perception, according to the Stagirite, is identical with a movement in the percipient. Here Aristotle seems to be using the term 'movement' in one of the 'non-physical' senses we discussed in chapter 3. However, as we noticed in that chapter, the distinctions which we now recognise were by no means so clear-cut in the fourth century BC. The psychological and physiological uses of the term were still to a large extent unanalysed.

Aristotle tends to side with those thinkers in the ancient world who, like Empedocles, believed that like is perceived by like. Hence he believed that a change in the sensory environment caused by the interpolation of a fresh object causes an analogous change in the sensorium. Thus, says Aristotle, the idea of 'hot or cold, or pleasant or fearful is like what the object would be, and so we shudder and are frightened at a mere idea' (32).

In general, the Stagirite writes, these movements of the sensorium take the form of various types of qualitative change. Although these alterations, perhaps minute alterations in moisture content or in temperature, are very slight, they occur at the centre and are hence able to cause large-scale activity at the periphery. Aristotle draws an analogy with the considerable change which occurs after a helmsman slightly alters the inclination of a ship's rudder: 'And it is not hard to see that a small change occurring at the centre makes great and numerous changes at the circumference, just as by shifting the rudder a hair's breadth you get a wide deviation at the prow' (33).

Aristotle thus believes that in this way he has satisfactorily accounted for the phenomenon of animal movement. We see, however, that Aristotle's psychophysics is not really psychophysics in the modern sense at all. The central problem of the modern subject, how an immaterial substance, the mind, acts upon a material substance, the body (to use the Cartesian formulation), does not trouble the Peripatetic, for he has circumvented its foundation. That the Cartesian problem existed in the Postsocratic world is

(30) *De Partibus Animalium,* 666a10 (my italics).
(31) *De Iuventute et Senectute,* 496a17.
(32) *De Motu Animalium,* 701b20.
(33) *Ibid.,* 701b1.

clear from what has been said about the doctrine of anamnesis and of Socrates' reaction to the Anaxagorean materialism. But for Aristotle the problem has already been solved in the *Metaphysics* and the *Physics*. For him there is no 'wall' between organism and environment, no barrier between body and soul. As we have already observed, Aristotle is the greatest of the hylozoists.

The insignificance of the brain

Our account of the Peripatetic physiology has shown it to be thoroughly cardiocentric. The heart is both the central motorium and the central sensorium of the body. It is somewhat ironic to reflect that the empirically minded biologist should have been wrong on this central topic while his literary and *'a prioristic'* teacher, Plato, should have been right. However, Aristotle thought he had good evidence for his belief. Some of this suppossed evidence we have already touched upon in the last section. In addition to these rather abstract reasons, the Stagirite brought forward evidence from anatomy, comparative anatomy and, especially, embryology.

Aristotle had observed that the heart is the first organ of the developing embryo to start moving. Recall here the ancient equation between spontaneous movement and the possession of a soul. Aristotle also observes that the heart is from the very beginning filled with blood. This, too, he believes to be significant: for he believes that the embryological process involves the 'crystallisation' of the various tissues and organs of the adult from this primordial pool of blood (34).

In the book he wrote about the movement of animals, Aristotle produces anatomical evidence to show that the heart is supplied with sinews which run out to connect with all parts of the body (35). This notion is repeated in *On Youth and Old Age* (36). The Stagirite's theory is clear. It is by means of these sinews that the slight movements of the heart are transmitted to all parts of the body. It is well positioned to cause the 'thrusting and pulling' movements which he believes to be basic to animal motion. Notice here the homogeneity of Aristotle's ideas on the roots of animal movement with the ideas about motion in the physical world discussed in chapter 9. 'Thrusting' and 'pulling' clearly inhabit the same universe of discourse as the movement in a resistive medium which he propounds in the *Physics*.

The brain, which we nowadays see as paramount in the initiation of animal movement, plays for the Stagirite a very minor role. In the *History of Animals* we read that 'the brain of all animals is bloodless, devoid of veins and naturally cold to the touch' (37). Aristotle placed great emphasis on warmth as a sign of the presence of life, and the importance of the vascular system in

(34) *De Partibus Animalium,* 651a15.
(35) *De Motu Animalium,* 703a25.
(36) *De Iuventute et Senectute,* 469a15.
(37) *Historia Animalium,* 495a5.

his view does not need further stress. It is thus not surprising to find that the brain had very little part to play in the Peripatetic physiology. In *The Parts of Animals* we read the following passage:

> That it [the brain] has no continuity with the organs of sense is plain from simple inspection and is still more clearly shown by the fact that when it is touched no sensation is produced . . . but Nature has contrived the brain as a counterpoise to the region of the heart with its contained heat and has given it to animals to moderate the latter . . . for this reason it is that every sanguineous animal has a brain; whereas no bloodless creature has such an organ. For where there is no blood there is but little heat. The brain then tempers the heat and seething of the heart (38).

In short, the brain merely existed to cool the blood. The Stagirite would have been astounded at our modern neurobiology. He would have been astonished to read that 'the biology of the future will centre on the human cerebrum' (39). For him the brain was but an auxiliary of the heart. It was the latter organ which constituted the central interest of both physiology and psychology.

Out-topping knowledge

Here we must leave our consideration of the Aristotelian system. It is hoped that an outline of the gigantic sweep and intricate coherence of the Stagirite's thought will have appeared through this necessarily short and oversimplified exegesis. It is hoped that the reader will have perceived how different was Aristotle's world from our own—how different, and yet in some ways how much the same: for to Aristotle, as to moderns, there was no real distinction between physics and biology. Both were aspects of a single science of nature; but for Aristotle the world of physics, like the world of biology, strove towards an end. Objects, for example, moved until they achieved their 'natural' place in a properly ordered cosmos. It is evident, therefore, that although biological and physical phenomena might be treated in similar terms, these terms were not those of post-Galilean physical science. The matter – form dichotomy, the cognate potentiality – actuality dichotomy, the concept of a final cause, play little or no part in modern physics. They seem to us today, as has several times been remarked, far more appropriate to a description of biological, if not psychological, phenomena.

The root which sustains this characteristic of the Aristotelian system is, as we saw in chapter 7, the Socrato-Platonic endeavour to counter the 'might is right' arguments of the Sophists. For, in order to confute this socially disruptive argument in depth, it was necessary to show (or so it appeared to Plato) that the happenings of the world are not the mere outcome of 'chance and necessity' (40). For, if such were the case in the great world, why might it not also be the case in the little world of human affairs? The Aristotelian

(38) *De Partibus Animalium,* 652b5.
(39) C. S. Sherrington, *The Integrative Action of the Nervous System,* Cambridge University Press (1906), p. 390.
(40) *Laws,* X, 889

science, stretching from meteorology to psychology, in consequence tries to show how the same principles apply everywhere: nature, as well as man, does nothing in vain. It is this deep and thorough-going refutation of the Democritean vision which makes Aristotelianism seem so alien to minds nurtured in the tradition of post-seventeenth-century physical science. It is the profundity of Aristotle's teleological vision which made Dante, as we noticed in the introductory quotation , regard him as 'the master of them that know'. And, finally, it is the refutation of this teleological position by modern science which throws into prominence the question which forms the subject of this book: 'What is life?'

11 ATHENIAN AFTER-GLOW

The Lyceum after Aristotle

Aristotle died in retirement in 322 BC. His one-time pupil, Alexander the Great, had died in the midst of his conquests the year before. The empires of both great systematisers soon fell into disrepair. In fact Aristotle was more fortunate than Alexander. He at least had had time to nominate a successor. His empire was thus preserved for a while from the dissension which tore apart that of Alexander. No warring satraps quarrelled over his great system of knowledge. Indeed, his successor Theophrastus carried on the great work of the Lyceum. We have already mentioned the two volumes on the plant kingdom with which he is credited. They complement in an exemplary fashion the Stagirite's great works on animals. However, Theophrastus' output, although encyclopaedic, did little to advance the system of his predecessor. It could hardly have been otherwise. The progressive method of modern science is a very modern acquisition. In the third century BC, as Gomperz points out (1), the Aristotelian system was quite probably the best possible. Thus, although Theophrastus possessed a fine critical intellect and attacked the Stagirite's philosophy in many places, he did not possess the ability to improve it greatly, let alone replace it; and with a closely organised *system* nothing less would really suffice. The intellectual revolution necessary to effect this change waited twenty centuries in the future.

Theophrastus lived to a great age and was succeeded as head of the school by Straton in 285 BC. Straton, unlike his two predecessors, was more at home with physics than with biology. What little we know of his views shows him to have reverted to the atomism of which, as we have seen, Aristotle was so convinced an opponent. Straton, however, believed that such a theory was necessary, especially the existence of a void, if he was to give a satisfactory explanation of the propagation of light, electricity and magnetism, the conduction of heat, and so on. It is interesting to note how, even in antiquity, physicists were drawn towards an atomic theory, whereas biologists, concerned as they were with organic design and fitness for purpose, regarded such hypotheses as little short of absurd.

(1) T. Gomperz, *Greek Thinkers,* John Murray, London (1912), vol. 4, chapter 39.

Alexandria

The career of Alexander marks a sea-change in the history of antiquity. The Greek world never recovered its concentrated parochial outlook after the glimpse of a world-state which Alexander offered. The marches and countermarches, the making and breaking of kingdoms, the mutual reaction of occident and orient which followed his death ushered in the so-called Hellenistic age—an age which, with its demarcation disputes and specialisms, bears, as many have remarked, a strange resemblance to our own.

Although Athens remained for centuries a centre of Greek culture and a university town, the central intellectual endeavour of the age shifted to Alexandria. Here the Ptolemies extended their patronage to the learned by founding the Museum and the Library. Within the research institute which was the Museum worked some of the intellectual giants of antiquity. Here the great astronomer Ptolemy worked out the elaborate system of cycles and epicycles which was to describe the motions of the skies until 1543. Technology and engineering were represented by Hero and Ctesibius, geography by Eratosthenes. At the Library, Euclid put together the book which many still regard as the supreme achievement of the human mind. Alexandria was thus a worthy successor to Athens; a worthy successor, in particular, to the empiricism of the Lyceum. Indeed, Straton who, as we have just seen became third head of the Lyceum, had much to do with the founding of the Alexandrian Museum.

Biological science was also well represented. It has been argued that the twin sciences of anatomy and physiology were both initiated in the Alexandrian Museum. According to Galen, Herophilus (*fl.* 300 BC) was the first man to practise human dissection, and many regard Erasistratus (*fl.* 260 BC) as the founding father of physiology.

Herophilus and the origin of human anatomy

If, as Galen asserts, Herophilus was indeed the first man to have had the privilege of making a systematic anatomy of the human body, it is small wonder that he made numerous discoveries. It is, however, principally through Galen's voluminous writings that these discoveries have come down to us. Several of the names which we nowadays commonly use to refer to parts of the body are straightforward latinisations of Herophilus' Greek. For instance, the first part of the small intestine is easily distinguished from the rest. The length of this initial segment is equivalent to the breadth of twelve fingers: that is, three hands laid together. In consequence Herophilus called it the *dodecadactylos*. We still know it as the duodenum. Similarly the sensitive layer at the back of the eye seemed to Herophilus somewhat like a spider's web. Hence his name for it was *amphiblestroides*. The Latin translation of this is 'retina'. Many other anatomical features were similarly first christened by Herophilus.

Herophilus not only recognised the difference between the arteries and the veins, alleging that the former were some six times as thick as the latter, but he also noticed the peculiarity of the pulmonary arteries and veins, calling the former the arterial vein and the latter the venous artery. He also did fundamental work on the nervous system. He strongly contested the Aristotelian idea that the heart was the centre and origin of the nerves. His dissections convinced him that this was the role of the brain. He described many parts of the brain for the first time and believed that the soul resided in the ventricles—the most important of which was the fourth. He was also the first to distinguish clearly between the nerves and the tendons. One other important result of his labours is best described in the words of Galen's commentary: '...the sensory nerves that descend from the brain to the eyes and which Herophilus called conduits because they...display visible channels' (2).

We shall see as this book continues that the notion of hollow nerves, perhaps acting as tubes for the transport of some agent, has a long and distinguished history in neurophysiology. Indeed, the idea remains very much alive today. We read in modern neurophysiological literature of axoplasmic flow, and endocrinologists provide evidence for the transport of neuro-secretions at quite rapid rates within certain axons. It is, however, clear that Herophilus had observed an artefact. It would have been impossible for him to have observed individual nerve fibres, and the cavity he speaks of must have been due to shrinkage.

One further opinion of Herophilus is worth noting. He believed that four and not three forces enlivened the organism. The first of these agents, a nourishing soul, inhabited the liver; the second, a thermal soul, was situated in the heart; the third, a perceptive force, was present in the nerves; and the last, a faculty of thought, was located in the brain. It is clear that the heart does not play the central role assigned to it in the Peripatetic physiology. The nerves and the brain are accorded equal, if not greater importance. Herophilus, however, agrees with Aristotle in seeing the heart as the fount of the body's heat. In the short Peripatetic treatise *On the Breath*, we may read, for example: 'The heart is the source of the innate heat. The soul is, so to speak, ...created in the heart of those animals which have blood' (3).

The Erasistratean physiology

Herophilus does not seem to have been very concerned to find an explanation of how the body 'worked'; it was sufficient to describe and give names to its parts. The mind of his younger contemporary, Erasistratus, however, was of a

(2) Galen, *On the usefulness of the parts of the body,* X, 12; quoted in E. Clarke and C.D. O'Malley, *The Human Brain and Spinal Cord,* University of California Press, Berkeley and Los Angeles (1968), p. 144.
(3) *De Respiratione,* 469b9.

more physiological bent.

One of the central concepts of the Erasistratean physiology was that each organ of the body was supplied by, indeed consisted of, three distinct and different 'vessels'. These we nowadays recognise as artery, vein and nerve. *Each* of these three vessels was, for Erasistratus, hollow. He believed that these vessels branched again and again below the level of vision, plaiting and twining together to form the substance of the tissues.

Each of three vessels contained, according to Erasistratus, a fluid. The veins contained blood, a nutritive fluid manufactured by the liver from the products of digestion. The arteries, however, contained a far more rarefied fluid—the vital spirit—derived from the environing pneuma. Finally the nerves contained a yet more refined and subtle fluid—the animal spirit—refined from the vital spirits of the arteries by the brain.

The surprising notion that the arteries contained pneuma and not blood has both physiological and metaphysical bases. The physiological basis is that when an anatomist dissects a recently killed animal he often finds that the arteries contain little blood compared to that contained in the veins. We nowadays understand the reason for this observation. The arteries, unlike the veins, possess muscular walls. On death these muscles enter a spasm of contraction. The blood in the arteries is thus forced across into the veins. The metaphysical basis, on the other hand, seems to have been a combination of pneumatology with the microcosm/macrocosm analogy which we discussed in earlier chapters of this book. It will be recalled from chapter 4 that Anaximenes maintained that, just as the pneuma enlivens the world, so does our breath enliven us. This living principle, according to Erasistratus, was taken into the lung and from there made its way to the left side of the heart via the pulmonary veins. From the left heart it was distributed all over the body by the arterial system. This idea, that the arteries contained 'air' and not blood, lingered long in medical and physiological theory. Indeed, the word artery itself originally meant an air duct. Arteries were for long considered to originate from the lungs and trachea rather than, as we perceive today, from the left ventricle of the heart.As late as the seventeenth century AD we find Francis Bacon writing, 'The lungs...through the artire, throat and mouth maketh the voice' (4).

But, it will be objected, surely a very elementary knowledge of anatomy would have shown Erasistratus the absurdity of his theory. For is it not the case that when an artery is severed blood escapes? Surely this must have been common knowledge in Alexandrian times and indeed throughout history. And does it not follow that the arteries must therefore contain blood in the living body? But Erasistratus, of course, had an answer to this obvious point. It was founded on his acceptance of the Peripatetic notion that 'Nature abhors a vacuum'. When an artery is cut, argued Erasistratus, the contained pneuma escapes. This leaves, or would tend to leave, a vacuum in the artery.

(4) F. Bacon, *Sylva Sylvarum* (1620), 199.

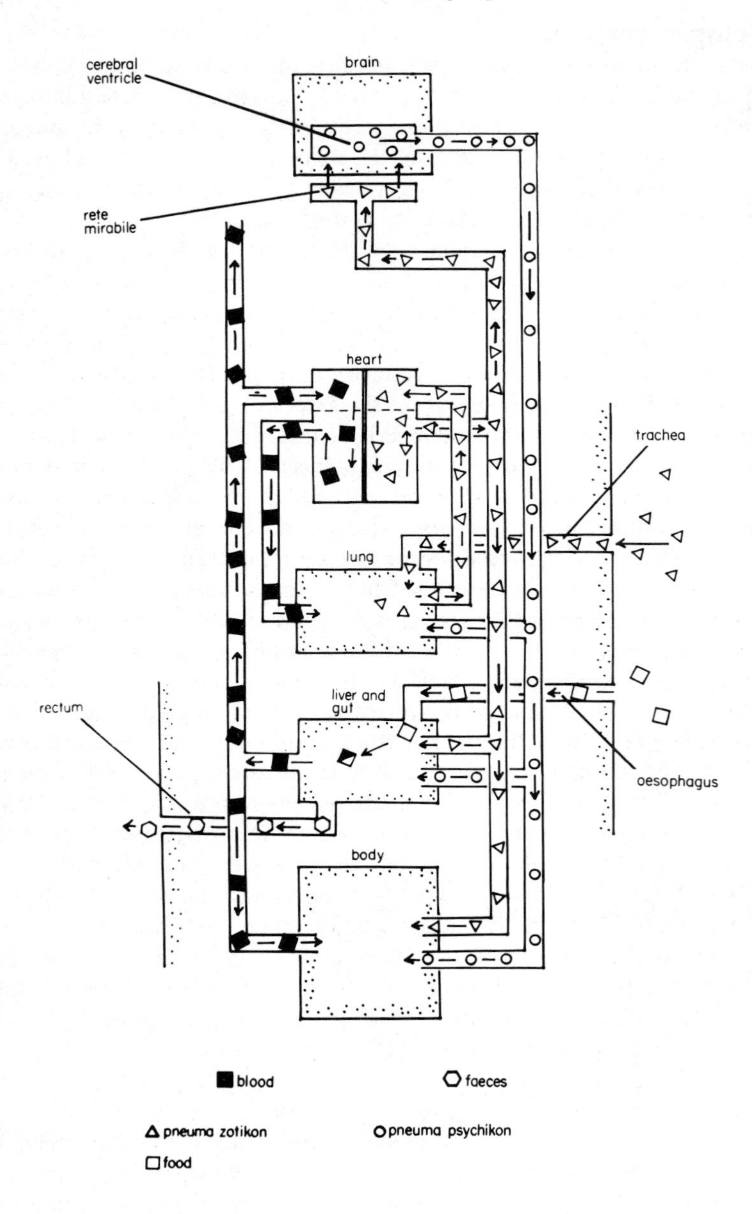

Figure 11.1 The Erasistratean physiology. The left and the right sides of the heart are quite separate. Both arteries and nerves contain pneumata: the former pneuma zotikon (vital spirits), the latter pneuma psychikon (animal spirits). The movement of both types of pneumata and of the blood was conceived to be rather sluggish, and all three substances were understood to be totally used up at the tissues.

But this cannot be. Instead blood from the veins immediately fills the vacant space. This is the blood which we eventually observe jetting from the cut. In parenthesis here note how close Erasistratus is to an understanding of the circulation of the blood. He is fully prepared to accept the existence of invisible connections between the arterial and venous systems. It is perhaps only the metaphysical climate of the time which prevents him following the example of his near-contemporary Aristarchus in astronomy and anticipating the foundation stone of the modern science of physiology by nearly two millennia.

Galen of Pergamon

So far as the biological sciences are concerned the last great name of classical antiquity is that of Galen (5). It was his system of medicine, his system of physiology, his psychophysiology, which was accepted by physicians right through the middle ages and lingered on, indeed, into the seventeenth and eighteenth centuries of our era. This need not altogether surprise us, for in some respects Galen's insights were very near to what we nowadays believe to be the truth. Often it is just his terminology which we find obscure. In other respects, however, Galen's biology is profoundly unmodern. He, like so many of the ancients, could not conceive that the exquisitely designed living organism could be explained in terms of a mindless interplay of atoms. He writes, for example, as follows:

> Is it by chance and not purposeful work what nature has accomplished? For when not a thousand or ten thousand people but only in ten thousand times ten thousand we observe malformations (as a sixth finger), will you still consider this accomplishment by nature accidental? (6)

Galen was born in AD 130 at Pergamon on the Aegean coast of Asia Minor. During the year AD 157 – 158 he was physician to the gladiators at Pergamon, but later removed himself to Rome. Here he had a great success, becoming eventually physician to the Emperor Marcus Aurelius. He died at the turn of the century.

Galen was an extremely prolific writer, not only on medical subjects but also on philosophical matters. Indeed, the title of one of his treatises, *Quod optimus medicus sit quoque philosophicus*, asserts that 'the best physician is also a philosopher'. Altogether Galen is believed to be responsible for about one hundred treatises, with a further sixty or so doubtfully his or almost certainly spurious. Indeed, Galen himself made an attempt to clarify the problem of which are and which are not genuine Galenical works by publishing a bibliography—*De libris propriis*—in which he lists 124 treatises.

This gigantic output was based not only on a wide medical experience but also on voracious reading. Galen's intellectual position is thus usually regarded as eclectic. He belongs completely to none of the principal schools

(5) Many of Galen's works are collected in French translation in C. Darenberg, *Oeuvres anatomiques, physiologiques et médicales de Galen,* Ballière, Paris (1854–6).
(6) Quoted by R.E. Siegel in *Galen's System of Physiology and Medicine,* Karger (1968), p.9.

of classical thought. He is neither totally a Peripatetic nor totally a Platonist, neither a Stoic, an Epicurean nor a Sceptic.

In this section an attempt is made to present in outline (it can be no more, for Galen's extant works consist of some ten thousand pages of Greek text) some of the central features of Galen's physiology.

We have already noticed something of Galen's animus against the atomists. 'Nature', for Galen, as for Aristotle, is a 'creative artist'. He pours scorn on those who would suggest that organisms come to be by the coming together of 'primary corpuscles'. He believes that 'nature' is 'prior' to the corpuscles, not a resultant of them, and that she attracts appropriate materials to the body and rejects those that are inappropriate (7). When pressed as to exactly how 'nature' 'attracts' appropriate materials, Galen replies: *'hypotheses non fingo'*. That she does is plain to him and must, he argues, be plain to any and every open-minded physician. Indeed, he goes on, it is not even strange to the atomists themselves. Epicurus, he points out, refers in his *Physics* to the fact that the lodestone attracts iron and that amber attracts chaff. Epicurus, he continues, develops a hypothesis to account for this observation, a hypothesis which he (Galen) proceeds to ridicule. Nevertheless, he says, the observation still stands, whatever the merits or demerits of the explanatory hypothesis. And if one of the major proponents of atomism admits that attraction occurs, how absurd, how doubly absurd, for would-be biophysicists to deny the evidence of their senses.

In parenthesis here it might be pointed out that we should be wrong to smile too quickly at Galen's shaping Demiurge. For what he says about organs and tissues we are nowadays only too inclined to say about cells. Consider the following passage:

> . . . all the observed facts testify that there must exist in almost all parts of the animal a certain inclination towards, or, so to speak, an appetite for their own quality and an aversion to, or, as it were, a hatred of a foreign quality . . . . But if there be an inclination or an attraction, there will also be some benefit derived; for no existing thing attracts anything else for the mere sake of attracting, but in order to benefit by what is acquired by attraction . . . (8).

How it is that cells attract materials of importance to themselves and reject unimportant and harmful matter is still a fundamental problem in biology. Biophysicists would nowadays seek an explanation in terms of membrane properties and the feedback circuits of cellular metabolism, but for many the Galenical account, shorn of its antique terminology, remains adequate.

However, Galen's chief importance in the context of this book lies in his treatment of cardiovascular and respiratory physiology. These two aspects of physiology were for him, as they are for us, closely linked and fundamental.

Galen's understanding of cardiovascular physiology was in one respect

(7) Galen, *On the Natural Faculties* (trans. A.J. Brock), The Loeb Classical Library, Heinemann, London (1952), 1, 12.
(8) *Ibid.*, 3, 6.

greatly in advance of the Erasistratean scheme. Galen was under no misapprehension about the presence or absence of blood in the arteries. Here was no subtle wind but, in the words of a later and equally great physiologist, 'the unequivocal gore'. His reasons for this belief derived from impeccable experimental and observational evidence. Galen had tied two ends of an artery and punctured the middle to show the indisputable presence of blood. He had also taken the trouble to observe the movement of blood in the mesenteric arteries. Galen's physiological system, as this first indication shows, was based not only on wide and eclectic reading but also on a great amount of careful observation and experiment. Nevertheless, and in spite of a great deal of invective directed at Erasistratus, Galen's overall system shares many features with that of the Alexandrian.

Galen, like Erasistratus, conceived that a vital principle was absorbed through the lungs from the circumambient air. He had made a close examination of the minute ramifications of the pulmonary tree and had concluded that 'under normal conditions pneuma passes from the trachea into the pulmonary veins in very small amounts' (9). Not only this, but he also understood that blood from the right side of the heart makes its way, although only very slowly, through the anastomoses of the pulmonary system to the left side (10). Galen thus regarded the blood which reached the left side of the heart and afterwards surged in the arteries as charged with a vital principle, the so-called *pneuma zotikon*, or vital spirit. This, stripped of its archaic terminology, has a surprisingly modern ring. Nowadays we should say that the blood in the left side of the heart and in the arteries is charged with a life-giving principle derived from the planetary atmosphere—oxygen.

Galen's cardiovascular physiology is, so far, close to our modern concept; but he veers from our present-day understanding when he asserts that the source of the body's heat is to be found in the left ventricle. This, as we saw in our earlier discussion of Herophilus, is an Aristotelian concept. Indeed it predates Aristotle and was widespread in the ancient world. It was only finally disproved in the eighteenth century AD when, with the development of the thermometer, it became possible to demonstrate that the heart was not in fact significantly warmer than the rest of the body's core. It is perhaps ironic to find Galen subscribing to this Peripatetic opinion when, as we shall see (page 154), his great seventeeth-century successor, Harvey, entered the lists against the Galenists wearing the Philosopher's colours! Thus, for Galen, the pneuma zotikon did not act as a fuel throughout the body's fabric as we nowadays recognise that oxygen does, but only, or at any rate chiefly, in the left heart: 'the hearth of the body's fire'. From this dark fire warmth was distributed all over the body by the arteries.

Galen also deviates from the twentieth-century understanding when he asserts as a central thesis of his system that a second type of pneuma, the

(9) Quoted in Siegel, *ibid.*, p.154.
(10) Galen, *On the Natural Faculties,* 3, 15.

pneuma psychikon, is associated with the nervous system. Here again we can, perhaps, detect the influence of the Alexandrian physiologists. The pneuma psychikon, or animal spirit, was secreted by certain vascular plexi which Galen had observed in the brains of the animals he had dissected. There were two important cases: the choroid plexi in the lateral and third cerebral ventricles, and the rete mirabile at the base of the brain. We now recognise that the choroid plexi secrete cerebro-spinal fluid into the ventricles, while the rete mirabile, ramifying around the pituitary in such animals as the domestic ox, forms a platform from which the arterial supply of the brain springs. The rete does not develop to the same extent in man, where its function is served by the circle of Willis.

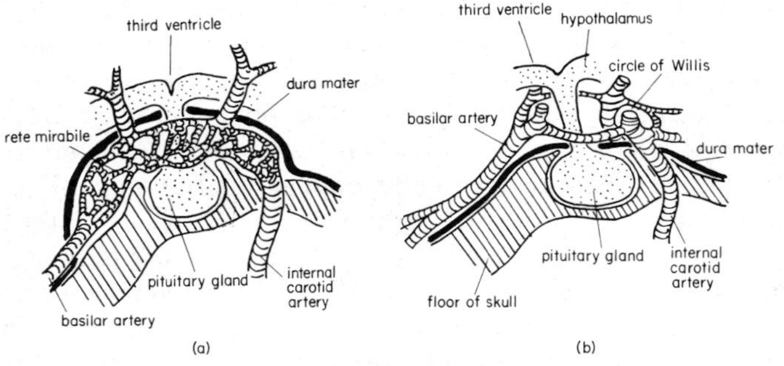

Figure 11.2. The rete mirabile and the circle of Willis. (a) The extensive rete mirabile situated above the dura mater may be found in the ox and certain other domestic mammals. (b) In humans an insignificant rete is found below the dura mater, and above it a strongly developed circle of Willis is to be found. Both rete and circle have the same function: to distribute the blood to the brain and to ensure that unilateral occlusion of one of the vessels leading up from the heart does not result in one side of the brain's being deprived of its blood supply. (After Siegel, *ibid.*, figure 6).

The pneuma psychikon was regarded by Galen as a second refinement of the compound of blood and vital principle derived from the surrounding air. The first refinement occurred, as we have seen, in the left ventricle of the heart, where pneuma entering through the lungs was concocted with the blood. The resulting pneuma zotikon was, in its turn, refined as it made its way slowly through the reticulations of the cerebral plexi. The highly refined pneuma which eventually percolates into the cerebral ventricles is ultimately reacted with fresh supplies of pneuma which Galen believed to enter via passages from the nasal cavity. This belief was probably based on the fact that many of the animals which Galen dissected develop long prolongation of the lateral ventricles towards the olfactory bulbs. The result of this final concoction, the pneuma psychikon, was believed not only to fill the cerebral ventricles but to permeate throughout the entire matter of the brain. It was conceived to be responsible for the phenomena of neurophysiology. Indeed Galen thought that the observed pulsations of the brain, which some have more recently considered to cause the electrical events of the EEG, forced the pneuma psychikon from the ventricles into the brain's substance rather as the movements of respiration forced pneuma from the atmosphere into the blood.

Once in the brain's substance, the pneuma psychikon made its presence felt at the periphery by means of the nerves. Galen appears to be in some doubt as to the precise mechanism involved. He proposes different solutions in different parts of his opus. Thus in some places we find him writing as if the pneuma psychikon acted by contact, the transmission down the nerve being analogous to a shock wave or chain reaction—a notion not so far removed from our present understanding. In other places he suggests that the pneuma strikes through the length of a nerve like 'the rays of sun through air and water' (11). It is interesting to notice that he does not in fact seem to have supported the Erasistratean idea that the nerves were hollow and that spirits flowed along them much as blood flows along a vein (12). The Erasistratean idea seems, however, to have been orthodox among Galenical physiologists and physicians until the seventeenth century. Vesalius, in the sixteenth century, although healthily sceptical of ventricular neurophysiology (see chapter 22), nevertheless accepted the existence of a psychic pneuma in the ventricles and its transmission via hollow nerves to the periphery.

Once again we see that Galen was closer to the truth than the later Galenists. Even with a concept so seemingly archaic as pneuma psychikon, Galen argued from more than merely metaphysical presuppositions; for he had observed that, whereas animals frequently recovered from surgery undertaken on superficial parts of the brain, this was seldom if ever the case when the knife cut into the third or fourth ventricle. Neurophysiologists today would say that the latter operations affect the 'reticular formation' which is believed to be responsible for 'setting' the level of consciousness of the cerebrum. Injury to this region is thus usually disastrous.

Galen's careful experimental approach led to yet further insights into neurophysiology. He was able to prove conclusively that the brain and not the heart was the source of sensations and motion. Furthermore he was able to show that different parts of the body are paralysed when different nerves are severed. Finally as an ultimate *coup de grâce* against the Peripatetics he was able to demonstrate that it was the brain and not the heart which controlled the voice. He is said to have stumbled over this result by accidentally cutting the recurrent laryngeal nerve of a squealing pig. However, he followed up this piece of serendipity by carefully investigating the effects of sectioning this nerve in a wide variety of mammals. Altogether Galen's experimental neurophysiology conclusively established the brain as the source and centre of sensation and motion and the nerves as its communication channels (13).

(11) Galen, *The Causes of Symptoms,* 1, 5, quoted in Clarke and O'Malley, *ibid.,* p.147.
(12) Siegel, *ibid.,* p.194
(13) See *Galen on the usefulness of the parts of the body* (trans. M.T. May), Cornell University Press, New York (1968).

However, one rather serious misconception marred the Galenical system, a misconception which, like the Peripatetic projectile, was to be revealed in the seventeenth century as the system's Achilles' heel. This misconception was the notion that the heart's interventricular septum was penetrated by minute and invisible pores.

We shall see in chapter 14 that one of William Harvey's most convincing pieces of evidence for the circulation of the blood lay in a quantitative evaluation of the volume of blood in the body. This, as we shall see, seemed a clinching argument when compared with the vague qualitative approach of the mediaeval Galenists. But when we return to the founding father of the tradition, Galen himself, we find, such are the ironies of history, that one of the principal considerations in his mistaken account of cardiovascular physiology was, also, a quantitative consideration. Galen had observed that the diameters of the pulmonary vein and artery were considerably less than the diameters of the vena cavae and aorta. It seemed, in consequence, impossible for all the blood on the right side of the heart to pass through the pulmonary circulation to the left side: hence the necessity for an alternative route to exist through the interventricular septum.

But in one quantitative matter Galen was wildly out. He grotesquely underestimated the quantity of blood passing out of the heart by the pulmonary artery and aorta or into the heart via the pulmonary veins and vena cavae. This, indeed, is the quantitative mistake which Harvey seized upon. For Galen, only drops of blood could squeeze through the anastomoses of the lung or the pores in the interventricular septum. It is a little strange that so ardent an experimentalist should make so elementary a mistake. One can only assume that here, at last, the great pressure of tradition and the great necessity to make sense of the body's operation overpowered the evidence of experiment, for Galen still worked within the ancient tradition which demanded the creation of complete and well-rounded syntheses. This pressure to provide a complete system was still too strong to allow the piecemeal approach of modern science with its loose ends and minute specialisms. No scientist nowadays feels professionally obliged to provide the whole picture!

Galen's quasi-static cardiovascular physiology ties in nicely, as we have implied above, with other parts of his system. For he believes that blood is manufactured in the liver from the chyle brought to it from the gut. He points out that only small quantities could be formed in this way in any short period of time. At the periphery of the 'circulation' the blood formed nutriment for the body's metabolism and was accordingly used up and removed from the veins. It was replaced by fresh blood manufactured in the liver. This somewhat sluggish physiological scheme is shown in figure 11.3.

One final feature of Galen's cardiovascular physiology should be noticed. The dark fire in the left ventricle of the heart resulted, as in other combustions, in the formation of wastes in the form of fumes. These

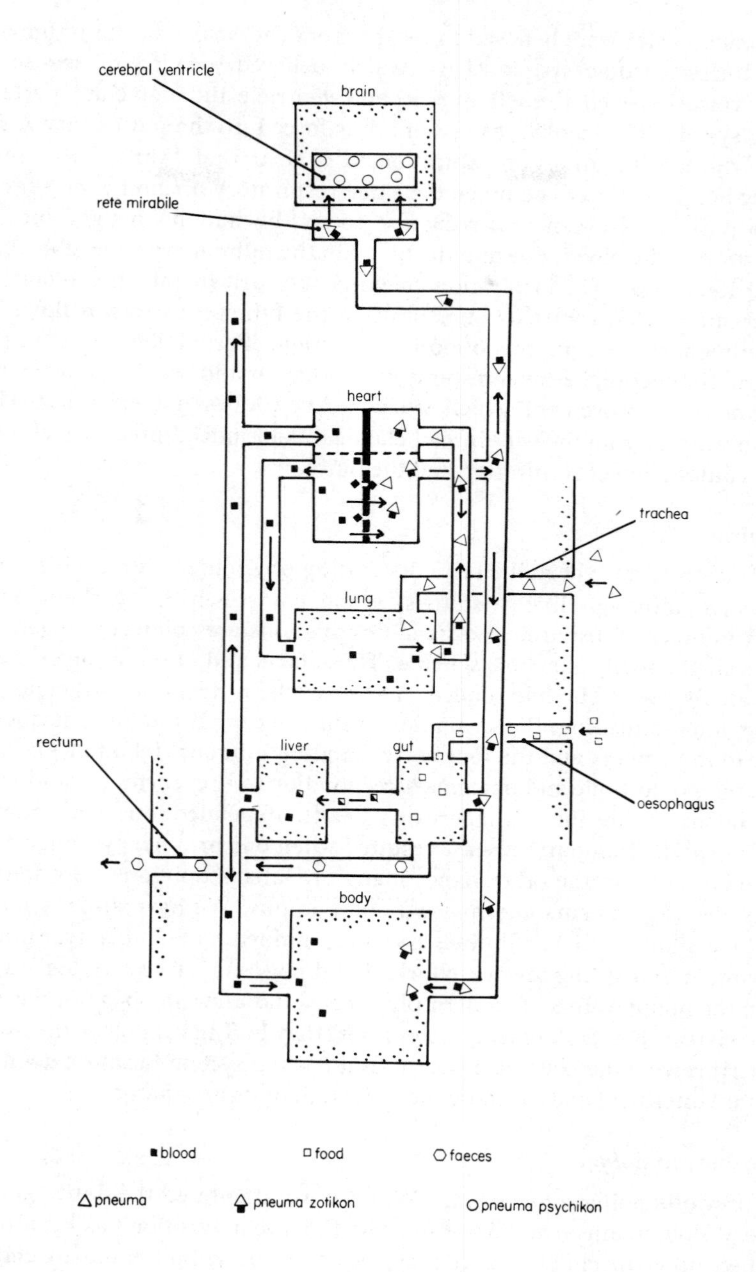

Figure 11.3. Galen's cardiovascular physiology

fumaceous wastes were believed to escape from the heart via the pulmonary veins during cardiac systole. This, Galen believed, occurred because the mitral valve between the left auricle and ventricle did not close perfectly during systole. The smoky wastes were thus forced up the pulmonary vein to escape through the lungs on expiration. It is clear that, while Galen in his archaic language described much the same respiratory phenomenon that we, in our post-Lavoisierian era, recognise today, he had no insight into the *circulation* of the blood, For him the blood in the pulmonary veins ebbed and flowed like a tide. On inspiration pneuma was drawn into the blood and passed into the left ventricle; on expiration the fumaceous wastes flowed up the pulmonary vein in the opposite direction. For Galen there was an intimate connection between respiratory movement and cardiac movement, a coincidence of movement which we now know does not exist but which probably arises from the widespread classical, and indeed preclassical, belief in the conjunction of respiration and the pulse.

Galenism

Galen seems to have been the last outstanding physiologist for a millennium. His system, although, like Aristotle's, wrong, compels by its synthetic power. A vast number of disparate facts and observations are connected together to form a single highly plausible scheme. These facts and observations were not all Galen's own. He had culled many of them from a voracious and indiscriminate reading. For example, he fully accepted the ancient doctrine of the four elements and the four corresponding humours (chapter 4): blood, black bile, yellow bile and phlegm corresponding to fire, earth, air and water. For Galen, as for the Pythagoreans, health depended on a harmonious blending of the humours: disease resulted when one or other tyrannised over the others. But, on the other hand, many of Galen's opinions were founded on his own experiments and his own observations. We have already noticed several instances above. It was, however, unfortunately this tradition of experiment and seeing for oneself which did not long survive Galen's death. When the inaugurators of modern physiology and anatomy began their work in the sixteenth and seventeenth centuries, they had first to clear the ground by overthrowing the Galenical system. But it was a system far more dogmatic, far more uncritical and far more theoretical than its originator.

The Stoic physiology

The prevailing philosophies as the Western world entered the dark ages were those of Epicureanism and Stoicism. The Epicurean position has been briefly touched upon in chapter 6. It may, however, be enlightening to end this chapter by sketching the outlook of Stoicism, as this philosophy coloured men's view of the world and of the nature of organisms until, insensibly, philosophy was displaced by religion.

Stoicism, like most post-Socratic Greek thought, was basically concerned with ethics: with what constituted a good life for human beings. The Stoics regarded man as something standing apart from nature: a thinking being. It is a commonplace that their doctrine maintained that an individual should strive to become as independent of the environment and thus as imperturbable as possible. They would have acclaimed Milton's hero when he scorns the burning lake and the sulphurous stench:

> The mind is its own place, and in itself
> Can make a Heav'n of Hell, a Hell of Heav'n.

If the chief object of Stoic thought is an attempt to discover what constitutes virtuous conduct it is clear that their interest in natural science, as such, would be minimal. However, virtue, according to the Stoics, consists in acting in harmony with the existing laws of the universe. Thus the Stoics did have a motive for studying the natural world: as a prolegomenon to their real concern—ethics. This motive, however, entailed that Stoic science would not be the detailed specialist study with which we are familiar today, but an enquiry into the general form of the laws of nature.

The Stoics resembled their rivals the Epicureans in believing that what was real was corporeal. In this they both made common cause against the two other major schools of classical thought: the Academicians and the Peripatetics. But to the modern mind the founders of Stoicism pushed their equation of reality with corporeality to rather absurd lengths.

> 'What do you mean by real?' asked the Sceptic. 'I mean solid and material. I mean this table is solid matter.' 'And God', asked the Sceptic, 'and the soul?' 'Perfectly solid', said Zeno [the founder of Stoicism], 'more solid, if anything, than the table.' 'And virtue or justice or the Rule of Three; also solid matter?' 'Of course', said Zeno, 'quite solid' (14).

This position seemed odd to many ancient thinkers, as it still does to us. Could 'summer' and 'intellectual concepts' really be material objects? The Stoics insisted that they were; they also insisted, as we saw in the quotation above, that the properties of things, virtue, justice, and so on, were also solid. This insistence resulted in a theory of 'intermingling' whereby the properties of a thing interpenetrate it. Now this concept is interesting in the context of the present book, as it throws light on one of the central concerns of our history: the interrelationship of body and soul. Chrysippus (280 – 205 BC), for instance, the third leader of the Stoa, used the theory of intermingling to explain the union of the psychical with the physical. He held that the soul permeated the body and held it together—a concept which we met earlier in Anaximenes.

However, whereas we have only a single sentence from which to reconstitute Anaximenes' thought, the Stoic theory is profusely documented. It is founded upon their concepts of matter and force. For although, as we

(14) G. Murray, *The Stoic Philosophy* (1915); quoted in B. Russell, *History of Western Philosophy*, Allen and Unwin, London (1946), p.176.

have just seen, the criterion of reality for the Stoics was corporeality, they were nevertheless acutely aware that this in itself was not enough. Matter on its own was merely passive and impressionable; in order to account for the flux of experience another concept, *force*, was indispensable. The interest of the Stoics in ascertaining the causes of things, an interest intimately connected with the central ethical concerns of their philosophy, resulted in their concentrating much attention on the nature of 'force'. Now, in contrast to the Academicians and the Peripatetics, the Stoics derided any cause other than the efficient cause. For the Stoics, efficient cause and force were synonymous.

Now this efficient cause, this force, was, of necessity, corporeal. The Stoics believed that this active substance, responsible for the world's flux, was, to cut a long story rather short, the pneuma. It was this substance which permeated all things and held them together. The Stoics believed, moreover, that this cohesive power of the pneuma was due to an inner tension. It is as if this continuous fluid which permeated and interpenetrated all things was stretched and would, if it could, contract. This elastic tension forced things caught in its field to cohere. Sambursky suggests that this Stoic concept prefigures in a rather remarkable way our modern concept of a field of force (15).

We have already seen that the Stoic concept of the pneuma, like Anaximenes', applied to both the microcosm and the macrocosm. It is another instance of that failure to make the modern distinction between the animate and the inanimate which is so marked a characteristic of ancient thought. Even in the times of Galen and Marcus Aurelius, the Stoics conceived that the macrocosm accurately reflected the major features of the microcosm. For this reason the Stoic philosophy is strongly deterministic. The Stoics asserted that the alleged ability of soothsayers to foretell future events accurately must depend on this interdependence of and correspondence between things. Thus physics and physiology were closely allied in the Stoic philosophy, but allied in a very unmodern way, for physiology was not conceived to emerge from physics: the two studies were treated analogously. Thus Sambursky, for instance, suggests that the best entry to Stoic dynamics is through a study of the Stoic theory of sense perception.

However this may be, the Stoic theory of sense perception is of interest in its own right. We have already seen that for the Stoics the soul penetrates throughout the body and holds it together. They believed, nevertheless, that there existed a coordinating centre or *hegemonikon*. In this centre were collated images of things transmitted from the sense organs. Most of the early Stoics located this centre in the heart (16). They argued with the Peripatetics that the voice, the expression of feeling and thought, emanates

(15) S. Sambursky, *Physics of the Stoics,* Routledge and Kegan Paul (1956), p. 136.
(16) E. Zeller, *The Stoics, Epicureans and Sceptics* (trans. O.J. Reichel), Longman, London (1892), p. 213.

from the region of the heart. *Vice versa*, it was conceived that volitions and feelings were sensed to occur in the heart: do we not still speak of someone in 'his heart of hearts' believing such and such? Do we not still speak of feelings as 'coming from the heart'? It would be easy to multiply these examples of linguistic fossils.

It was alleged that the hegemonikon communicated with the exterior by way of the pneuma. We have already emphasised that for the Stoics natural science, including physiology, was studied merely as a prelude to the central concern of philosophy—ethics. Consequently it is not surprising to find that the major Stoic philosophers failed to explain in anatomical or physiological detail exactly how the pneuma allowed the central hegemonikon to communicate with the exterior. There is evidence, however, that they accepted the Erasistratean idea that, while blood is present in the veins, the arteries contain pneuma (17). Hence it is not unreasonable to suppose that the arterial system, distributed as it is throughout the body and centred on the heart, might have provided the communication system required by the Stoic theory.

Now we have already noticed that the Stoics conceived the pneuma as an elastic fluid under a certain tension. Such a fluid, they understood, could conduct a wave. Chrysippus, for instance, is reported (18) to have explained that 'sight is due to...the signal being transmitted to the observer by means of stressed air'. A similar concept seems to have accounted for the transmission of information between the hegemonikon and the body's periphery. Chrysippus writes as follows:

> In the same way as a spider in the centre of a web holds in its feet all the beginnings of the threads, in order to feel by close contact if an insect strikes the web, and where, so does the ruling part of the soul, situated in the middle of the heart, check on the beginnings of the senses, in order to perceive their messages . . . (19).

There is no suggestion of hydrodynamics, no indication that the pneuma *flows* in the arteries: rather that shock waves travel to and fro in an elastic medium. We have already seen that Galen, to whose eclecticism Stoic ideas must have been quite familiar, proposed a somewhat similar mechanism for the transmission of messages along nerves. Chrysippus believed that a similar mechanism accounted for the soul's control of bodily movement. Like Aristotle he believed that the psychological impulse was the first cause of motion. He believed that this impulse was transmitted via the pneuma to the effector organs (20).

Finally, to end this chapter, it is not difficult to see that the Stoic concept

(17) See, for instance, Seneca, *Naturales Quaestiones* (trans. T.H. Corcoran), The Loeb Classical Library, Heinemann, London (1971), 3, 15, 1: ' . . . the earth is governed by nature and is much like the system of our own bodies in which there are both veins (receptacles for blood) and arteries (receptacles for air).'
(18) Diogenes Laertius, *Lives of Eminent Philosophers*, 7, 157.
(19) Sambursky, *ibid.*, p.24
(20) Zeller, *ibid.*, p.244.

of 'soul' and 'body' grades easily into the theories held by theologians in mediaeval Europe; for it is, in an important sense, far more open to a 'two-substance' interpretation than the Aristotelian 'soul is form' ... The theory of 'commingling' mentioned above implies that, for many Stoics, the soul or pneuma was a shadowy replica of the body: a ghost which energised the machine. Thus, to quote the great nineteenth-century authority on late classical thought:

> The statements of Seneca, that this life is a prelude to a better; that the body is a lodging house, from which the soul will return to its own home; his joy in looking forward to the day which will rend the bonds of the body asunder, which he, in common with the early Christians, calls the birthday of eternal life, his description of the peace of eternity there awaiting us, of the freedom and bliss of heavenly life . . . all contain nothing at variance with the Stoic teaching, however near they may approach to Platonic or even Christian modes of thought (21).

(21) Zeller, *ibid.*, p.220.

12 ALCHEMY

Aristotle's Physics *applied*

To the majority of modern readers, especially those with a scientific training, the word 'alchemy' conjures up only ideas of charlatans engaged in an obscurantist search for impossible objects. We have all seen pictures of the alchemist toiling at his retort amidst a chaos of clumsy, seemingly dusty, apparatus. All scientists hope that this picture is the antithesis of one of themselves at work in their laboratory. With the establishment of chemistry towards the end of the seventeenth century, the vigorous reaction against alchemy which had been gathering momentum for several centuries finally burst into the alchemical laboratories and swept the alchemist and his theories into near oblivion. This, anyway, is how it seems from our twentieth-century vantage point. In fact the reality was more gradual: there is no doubt that the science of chemistry evolved from the alchemist's art.

Alchemy may be seen as the last and greatest of those conflations of the animate with the inanimate which, as we have noticed in earlier chapters, is so marked a character of ancient science. As such it is of peculiar interest in the context of this book. The *leitmotiv* running through the obfuscations and complexities of alchemical literature is an attempt to treat inorganic nature in terms which we now regard as appropriate only to animate creation. In this conflation alchemy is, as Hopkins observes, the true child of Greek thought (1).

Greek thinkers had long recognised that one of the most striking features of an organism was its metabolic power to transform food materials into its own likeness. This ability had riveted the attention of both Aristotle and Galen. Does this metabolism also occur in what we now regard as the inorganic world? The alchemists believed that it did. They treated the happenings of the inorganic world in terms more appropriate to the happenings within a living organism. To this extent alchemy is the offspring of the Peripatetic philosophy. It is also, as we shall see, strongly influenced by mystical currents originating in neo-Platonism.

We realise today that the paradigm was wrong. Although geologists and geochemists will nowadays confirm that the lithosphere does have a chemistry, does undergo cycles of transformations, these changes are never

(1) A. J. Hopkins, *Alchemy, Child of Greek Philosophy*, Columbia University Press, New York (1934).

Figure 12.1. Alchemy. This late sixteenth-century painting by Johannes Stradanus shows many of the features of an alchemical laboratory. At the centre of the picture is depicted a furnace and a cluster of stills. Singer (1956) believes that the majority of workers depicted are involved in the preparation of herbal medicines.

viewed in the framework of organism. Crustal chemistry, in addition to being almost infinitely slower than physiological chemistry, strives to no end. The reactions, at root, are treated as expressions of inert matter tending to its lowest energy state. There is no suggestion of matter arranging itself for the good of the whole or of striving to achieve perfection. It is interesting to notice, once again, that in modern science the pendulum has tended to swing to the opposite extreme. Biochemists and biophysicists strive to account for the transformations of matter within organisms in terms derived from the sciences of inorganic matter.

The artisan's Aristotle

We have already observed that after the death of Aristotle the centre of intellectual activity had moved from Athens to the nascent city of Alexandria. In this polyglot city on the estuary of the Nile the Peripatetic tradition flourished but, outside the Museum, tended to become alloyed with a narrower vision, a more superstitious outlook. The Aristotelian theory of matter seemed, indeed, well adapted to explain the experiences of metallurgists and jewellers, tanners and perfume makers. The mind of the reflective smith was at home in a universe of qualitative change and of progress from potential to actual. But the broader vision of the Aristotelian system was of no interest to a practical-minded craftsman.

Holmyard points out (2) that the word 'alchemy', although Arabic, has its etymological roots in ancient Egypt or perhaps Greece. The fact, however, that it is essentially an Arabic word reminds us that the great flowering of

(2) E.J. Holmyard, *Alchemy,* Penguin, Harmondsworth (1957).

alchemical art occurred in the Moslem world in the eighth and ninth centuries AD. Islamic alchemy seems to have developed partly from contact with the artisans of Alexandria and partly from contact with the large number of philosophical and scientific treatises, many originating in the Lyceum, translated in the early centuries AD from their Latin and Greek originals into Arabic. Islamic workers thus developed and systematised the works of their forerunners in Greece and Egypt. This development amounted, in effect, to the elaboration of a theory of matter.

This theory of matter accounted for the manifold phenomena of the jeweller's shop, the smith's furnace and the dyer's vat. But the theory was not merely a descriptive device. Like all scientific theories it aspired to being predictive. Understanding, as Francis Bacon was later to point out, brings power; understanding of what happened when matter metabolised was a prerequisite for technological – economic gain. Hence the alchemist's belief in and search for the object which Geoffrey Chaucer calls 'the privy stoon'. He who possessed this possessed the power of transmutation, possessed, in other words, the metabolic power which (so they believed) all nature possessed. In many cases also they hoped in addition to obtain the power of prolonging human life. These two sides of alchemy, the theoretical or speculative and the practical, are well brought out by Roger Bacon (3). The former

> treats of the generation of things from the elements and of all inanimate things and of common stones, gems, marbles, of gold and other metals, of sulphurs and salts and pigments, of lapis lazuli and minium and other colours, of oils and burning bitumens and other things without limit...

while the latter

> teaches how to make noble metals...and also how to discourse on such things as are capable of prolonging human life....

As in modern science, the distinction between the two, between the pure and the applied, was always somewhat arbitrary: the two aspects of the subject were inextricably interwoven. Both aspects have, however, considerable interest for the historian of the concept of organism; for in both aspects we can follow the working out of the Aristotelian 'biological' physics to its ultimate limits.

Alchemy had, however, as we have already mentioned and as we shall see more fully below, many other sources than the comparatively straightforward Peripatetic tradition. From its earliest days in pre-Christian Alexandria it was strongly influenced by mystical and later, after the work of Plotinus, by neo-Platonic thought. Thus the description of an alchemical reaction may stand for man's evolving relation with the deity, or the soul's upward gyres toward the beatific vision. On the other hand a poem ostensibly about the struggles of fabulous beings in an apocryphal world may in fact symbolise the chemical procedures recommended for a transmutation.

The meaning of many alchemical writings is thus at best obscure and open to many different interpretations; at worst it is totally unintelligible.

(3) Quoted in *ibid.*, p.117.

The irritation which such ambiguity and obscurity induced is well expressed in the book which founded the modern science of chemistry: '...such captious subtleties', writes Robert Boyle,

> do indeed often puzzle and sometimes silence men, but rarely satisfy them. Being like the tricks of jugglers, whereby men doubt not that they are cheated, though oftentimes they cannot declare by what slight they are imposed on (4).

Thus all that can be attempted here is to extricate some of the recurrent themes in the thousand-year history of alchemy and to use them as lenses to illuminate the mediaeval concept of 'inorganic' matter.

The psychobiology of matter

Chaucer's assistant alchemist, when taxed with his ruined complexion, replied that his accustomed place was at the furnace and that

> I am nat wont in no mirour to prye,
> But swinke sore and lerne multiplye (5).

His avowed aim is thus 'to lerne multiplye'. In the twentieth century we are only too well acquainted with the notion of a 'chain reaction'. We are well aware that in this way one neutron may be multiplied a millionfold. Something of the same sort was striven for by the mediaeval alchemist—only in his case, what he was concerned to multiply was gold. The analogy between the achievements of the modern radiochemist and atomic physicist and the objectives of the alchemist is, as many have pointed out, quite close. Perhaps, however, a biological analogy corresponds even more closely to the alchemist's dream. Seeding a microorganism into a nutrient broth quickly leads to a vast multiplication of living matter. The alchemist hoped to achieve something very similar in the world of minerals.

For, to the alchemist, all nature was alive. A passage from a very influential work published by Petrus Bonus about AD 1330 entitled *The New Pearl of Great Price* may help to clarify this alchemical vision. Bonus writes as follows:

> Something closely analogous to the generation of alchemy is observed in the animal, vegetable, mineral and elementary world. Nature generates frogs in the clouds or by means of putrefaction in dust moistened with rain....Avicenna tells us a calf was generated in the clouds amid thunder....In the dead body of a calf are generated bees, wasps in the carcass of an ass, beetles in the flesh of a horse and locusts in that of a mule. ...The same law holds good in the mineral world, though not to quite the same extent (6).

Here we have a very plain indication that the modern idea of inorganic matter was alien to the mediaeval mind. 'Metals', writes Cassirer (7), 'were for Cardanus only buried plants leading an underground existence, stones have their development, their growth, their maturity....' The phenomena of

(4) R. Boyle, *The Sceptical Chymist,* Cooke, London (1661), p. 15; facsimile reprint by Dawsons, London (1965).

(5) G. Chaucer, *The Canon's Yeoman's Tale,* 668–9, in *The Complete Works of Geoffrey Chaucer* (ed. W.W. Skeat), Oxford University Press, London (1912).

(6) Holmyard, *ibid.,* pp.140–1.

(7) E. Cassirer, *Individual and Cosmos*, Oxford University Press, London (1963), p.149.

embryology were not confined to the biological world. They are also to be observed, though on a rather longer time scale, in the world of minerals. For Petrus Bonus the generation of metals may take thousands of years and occurs in the bowels of the earth. It is helped by a gentle heat both from the sun and from the earth's centre. It is the alchemist's ambition to repeat and speed up this geological epigenesis in the laboratory. Hence the centrality of the furnace.

The belief that minerals fructify in the womb of the earth is thus one of the basic tenets of theoretical alchemy. In order, however, to understand why the alchemist thought it possible to achieve a transmutation, a further idea has to be added. This is the idea of perfection. It was emphasised above that alchemy grew up within the framework of Peripatetic theory which, as we noticed in earlier chapters, accorded great significance to qualitative change. The journey from imperfection to perfection is most certainly a qualitative change. But we also emphasised above that alchemy had strong connections with neo-Platonic thought. One of the major themes of the latter tradition was an equation of moral worth with different levels of 'being'. Starting with inert, formless, inchoate matter, the great chain of being pressed upward through minerals, plants, animals, men, principalities and powers to God. The scale of moral worth ran alongside, in parallel. Thus we can see how it was possible for alchemists to regard their operations as techniques designed to accelerate the otherwise slow ascent of matter towards perfection. We can find, for example, in one of the most important texts in the whole of mediaeval alchemy the following definition of the subject: 'This science [alchemy] treats of the *Imperfect Bodies* of *Minerals* & teaches how to perfect them; we therefore in the first place consider *two things, viz. Imperfection* and *Perfection*' (8).

Gold was valued for its constancy and incorruptibility. We are at once reminded of the stress placed by the Presocratics on unchangeableness. Hence gold was taken to be the 'noblest' of metals to which the baser metals aspire. In the earth's womb, gold is formed by an almost infinitely slow embryology. Geber recounts one of the observations which may have sustained this mistaken belief. He describes how water flowing out of copper mines carries with it 'scales of copper' which, after much washing and digestion in the heat of the sun, over a period of years is eventually transformed into pure gold. Holmyard (9) provides a modern explanation for this opinion. He points out that many copper ores contain small quantities of gold, and that this may be revealed after weathering by water and sun.

Thus we can discern the pattern behind the apparent obfuscations of the alchemist. Matter, for him, was not the inert, inanimate stuff we believe in today. Newton and his blindly running atoms were still far in the future.

(8) Geber, *The Investigation of Perfection,* translated by Richard Russell in 1678. Geber is the latinised name of a very eminent, though perhaps mythical, Islamic alchemist, Jabir ibn Hayyan (A.D. 721–813)
(9) Holmyard, *ibid.*

Epicurus lay neglected in the past. Matter, for these early experimentalists, was subject to qualitative change and development, naturally strove towards perfection, could be helped on its way and was, moreover, affected by influences emanating from the stars and planets. The world, in short, was still a living organism, a system held together by numberless sympathies. The microcosm and the macrocosm, as the Stoics saw, reflected each other back and forth. No alchemist began his work without first taking the positions of the heavenly bodies.

The stuff with which the alchemist laboured possessed (perhaps projected from the alchemist's own psyche (10)) its own rich personality. The allegories through which the alchemist reported his results and methods powerfully sustained him in this attitude towards matter and material change. The boundary which became so apparent in later centuries, the boundary which divides the quick from the dead, the organism from the inorganic, seemed hardly to exist in the alchemical laboratory. Matter, like a foetus, underwent its own peculiar embryology, albeit immensely more drawn out. In both cases there was progress from imperfection to perfection. Matter, too, had for the alchemist its own characteristic spirit. Many alchemical procedures were indeed designed to facilitate the achievement of perfection by separating body from spirit. Thus the development of the still, fairly late on in the history of alchemy, was one of its major theoretical and technological advances. The spirits which it then became possible to distil off from the 'crude' must have seemed to confirm all their theories. This 'essence' both strove upwards, tending to burst like a genie from any container, and, in the case of the alcohols, concentrated a 'vitality' unknown to the original. The terms still remain in the vocabulary of modern chemistry. We still refer to methylated 'spirits'.

None the less, mistaken though their theoretical structure was, it would be wrong to suppose that in a millennium and a half of painstaking effort the alchemists did not stumble across a fair number of 'genuine' chemical facts and techniques. We have already noticed the important development of the still, and it is sometimes possible to 'see' in their flowery and allegorical writings the outlines of a 'hard' laboratory procedure. That in a theoretical sense they were unable to make anything lasting out of their discoveries was perhaps not altogether their fault. It requires genius far out of the common run to perceive in nature something completely alien to our accustomed thought process. Max Planck put it well when he wrote '...a new scientific truth does not triumph by convincing its opponents and making them see the light, but rather because its opponents eventually die and a new generation

(10) C. Jung, *Psychology and Alchemy* (trans. R.F.C.Hull), *Collected Works,* vol. 12, Routledge and Kegan Paul, London (1953): 'The alchemy of the classical epoch (from antiquity to about the middle of the seventeenth century) was, in essence, chemical research work into which there entered, by way of projection, an admixture of psychic material' (p. 476).

grows up that is familiar with it' (11). And the facts and truths which the alchemists turned up were particles of the modern world displaced far in time out of their proper context. The alchemists were not 'set' to weigh and measure and look for functional relationships; their 'set' was qualitative, based on the interchangeability of the ancient elements, Peripatetic and neo-Platonic. Indeed, the alchemists had set themselves to discover something which just stubbornly did not exist. If persistence and resolution are virtues, then many alchemists were virtuous: with unremitting wearisome labour they tried to force the material world into a mould which it just did not fit. This Procrustean exercise was doomed; centuries later Francis Bacon wrote an epitaph for alchemy and alchemists: It is not for man to narrow down the world to the confines of his intellect but rather for him to enlarge his vision to encompass the world (12). Thus to the tenacious and, as Geber put it, the 'stiff necked' alchemist, the best words of advice available were those of the sceptical Chaucer:

> sith god of hevene
> ne wol nat that the philosophres nevene
> How that a man shal come un-to this stoon,
> I rede, as for the beste, lete it goon.
> For who-so maketh god his adversarie
> As for to werken any thing in contrarie
> Of his wil, certes, never shal he thryve (13).

(11) M. Planck, *Scientific Autobiography and Other Papers* (trans. F. Gaynor), New York (1949), p. 33.
(12) Of alchemists in particular, Bacon has a characteristically witty apophthegm: '...we may well apply to them the fable of the old man, who bequeathed to his sons somē gold buried in his garden, pretending not to know the exact spot, whereupon they worked diligently in digging the vineyard and though they found no gold, the vintage was rendered more abundant by this labour'. *Novum Organum* (1620), 85.
(13) Chaucer, *ibid.*, 1474–8.

13 MUTATION OF THE PARADIGM

The nature of Kuhnian paradigms

T S Kuhn, in his discussion of the nature of scientific paradigms (1), shows that they first gain a foothold by being more successful than their competitors in accounting for certain critical problems. Scientific advance, Kuhn goes on, then consists in the demonstration that the paradigm is appropriate to wider and wider classes of phenomena. Eventually a successful paradigm articulates a wide field of observations into a consistent logical structure.

Now in our discussion so far we have stressed that the predominant paradigm has all along been the concept of organism. Subject matters which we nowadays believe to be entirely non-organismic in character, chemistry, for example, and physics, were treated, as we have seen, in terms which we now consider appropriate only to organisms. We saw, in the last chapter, a good example of this Procrustean practice. Material change was treated as if it were occurring within a metabolising, developing 'world' organism.

In this chapter we shall see the organismic paradigm begin to be replaced by one generally known as 'mechanistic'. It is not, of course, true that the latter paradigm sprang fully armed from the head of Galileo Galilei. We shall note, briefly, that it had a long gestation in mediaeval Europe. Fragments of it also existed in the classical world. We have seen how Leucippus, Democritus and their followers proposed an alternative to the Peripatetic vision. But we have also noticed the response of biologists and theologians to this premature biophysics. Epicureans, moreover, as self-confessed atheists, could hardly hope to be influential during the ages of faith. Yet, such are the paradoxes of history, it seems that it was partly the reorientation of thought following the donation of Constantine that lit the fuse leading to the seventeenth-century explosion of science.

(1) T.S. Kuhn, *The Structure of Scientific Revolutions,* University of Chicago Press, Chicago and London (1962). Kuhn has somewhat revised his ideas in his later work, but the notion of underlying paradigms or, as Holton calls them, *themata* – 'unverifiable, unfalsifiable and yet not-quite-so arbitrary hypotheses' which arise from 'our general intuitions and beliefs about how we expect nature to behave' (*Nature,* **247** (1974), 317) – retains its value.

The evolution of machinery

Why, then, did the mutation occur? And why did it occur in western Europe rather than in other, perhaps more civilised, parts of the world? Why did a paradigm which, at root, conceives the relations between parts of a complex whole as 'external', as if in a piece of clockwork, displace a paradigm which saw such relations as 'internal', as consisting of a system of mutual sympathies? These are questions which have occupied the working lives of many historians of science. To mention just one famous contemporary instance: Joseph Needham left a brilliant career in biochemistry to devote three decades to attempting to understand why the change did not occur in the great civilisations of the orient (2). It follows that in this section we can do no more than look at some of the hints these great scholars have provided.

We may perhaps notice first of all that throughout the mediaeval period there was a slow but continuous development of technology. We saw, indeed, in the last chapter that the practical-minded alchemist could hardly help stumbling across some 'hard' facts even when working within an alien paradigm. The implication that the modern metaphysics is partly tied up with the development of technology is interesting in the context of the present book. It will be recalled from chapter 3 that the origin of symbolic thought itself is believed to have been very closely associated with the development of tools and weapons.

Some of the significant dates in the development of technology during the mediaeval period are shown in table 13.1. Perhaps the most important innovations in the context of the present chapter are those to do with power and mechanical engineering. In classical civilisations the term *mechanē* seems to have been applied to any contrivance which interposed a number of moving parts between the motive force and the acting part. In classical times the motive force was in practically all cases either human or animal muscle. Figure 13.1 shows such a mechanical contrivance. A crane used for the building of Rome is energised by a treadmill worked by human muscle. We should thus not be surprised at the widespread classical identification of prime movers with living souls (chapter 10). The ideas for many simple machines were, moreover, derived from living nature. The shells of bivalves suggested pincers, the bones of human anatomy suggested levers. During the centuries following the birth of Christ, however, a progressive development of machines in their own right and motivated by non-living forces gathered momentum. First water and then wind were harnessed as sources of motive power.

Perhaps, indeed, the origin and growth of Christianity and the contemporaneous growth of technology are not entirely coincidental. Several

(2) J. Needham, *Science and Civilisation in China,* Vols. 1, 2, 3, 4 (continuing), Cambridge University Press (1954–74).

a

b

Figure 13.1. Treadmill cranes. (a) Bas-relief of a crane used during the building of Rome in classical times. (b) the large treadmill crane at Bruges was still in use in the sixteenth century AD. Both these illustrations emphasise the significance of animate prime movers in ancient and mediaeval technology.

Table 13.1. Development of technology in Europe in the dark and middle ages

A.D. 300	Water wheels in common use with gearing to allow the grinding of corn[a]
A.D. 400	Inception of the 'overshot' water mill and the development of mill ponds, mill races etc[a]
A.D. 500	Introduction of the stirrup, probably from China[b]
A.D. 600	
A.D. 700	
A.D. 800	Earliest Chinese reference to gunpowder, 850[c]
	Earliest known printed work, Chinese, 868[d]
A.D. 900	Earliest European reference to a collar in the harness of horses, 920: allowed more effective use of the horse as a draught animal[c]
A.D. 1000	Water-powered hemp and fulling mills in France, 1050[c]
	Tidal mill recorded in Dover harbour, 1066[a]
	Domesday book records 5624 water mills for grinding corn in Southern England and several water mills for other purposes, 1087[a]
	The introduction of mould boards on ploughs allowed the more effective tilling of heavy north west European soils[e]
A.D. 1100	Breast-strap harness for horses developed, 1130[c]
	Translations from Arabic into Latin included the introduction of Arabic numerals, 1165[c]
	First mention of windmills in Western Europe (Normandy), 1180[c]
	First reference to stern-post rudder instead of steering oar, 1180[c]
	Introduction of paper through Spain and Southern France[f]
	Construction of Gothic cathedrals in France and England involved a practical understanding of statics[f]
A.D. 1200	Villard de Honnecourt's notebook (33 parchment sheets), 1235, shows many engineering and architectural devices including flying buttresses, water-powered sawmills, screw-jacks, etc[g]
	First illustration of wheelbarrow in Europe, 1250[c]
	First record of spinning wheel, 1280[c]
	Compass in use by European seamen, 1283, although recorded in China from 1117[c]
	First record of spectacles, 1284[c]
A.D. 1300	Berthold Schwarz invents gunpowder, 1313, although known in China from 850[c]
	Earliest record of European cannon, Metz, 1324[c]
	Florentines use cannon in war, 1326[c]
	Mechanical clock at Dover castle, 1348[c]
	Clockwork automata striking the hours etc., Orvieto, 1351[h]
	Earliest reference to a tower windmill, i.e. can be turned to face the wind, 1390[a]
	Movable metal type printing developed at Limoges, 1381[f]
A.D. 1400	Windmills geared to Archimedean screw to raise water, Holland, 1404[c]
	First illustration of a crank mechanism for converting rotary to reciprocating motion, 1421[i]
	Brunelleschi (1377–1446) completes dome of Cathedral, Florence, 1420–1434
	Alberti's *Architettura*, 1450, shows good understanding of mechanics
	Leonardo da Vinci, b. 1452

[a] J.R.B. Forbes, Power, in *A History of Technology* (ed. C. Singer *et al.*, vol. 2, Oxford (1956) pp. 589–622.

[b] E.M. Jope, Vehicles and harness, in Singer *et al.*, *ibid.*, pp. 537–62.

[c] R.L. Storey, *Chronology of the Medieval World,* London (1973).

[d] G.O. Walter, Typesetting, *Sci. Amer.*, 220 (1969) (5), 60–9.

[e] J.M. Jope, Agriculture, in Singer *et al.*, *ibid.*, pp. 81–102.

[f] A.C. Crombie, (ed), *Scientific Change,* London (1963).

[g] F. Klemm, *A History of Western Technology,* trans. D.W. Singer, London (1959).

[h] M. Daumas, *Histoire générale des Techniques,* Paris (1964).

[i] B. Gille, Machines, in Singer *et al., ibid.,* pp.629–57.

authorities (3) have suggested that the interdict placed on animism by the early fathers of the church altered the particular relationship between man and nature which had been so prevalent from the earliest times. Instead of 'seeing' natural events as the manifestation of a living force, a hidden will, the peasant and artisan slowly began to 'see' that the happenings in their environment were there to be controlled and manipulated, if not for the greater glory of God, then for the alleviation of their own daily labour.

Whatever may have been the sociological and psychological compulsions, the fact remains that throughout the mediaeval period there was a slow but steady growth of technological knowledge and know-how. It has been said that craftsmanship is one of the most conservative of human activities. This conservatism must have been more marked during the middle ages than in more modern times for, in general, the mediaeval craftsman could hardly have understood, in our sense of that term, exactly what he was doing. For, firstly, he was frequently illiterate, out of touch with the world of learning which, literally, spoke and wrote in a different language: Latin. And, secondly, even had he been *au courant* with the world of letters, he might not have been greatly enlightened; for, as we have been stressing, mediaeval thought did not move within a paradigm appropriate to engineering technology (4). In some respects, we may suppose the craftsman and artisan was fortunate in his position: he was left to get on with his craft as best he might, unworried by theoreticians. He learnt his craft from his seniors and passed it on to his juniors, generation after generation. Jenkins, for example, mentions ten, perhaps a hundred, generations of Welsh coracle builders (5). Yet, in spite of this deep-rooted conservatism, innovation could and did occur. Such innovations must have appeared and spread rather like beneficial mutations in the modern theory of evolution. To continue the analogy, deleterious changes in technique would have been quickly eliminated; indeed they might well have been literally lethal. Thus the craftsman's stock-in-trade, like the genetic make-up of a population, must have slowly changed as the centuries passed—and always changed in one direction: towards a greater control of the environment, towards greater manipulative ability.

(3) See, for instance, L. White, Technology of the middle ages, in *Technology and Western Civilisation,* vol. 1 (ed. M. Kranzberg and C.W. Pursell), Oxford University Press, London (1967), p. 68. Lynn White makes much the same point about the exploitative nature of the Judaeo-Christian ethic in his influential paper, The historical roots of the ecologic crisis, *Science,* **155** (1967), 1203–7.
(4) We shall see in the next section of this chapter, however, that scholastics frequently interested themselves in technology and sought its theoretical basis.
(5) T.G. Jenkins. The ancient craft of coracle building, *Country Life,* **125** (1959), 716–17.

In addition to new prime movers in the form of waterwheels and windmills, the middle ages also saw the development of machinery in the classical sense mentioned above: moving parts interposed between the prime mover and the working part. Thus, in the eleventh and twelfth centuries, waterwheels powered not only mills for grinding corn but also mills for sawing timber, for

Figure 13.2. A large cannon being raised onto a gun carriage. Leonardo's drawing (*c* 1485–8) shows several features of fifteenth-century technology. It again emphasises the dependence of machinery on the contractile power of muscle. The stacks of cannon in the background are, however, prophetic of a new source of energy.

grinding cutlery, for tanning, for turning wood, for making paper and for many other purposes. With the development of metallurgy, water-powered mills were also adapted to pump bellows at primitive blast furnaces.

The accelerating development of metallurgy also gave a powerful impetus to military technology. The selective force acting upon this technology was, of course, always particularly intense. Then, as now, military engineers frequently showed the way forward to the rest of technology. Gunpowder, probably derived from China, appeared in Europe at the beginning of the fourteenth century, and this very rapidly led to the development of many different sorts of firearm. Warfare was revolutionised. The invention and development of cannon reacted back on metallurgy and forced the development of metals technology. It also increased the interest of theoreticians in the science of ballistics: a science of very considerable significance in the origins of the seventeenth-century scientific revolution. Last, but far from least, several authorities see in the explosive action of cannon an early invitation and stimulus to the study of pneumatic chemistry.

Towards the end of the fifteenth century the notebooks of Leonardo da Vinci (1452 – 1519) are filled with the designs of intricate machinery (figure 13.2). It is interesting also to note that when he sued for employment by the Duke of Milan he recommended himself not so much as an artist but more as a highly competent military engineer. He claimed to have 'fully studied the works of all those who claim to be masters and artificers of the instruments of war' (6), and went on to enumerate the various different war machines he felt himself competent to design and construct.

The universality of Leonardo's interests are nowadays thought to be less the index of outstanding originality than the sign of voracious reading in the literature circulating in fifteenth-century Italy. This, of course, increases rather than detracts from the interest of the notebooks. The crabbed mirror-writing and the superb drawings preserve the insights, the know-how, in short the state of the technological art in the centuries immediately preceding the scientific revolution. Although Leonardo is not generally regarded as one of the great instaurators of the Renaissance revolution in physical science—though a harbinger he certainly was—his place in the early history of biology is more secure. His anatomical drawings have never been surpassed. He is said to have been the first man since classical antiquity to have seriously questioned the inheritance of acquired characters (7), and his understanding of geology antedated Lyell by some four hundred years.

Scholastic disputation

Francis Bacon (1561 – 1626) characterised the efforts of the schoolmen as the work of spiders endlessly spinning webs to catch the man of common sense and, in another place, he refers to them as *cymini sectores*—hair-splitters (8). 'This kind of degenerate learning', he writes (9),

> did chiefly reign among the schoolmen: who, having sharp and strong wits, and an abundance of leisure, and small variety of reading, but their wits being shut up in the cells of a few authors (chiefly Aristotle, their dictator) as their persons were shut up in the cells of monasteries and colleges, and knowing little history, either of nature or time, did out of no great quantity of matter and infinite agitation of wit spin unto us those laborious webs of learning which are extant in their books. For the wit and mind of man, if it work upon matter, which is the contemplation of the creatures of God,

(6) K. Clark, *Leonardo da Vinci,* Penguin, Harmondsworth (1958), p. 45.

(7) C. Zirkle, The knowledge of heredity before 1900, in *Genetics in the Twentieth Century* (ed. L.C. Dunn), Macmillan, New York (1951).

(8) F. Bacon, Of study, *Essays,* Charles Knight, London (1840).

(9) F. Bacon, *The Advancement of Learning* (1605), in *Philosophical Works of Bacon* (ed. J.M. Robertson), London (1905), p. 55.

worketh according to the stuff and is limited thereby: but if it work upon itself, as the spider worketh his web, then it is endless, and brings forth indeed cobwebs of learning, admirable for the fineness of thread and work, but of no substance or profit.

Although Francis Bacon is probably better taken as one of the heralds or, to use his own term, 'trumpeters' of modern science rather than as one of its 'instaurators' (10), his opinion of the schoolmen was widespread among the early scientists. The endless disputations of the schools, the bringing forward of argument and authority to be countered by the production of counter-argument and counter-authority, seemed to the positive spirits of the scientific renaissance a singularly futile process. The distaste which the Renaissance thinkers felt for the alien procedures of the schools still colours our vision; it is still difficult for the modern mind to penetrate the smokescreen of adverse propaganda generated by the sixteenth- and seventeenth-century revolution and discover the genuine interests of the scholastic. Yet if the attempt is made we frequently find that the debates of the fathers of the church, although ostensibly about very different matters, in fact formed part of the complex crucible of interests out of which, as an amalgam, arose Galilean science.

The chaotic conditions prevailing in western Europe after the fall of Rome (11) in the fifth century AD led, until about the twelfth century, to an almost complete loss of classical knowledge. It is possible to detect, however, beginning in the tenth century and slowly gathering strength throughout the succeeding two centuries, the first stirrings of a rebirth of learning. These stirrings initially took the form of translations of numerous Arabic works of medicine and science (12). These Arabic treatises and encyclopaedias were themselves translations, often via the intermediate language of Syriac, from the Greek works of classical antiquity. In this way classical thought, often in strange disguise, percolated back into Christendom.

Thus at the beginning of the thirteenth century AD we find that most of Aristotle's thought is once more available in the universities and monasteries of the west. And it was soon clear that the Stagirite's system was, as we ourselves saw in chapters 8, 9 and 10, all-embracing and self-consistent. On the other hand it was not, perhaps, at first sight fully consistent with the truths of revealed religion. Hence Aristotelianism had, in the beginning, a somewhat mixed reception. The philosophy was banned by several Popes, and interdicts were placed upon its exposition in the Schools. However, by the end of the thirteenth century, three great theologians had shown that

(10) Francis Bacon has sometimes been styled 'The Great Instaurator' from the title of one of his books, *The Great Instauration,* in which he sought to lay the foundations of a post-Scholastic inductive science.

(11) Rome was sacked by Alaric the Goth in AD 410.

(12) During the centuries after the fall of Rome the intellectual and cultural ascendency passed to Islam. From the Pyrenees almost to the banks of the Indus, Saracenic culture was dominant. Throughout this vast tract of the world, Arabic was the language of the intellect. Just as nowadays scientists strive to learn English, so in those days, and for centuries, a knowledge of Arabic was vital to the would-be scientist, philosopher or physician.

Aristotle and revelation did not disagree in essentials, and the Athenian philosopher became for the late middle ages 'The Philosopher'.

The three theologians principally responsible for effecting this near-apotheosis were firstly Alexander of Hales, secondly Albertus Magnus, thirdly, and greatest of all, Thomas Aquinas. The Thomist synthesis, the philosophical parts of which are still acceptable to the Roman church, used the Aristotelian system to bind theology and science together into a single whole. It seemed, consequently, that an attack on the scientific areas of the Stagirite's system was also an attack on the theological areas. We can thus, remembering the interlocking nature of the Aristotelian synthesis, understand the vigour with which the church defended what seem to us preposterous aspects of his science.

Thus the beatification of the Aristotelian system had nearly as many 'cons' as it had 'pros' for the advancement of science. However, it did have one very definite advantage, at least in the work of Albertus Magnus (13) and St Thomas, and this is that Aristotle, as against Plato, was, as we saw, preeminently a philosopher who believed that 'knowledge enters through the doors of the senses'. The Plato of mediaeval times was too often confused with Plotinus, and both of them with the notion that truth is ultimately to be found in the mystic ecstasy of the enthusiast. The works of the Stagirite have no element of this mystagogy. It will be recalled that the *Metaphysics* opens with a paean of praise to the senses (chapter 8), and through all his works we feel that he is struggling to make sense of the observational world.

But when men began again to take Aristotle seriously, and to compare what he said with what their own senses told them, serious difficulties began to emerge. It is partly through an analysis of these difficulties that thinkers groped their way towards the seventeenth-century revolution. Thus once again we may marvel at the twists and paradoxes of the history of scientific ideas. It was Aristotle, the most empirical of the great classical philosophers, whom the founders of modern science had to combat along the long road leading to the seventeenth century.

This iconoclasm was, of course, often dangerous. In consequence one of the techniques much used in scholastic argumentation was to treat the whole debate as a game. Positions and counterpositions were endlessly advanced, not, so the protagonists asserted, because they were in any sense true, but as intellectual exercises to test the sharpness of the opposition's wits. In a way this divorce of the theoretical intellect from the everyday practicalities of truth and falsehood reminds us of an aspect of the scientific method. The hypothetico-deductive method involves a rather similar attitude: a *hypothesis* is made about the way things are, and *at this stage* the truth or falsity of the proposition is not in question. We, however, nowadays insist that the consequences of the hypothesis be deduced and tested, not

(13) A common refrain in Albertus Magnus is the expression, *'fui et vidi experire'*, 'I was there, I saw it happen'.

against the opinions and wits of other disputants, or against the writing of a classical or theological authority, but against the appropriate sense data.

From another angle again we can perhaps also detect in these thirteenth-century analyses a cast of mind which still forms a significant strand in the scientist's approach. This cast of mind is represented by the movement of thought called nominalism, and in particular by the work of William of Occam (14). In essence nominalism was a reaction against what seemed and still seem the absurdities of the Platonic metaphysics of universals. It seems to us absurd, and by definition it seemed to the nominalists absurd, to suppose that in addition to the many individual horses existing in the world there also existed, in the same sense of that verb, a universal horse. 'Entia non sunt multiplicanda praeter necessitatem.' Nominalists wished to assert that universals were concepts restricted to the human mind; that the word 'horse' was merely a sign applied by humans to a group of rather similar animals.

Now the point about all this which is of interest to us in the context of the present book is that the nominalist was not *per se* interested in what things 'really are in themselves', which he was content to leave to the authorities of the church, but only in the relations and internal consistencies of his systems of signs. Thus we can see in this movement within fourteenth-century scholasticism the forerunner of our modern formalisms. Indeed Dijksterhuis (15) goes so far as to assert that 'whenever in the work of a Schoolman of the fourteenth or fifteenth century we are struck by an utterance harmonising with present day physical conceptions the author proves to be a pupil or a follower of the *Venerabilis Inceptor*'. And, as we noticed in chapter 1, it was no less a scientist than Albert Einstein who believed that 'scientific concepts are free creations of the human mind and are not, however it may seem, uniquely determined by the external world'. The contents of modern science, unlike the contents of Peripatetic science, are calculating devices, predictive machinery: not descriptions of the 'subjective' states of matter.

In particular we can see in the nominalist movement the beginnings of the evaporation of the concept of 'soul' from the world of science; for 'soul' is *a fortiori* something felt privately and not something which can be caught in a system of descriptive signs. The latter, for the nominalist, describe only the behaviour of observables in the public domain. No hypotheses need be made as to what *causes* the behaviour: that is a theological or metaphysical question. But this conclusion, as it applies to human behaviour, was not of course explicitly drawn in the fourteenth century. It was not vigorously pointed out until the rise of Watsonian behaviour theory some six centuries in the future.

(14) William of Occam, the Venerabilis Inceptor (that is, he did not hold a chair in Theology), was an Englishman born at Occam in Surrey between 1280 and 1290.
(15) E.J. Dijksterhuis, *The Mechanisation of the World Picture* (trans. C. Dikshoorn), Oxford University Press, London (1961), p. 166

The projectile again

The arguments of the scholastics impinged on the physical world at one crucial point: for one of the questions which received endless discussion was the question of the projectile. This was a question which, it will be remembered, was left dangling rather unsatisfactorily by the Aristotelian system. It was the weakest link in the physics which the fourteenth century had received from Islam and, increasingly as time went on, from direct translation of the Greek originals. For, as we noticed in chapter 10, the Peripatetic theory of the projectile was contradicted by simple everyday practice: the practice of throwing. And, now, it was contradicted *a fortiori* by the action of the newly invented artillery.

It will be recalled from our discussion of Aristotelian physics that for the Stagirite all motion of a *corpus inanimatum* must, by definition, be caused by something else. In the absence of an external *motor conjunctus* the motion of an inanimate body would soon cease. Moreover this external *movens* must all the time be in contact with that which is being moved, for action at a distance is inconceivable: 'A body cannot act where it is not'.

We have already seen that this Peripatetic notion was from the first regarded with dubiety. We saw how Aristotle attempted to get around the question of the projectile by recourse to the rather implausible theory of antiperistasis. A second difficult case for his general theory was to find an explanation for the motion of a falling body. The final cause was clear: the tendency of a heavy body to seek its 'natural' place. But what could be the efficient cause? What could serve as the *motor conjunctus*? Some scholastics supposed that there was an inherent principle which actively caused a heavy body to move towards the earth. This, as Dijksterhuis points out (16), implies a softening of the Aristotlelian distinction between animate and inanimate. It would seem that the heavy body is in some strange way able to move itself. This must have struck many practical men as an odd conceit: and, indeed, there were many other theories in the fourteenth and fifteenth centuries to account for the observed facts of gravitational movement.

The idea of an internal motive force was, however, taken up by Scholastic thinkers interested in the related problem of the flight of a projectile. An alternative to the Aristotelian position had in fact been elaborated in antiquity by Philiponus. It was this alternative that was taken up and developed by the fourteenth-century Scholastics. In essence it was suggested that the projectile remains in motion after it has left the hand because the thrower has imparted an 'impetus'. The stone flying through the air is thus different from the same stone stationary on the ground in that it is endowed with a *vis impressa, impetus,* or *virtus movens.* This impetus gradually ebbs away, like heat from a hot body, after the projectile has left the projicient.

Many authorities have seen in this fourteenth-century impetus theory a

(16) *Ibid.*, p.177.

forerunner of the modern concept of momentum; for it was not long before, in the mind of John Buridan and others, the impetus theory was refined so that the quantity of impetus possessed by a body was calculated as the product of its velocity and mass. Furthermore, Buridan was of the opinion that impetus was lost when a body met a resisting force, whether viscous drag or gravitational attraction.

Yet, although the Scholastic impetus theory is superficially similar to the concepts of modern dynamics, it is, at a deeper level, greatly different. The metaphysical background, the world picture, remains alien. Fourteenth-century Scholastics followed Aristotle in 'seeing' motion as a *process*, as being initiated by someone or something. The influence of the Peripatetic philosophy remained paramount. Post-Newtonian physicists regard motion as a state and rest as the limiting case of that state. As a great Newtonian scholar puts it, 'In the new physics motion and rest were equivalent states neither of which require the continuous action of an external force in order to maintain themselves' (17). In contrast the fourteenth-century physicist's paradigm was still the ship hauled through the waters of the Aegean (chapter 9): motion in a resistive medium.

There was one exception to this pervasive paradigm, and this exception was heavenly. The ancients and the earlier Scholastics believed that the heavenly bodies moved in their courses under the impulsion of intelligences. In the fourteenth century, however, it began to be said that the stars and planets kept their *perpetuum mobile* because they moved *in vacuo*. There was, in consequence, no way for their inherent impetus to be dissipated.

It required the unification of heavenly and earthly dynamics to achieve the modern understanding. As has been often pointed out this unification was Sir Isaac Newton's unique contribution. After Sir Isaac the same mathematical system comprehends the movements of apples, planets, tides and projectiles. The projectile problem thus disappears. Its motion is no more (and no less) mysterious than the state of rest. Nevertheless even Newton was unable to free himself of all trace of mediaeval thought. For in order to make acceptable his proposition that a body (projectile) *naturally* continues in rectilinear motion or at rest, he found it necessary to import the notion of a *vis inertiae*, a fundamental laziness in things. This importation of an 'occult' force to account for the observations summed up in his first law stands in contrast to the clarity of his refusal to make hypotheses about the nature of the gravitational force: *hypotheses non fingo*.

Hypotheses non fingo was also, as we saw above, a characteristic of the Ockhamist school of mediaeval thought. It is strongly emphasised in the Newtonian tradition. Sir Isaac writes to Bentley, for instance, 'for the cause of Gravity is what I do not pretend to know' (18), and his great French

(17) I.B. Cohen, 'Quantum in se est': Newton's concept of inertia in relation to Descartes and Lucretius, *Roy. Soc. Notes and Records,* 19 (1964), 132–55.

(18) 17 January 1693: in *Isaac Newton's Papers and Letters on Natural Philosophy* (ed. I.B. Cohen), Cambridge University Press (1958), p. 298.

populariser, Voltaire, generalises this *docta ignoranta*:

> We ought to suppose we know no more of the cause of impulsion than we do of that of attraction. We even have not a greater idea of the one than the other of these powers; for no-body can conceive why a body has power by them to move another from its place. Neither, indeed, do we conceive better why the parts of matter gravitate towards each other. Newton himself did not pretend to know the reason of this attraction. He has only proved its existence... (19).

We shall see in the next chapter that the Galilean revolution was based on an identical shift of attention. Famously, Galileo directed his attention away from a search for the nature of essence and substance and proposed instead to concentrate on the correlation of certain 'accidents' or contingent properties. This redirection of interest thus lies very close to the roots of modern science and remains to this day one of its predominant characteristics. Attempts to visualise, to imagine, to 'feel on one's pulses' what is actually happening at the crucial points in the new dynamics are doomed to failure. We have, as we have already noticed, a magnificent calculating device, a system of signs which provides gigantic predictive power, but we have lost the Aristotelian 'feel' for what is going on: in short *hypotheses non fingo*; *entia non sunt multiplicanda praeter necessitatem*.

The mechanistic paradigm

Physics, and its most fundamental branch, dynamics, form a paradigm to which all the other post-Newtonian sciences aspire. In the nineteenth century, for instance, it was quite clear to Hermann Helmholtz that 'the task of physical science is finally to reduce all the phenomena of nature to forces of attraction and repulsion, the intensity of which is dependent upon the mutual distance of material bodies' (20), and in another place he writes, '...the phenomena of nature are to be referred back to the motions of material particles possessing unchangeable moving forces, which are dependent upon conditions of space alone ... (the vocation of theoretical natural science) will be ended as soon as the reduction of natural phenomena to simple forces is complete...'. And this is still the avowed aim of 'mainstream' biologists. The Aristotelian organismic paradigm has been replaced by one deriving from the seventeenth-century revolution in physical science. The Aristotelian notion of 'soul' either as self-moving mover or as form disappears from biological thought. Its departure was not, as we shall see, easy or sudden. It lingered in the minds of certain biologists well into the twentieth century. But it is orthodox nowadays to 'see' an organism as an exquisite piece of chemical engineering designed by the trial-and-error forces of natural selection' through several billion years. Its features are all considered to flow from the 'blind running' of Newtonian atoms, or their

(19) F.M. Voltaire, *The Elements of Sir Isaac Newton's Philosophy* (trans. John Hanna, 1738); republished by Cass, London (1967).

(20) H. Helmholtz, *Ueber der Erhaltung der Kraft,* Berlin (1847), trans. J. Tyndall in *A Source Book in Physics* (ed. W.F. Magie), McGraw-Hill, New York and London (1935), pp. 212–20.

modern successors. Thus we see that the Abderan conception which seemed so absurd to Aristotle is triumphant today. But, as we shall see, its triumph was not by straightforward conquest. It also involved a very significant development in matter theory itself. This development, together with neo-Darwinism, shows us, however, that Aristotle's scepticism was unwarranted.

14 ESCAPE FROM THE LABYRINTH

Plato and Plotinus

A common summing up of what happened during the scientific renaissance is to say that it was occasioned by a victory of Plato over the church's Aristotelianism. We may perhaps recall Walter Pater's compelling image of the young Pico della Mirandola, fresh from wide-ranging studies in the north, standing at the threshold of the room in Florence where Marsilio Ficino had just completed his Latin translation of Plato. The enthusiastic Pico was later to undertake the defence of nine hundred theses in Rome, an undertaking which was in the event proscribed by the Pope, and Ficino was to devote the rest of his life to the translation of Plotinus, 'that new Plato, in whom the mystical element in the Platonic philosophy had been worked out to the utmost limit of vision and ecstasy' (1). The Florentine Academy in which Ficino had been dedicated from his youth to the study and translation of Plato was indeed a focal point in the focal city of the Italian Renaissance.

These images suggest that the Florentine appreciation of Plato still retained a thoroughly mediaeval flavour; that, indeed, it saw Plotinus as Plato's heir and *terminus ad quem*. This was indeed a labyrinth; to some minds, to the ardent mind of della Mirandola, for instance, an even more bewitching labyrinth than Aristotelianism. But this emotive and mystical Plato was not the Plato who interested the founders of the scientific revolution. For sixteenth- and seventeenth-century physicists it was rather the Plato who caused to have inscribed on the lintel to the Academy, 'let none enter here who know no geometry'. This can be seen, for example, in figure 14.1, the frontispiece to Tartaglia's *Nova Scientia* (1537), where Euclid acts as the usher and gatekeeper to the scientific world. The new science was also prefigured by the Plato who had invented the 'myth of the cave' as a representation of the human condition. The burden of this metaphor, it will be remembered, was that man's senses are faulty and provide only distorted information about the 'real' world. The prisoners in the cave can only see the shadows cast by the events happening above them in the broad light of day. These they have to interpret if they wish to understand what is happening in the 'real' world. Both these aspects of Plato, both his insistence on

(1) W. Pater, *Studies in the History of the Renaissance*, Macmillan, London (1873), pp. 24–6.

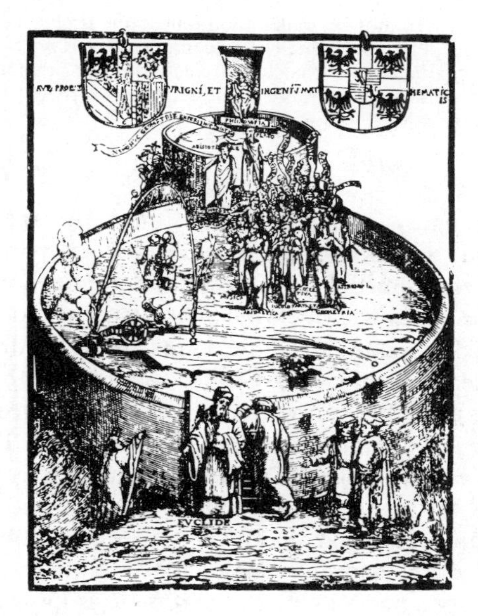

Figure 14.4. Frontispiece of *Nova Scientia Inventa da Nicolo Tartalea.* Euclid guards the gate leading into the castle of the sciences. That there is no other entry is shown by the individual on the left of the gate who finds his scaling ladder far too short. Within the walls stand Nicolo Tartaglia and his companions. The latter are represented as a chorus of muses and are identifiable as Aritmetica, Musica, Geometria, Astronomia and many others. Before them a mortar and a cannon demonstrate the trajectories of their shot. The latter problem forms the kernel of Tartaglia's *Nova Scientia.*

Behind Tartaglia and his colleagues rises an inner keep. At its gate stand Aristotle and, behind him, Plato. Beyond them both is enthroned Philosofia herself. The position of Aristotle suggests that he is closer to the level of the sciences than is the 'divine' Plato. Plato, however, doubles the role of Euclid at the outer gate and bears in his right hand a banner inscribed *Nemo huc geometrie expers ingrediatur:* let none enter here who know no geometry.

Two small, apparently nude, figures are to be observed clambering amongst the foliage on the vertiginous cliff below Philosofia's fastness. It is just possible that with the help of the shrub they will escape falling to their deaths and enter the sanctum.

mathematics and his belief that observational data have to be interpreted as distorted representations of what is 'really' going on, are to be found in the greatest of seventeenth-century physicists: Galileo Galilei (2). Galileo's approach is epitomised in the following famous passage:

(2) *Il Saggiatore,* Roma (1623). *Dialogue on the Great World Systems.* (trans. Thomas Salusbury, 1661) (ed. G. de Santillana), Chicago University Press (1953). Facsimile reprint by Culture et Civilisation, Bruxelles (1966): *Dialogo di Galileo Galilei Linceo dove ne i congressi di quattro giornate si discorse sopra i due massimi sistemi del mondo, telemaico e copernicano,* Fiorenza (1632). *Dialogues concerning Two New Sciences* (trans. H. Crew and A. de Salvio) Dover, New York (1953). Facsimile reprint by Culture et Civilisation, Bruxelles (1966) : *Discorsi e dimonstrazioni matematiche intorno a due nuove scienze,* Leida (1638).

> . . . the book[of nature] cannot be understood unless one first learns to comprehend the language and read the letters of which it is composed. It is written in the language of mathematics and its characters are triangles, circles and other geometrical figures, without which it is humanly impossible to understand a single word of it; without these one wanders in a dark labyrinth(3).

The correlation of accidents

Galileo was thoroughly convinced that only by the application of mathematics could an escape from the dark labyrinth of scholastic disputation be effected. Furthermore, it was an application of mathematics to certain properties of things which the Scholastics would have regarded as unimportant, indeed trivial. As Crombie points out in his detailed account of the origins of Galilean dynamics (4), Galileo sought only to correlate certain 'accidents', leaving aside the conventional Scholastic concern with 'essence' as something 'to be left to a higher science than ours' (5). This is made quite explicit in Galileo's exposition of Copernican astronomy. In the person of Salviati he asks the Aristotelian, Simplicius, what causes a heavy body to fall; for, says he, if he can be satisfied on this point then he will be able to account for the movements of the heavenly bodies too. Simplicius, of course replies tartly that the cause of a body falling to the ground is known to everyone and it is called gravity. Salviati responds, also tartly, to the effect that Simplicius should have said that everyone knows that the cause is *called* gravity; but what 'principle or virtue moves a stone downwards' we no more understand than that which 'moves it upwards' (6). This attitude towards the observational world reminds us of the nominalism which we discussed in the last chapter.

This clear-sighted appreciation of the distinction between words and things is an important strand in Galileo's thought. In parenthesis we might remind ourselves that Galileo's tartness could have been profitably applied to some of the biological theorists of the first part of the twentieth century. All too often terms like *élan vital* and 'entelechy', coined to describe certain biological phenomena, have been mistaken for actual existents somehow located within organisms.

It was suggested above that Galileo was a follower of Plato rather than of Aristotle not only because of his insistence on the crucial significance of mathematics but also because of his overall attitude towards nature. Indeed, as George Sarton points out (7), it is incorrect to suggest that Plato was himself a great creative mathematician. Archimedes was, in fact, a far more

(3) *Il Saggiatore,* 6, 232.
(4) A.C. Crombie, *Augustine to Galileo: the history of science AD 400–1650*, Falcon Press, London (1952). See also A.C. Crombie in *Critical Problems in the History of Science* (ed. M. Claggett),University of Wisconsin Press (1969), p. 86.
(5) *Two New Sciences*, Day 3, Cor. 3.
(6) *Great World Systems,* Day 2.
(7) G. Sarton, *Introduction to the History of Science,* Vol. 1., Harvard University Press (1952). Sarton is well qualified to judge this point, for in addition to being the doyen of the historians of science he started his career as a professional mathematician.

powerful influence on Renaissance scientists and mathematicians. It was, however, the Platonic metaphysical vision which Galileo and his contemporaries resurrected. Behind the observational world of shadowy approximations roughly, but never exactly, this or that, lay an ideal reality. It was to this supra-sensible world that the mathematics could be applied. It was this world which could be described in a language whose characters were 'triangles, circles and other geometrical figures'.

The difference between this vision and the Aristotelian immediacy needs no further emphasis. One consequence of considerable significance to biology flows from it. This is the notorious distinction between 'primary' and 'secondary' qualities. For Galileo the 'primary' qualities (as we noticed in chapter 6) are those which are amenable to mathematical manipulation. 'I hold' he writes (8), 'that there exists nothing in external bodies for exciting in us tastes, odours, sounds except sizes, shapes, numbers and slow or swift motion.'

This sentence is pregnant with the modern concept of man's place in nature. In the seventeenth century it was revolutionary. It set against the qualitative Aristotelianism of the schools a stark mathematical philosophy; a philosophy which, as Burtt emphasises again and again (9), ultimately reads human reality right out of nature. What is 'really real' are the mathematical abstractions, the 'triangles, circles and other geometrical figures'; our feelings, emotions, sympathies, and so on, appear to be merely epiphenomena.

It will by now be apparent that the Galilean vision of nature is remarkably modern. It wants, however, one final ingredient—the atomic theory; for without it the Galilean mathematical abstractions remain only calculating devices, pictures in the mind. It is salutary to remember that it was precisely for not admitting this point with respect to the Copernican hypothesis that Galileo was censured and imprisoned by the Roman church.

Atomism, however, plays only a very minor role in the Galilean corpus. Indeed, for Galileo the world is still a perfectly ordered cosmos (10), a concept which, it will be recalled, was difficult and is still difficult for atomists to accept. From this basic premise, indeed, Galileo deduced some of the most important components of his system. Believing, for example, that circular motion is perfect motion he deduces that all natural motion, whether supra or sublunary, is circular. Indeed we have to wait for Descartes before the ancient and all-pervasive notion that circular motion, because perfect, is therefore natural, is broken. And, as we shall see later in this chapter, William Harvey, steeped in Aristotelianism from his youth, brilliantly

(8) *Il Saggiatore,* Q.48.

(9) E.A.Burtt, *The Metaphysical Foundations of Modern Physical Science,* Routledge and Kegan Paul, London (1950).

(10) *Great World Systems,* Day 1. Salviati (Galileo): '... but we suppose them (the parts of the world) to be perfectly ordered....' '...it is necessarily most ordered, that is, having parts disposed with exquisite and perfect order.'

succeeded in discovering the same epitome of perfection in the microcosm.

In addition to contradicting the vision of a perfectly ordered cosmos, an inkling of the ecclesiastical wrath to come may have prevented Galileo from investigating, or at least publishing, in this field. It will be remembered that, although never entirely forgotten, the ancient atomists had long laboured under the damning stigma of atheism. The credit for finally exorcising this stigma, for baptising the Abderan theory, belongs to a Roman Catholic priest: Pierre Gassendi (1592 – 1655).

Atomism revived

The atomism revived by Gassendi and published in 1649 (11) was identical to that propounded by Leucippus and Democritus and popularised by Epicurus and Lucretius. Indeed, the book which Gassendi published in 1649 was an annotated translation of the tenth book of Diogenes Laertius' *Lives* which, it will be recalled, is one of the best ancient sources for the atomic doctrine. The annotations which Gassendi added were chiefly designed to cleanse the work of all taints of atheism. Thus, instead of the atoms falling through the void from infinity to infinity, Gassendi conceived that they were in the beginning created by God and endowed with a certain impetus or 'pondus' by which they were thenceforward compelled to move. In a later work, the *Syntagma*, he also writes of their having a *sensus naturalis*, a species of awareness. In essence, however, his is a work of exegesis, and the Democritean theory remains thereafter at the centre of science. And it remains, as it did in the ancient world, as a persistent challenge, not only to the theologian but also to the biologist. In both cases the implication was clear: there was no creative intelligence guiding the world, whether macrocosm or microcosm. The theologians were forced to recognise that the atoms, having been 'bowled' into the void on the first day, retained their impetus until the end of time. Thus, by easy stages, arose the concepts of the *deus ex machina*, the superannuated engineer of the mechanists, and the *deus absconditus* of Pascal and the moderns. Laplace, when asked by Napoleon about the place of God in his system, replied, famously, that he had no need of such a hypothesis. Similarly with the microcosm: the ancient notions of a guiding entelechy, of formal and final causes, to account for the development and integration of a living organism, although lingering for centuries, were doomed by the publication of the *Animadversiones* in 1649.

(11) P.Gassendi, *Animadversiones in decimum librum Diogenes Laertius* (Observations of the tenth book of Diogenes Laertius), Lyon (1649). Although Gassendi is normally regarded as the inaugurator of atomism in the modern world, he was in fact preceded by an Englishman, Nicholas Hill, who published a tract actively supporting the atomism of Democritus and Epicurus nearly fifty years earlier. The *Philosophia Epicurea,* Paris (1601), attempted to make the theory theologically palatable by maintaining that the atoms were in the beginning made by God. The *accidental* coming together and falling apart of the atoms accounting for generation and dissolution remained, however, to perplex the faithful. See G. McColley, Nicholas Hill and the *Philosophia Epicurea, Ann of Sci.,* 4 (1949), 390–405.

Art and anatomy

The sea-change in the foundations of physics, which soon became the paradigmatic science, quickly made itself felt throughout the sciences of living matter. Indeed, several of the immediate disciples of Galileo attempted to apply the new method to investigate the old science of biology. The best known name amongst these would-be biophysicists or iatrophysicists is that of Borelli. A description of his work is, however, postponed until chapter 16, for we are running ahead of the main story. Long before Galileo published his discourses, the ancient science of anatomy had been revitalised, and it is to this event that we must now turn.

In a famous passage Burckhardt sums up his interpretation of what happened in the Renaissance thus:

> In the Middle Ages both sides of human consciousness – that which was turned within us as that which was turned without – lay dreaming or half awake beneath a common veil. The veil was woven of faith, illusion, and childish prepossession, through which the world and history were seen clad in strange hues. Man was conscious of himself only as a member of a race, people, party, family, or corporation – only through some general category. In Italy this veil first melted into the air; an *objective* treatment and consideration of the State and of all the things of this world became possible. The *subjective* side at the same time asserted itself with corresponding emphasis; man became a spiritual *individual* and recognised himself as such (12).

This redirection of activity can be seen in all spheres of human activity. Italy, as Burckhardt goes on to point out, began, at the end of the thirteenth century, 'to swarm with individuality'. One has only to glance at the crowded pages of the *Divina Commedia* to confirm that this is the case; one has only to compare the personable madonnas of the Italian quattrocento with the anguished, soulful, but essentially abstract icons of earlier centuries to recognise that an alteration of consciousness has occurred.

This shift of consciousness also made itself felt in anatomy. After the death of Galen at the end of the second century AD the practice of dissection, especially human dissection, seems to have fallen into desuetude. For centuries anatomy seems to have been taught to aspiring physicians by means of five traditional figures: of the skeletal, muscular, nervous, arterial and venous systems. These systems were represented in crude human figures drawn in a squatting position (figure 14.2). There is evidence to suggest that these manikins originated in antiquity and hardly altered through successive centuries. There is no need to stress the difference between these grotesques and the magnificent Leonardo drawings of the turn of the fifteenth century (see, for example, figure 14.3).

Anatomy showed the first stirrings of a new life at the beginning of the fourteenth century, when the first records of original human dissections begin to appear. In 1316 Mondinus published his *Anathomia*, which nudged the subject out of the Arabian twilight. There is little doubt that this

(12) J. Burckhardt, *The Civilisation of the Renaissance* (trans. S.G.C.Middlemore), Phaidon Press, London (1945), p. 81.

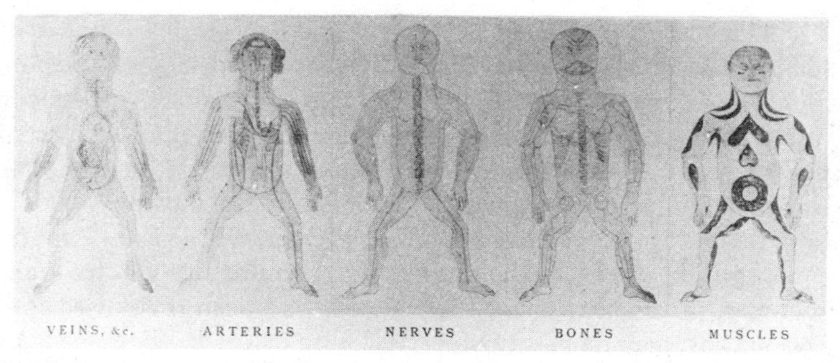

Figure 14.2. The five-figure diagram. From a Bodleian MS (*c* 1292). Note how the trachea is connected to the heart in the artery figure.

work is partly based on the author's own dissections. It was not, however, until the middle of the century that the practice of dissection received official sanction (see table 14.1). Even then the opportunity of the average medical student watching an anatomy, even from a position far back behind the crowd of notables surrounding the dissection table, was customarily restricted to no more than once a year.

Medical students were not the only individuals of the Italian quattrocento interested in anatomy. An avid interest was also shown by the painters and sculptors. It has been suggested that the initial point of contact between these two otherwise rather disparate groups was the apothecary's shop (13). Here the physician would come for his specifics and the artist for his pigments. The interest of the artist in surface anatomy would lead him to accompany his new acquaintance to the still rather infrequent public anatomies. Later, in their ardent pursuit of naturalism, the artists themselves began to make dissections. So far as is known, the earliest instance of an eminent artist – anatomist is Donatello (1386 – 1466). Of the slightly later Pollaiuolo (1432 – 1498), Vasari writes: 'He understood the nude in a more modern way than the masters before him; he removed the skin from many corpses to see the anatomy underneath' (14). From Pollaiuolo a direct line runs through Verrocchio to Leonardo da Vinci.

Perhaps too much should not be made of this conjunction of anatomists and artists. For, with the gigantic exception of Leonardo himself, the painters, almost by definition, were interested only in superficial anatomy. With Leonardo, however, we meet an altogether different phenomenon. Without necessarily accepting his reported statement (15) that he had dissected more than thirty bodies, there can be no doubt at all that he was no

(13) E.C. Streeter, The role of certain Florentines in the history of anatomy, artistic and practical, *Johns Hopkins Hosp. Bull.*, 27 (1916) 113–18.

(14) G. Vasari, *Lives of the Artists,* Florence (1550); trans. George Bull, Penguin, Harmondsworth (1965).

(15) Reported by Antonio de' Beatis, secretary to Cardinal Louis of Aragon, who visited Leonardo on 10 October 1517 (in Clark, *Leonardo da Vinci,* Penguin, Harmondsworth (1958), p. 157).

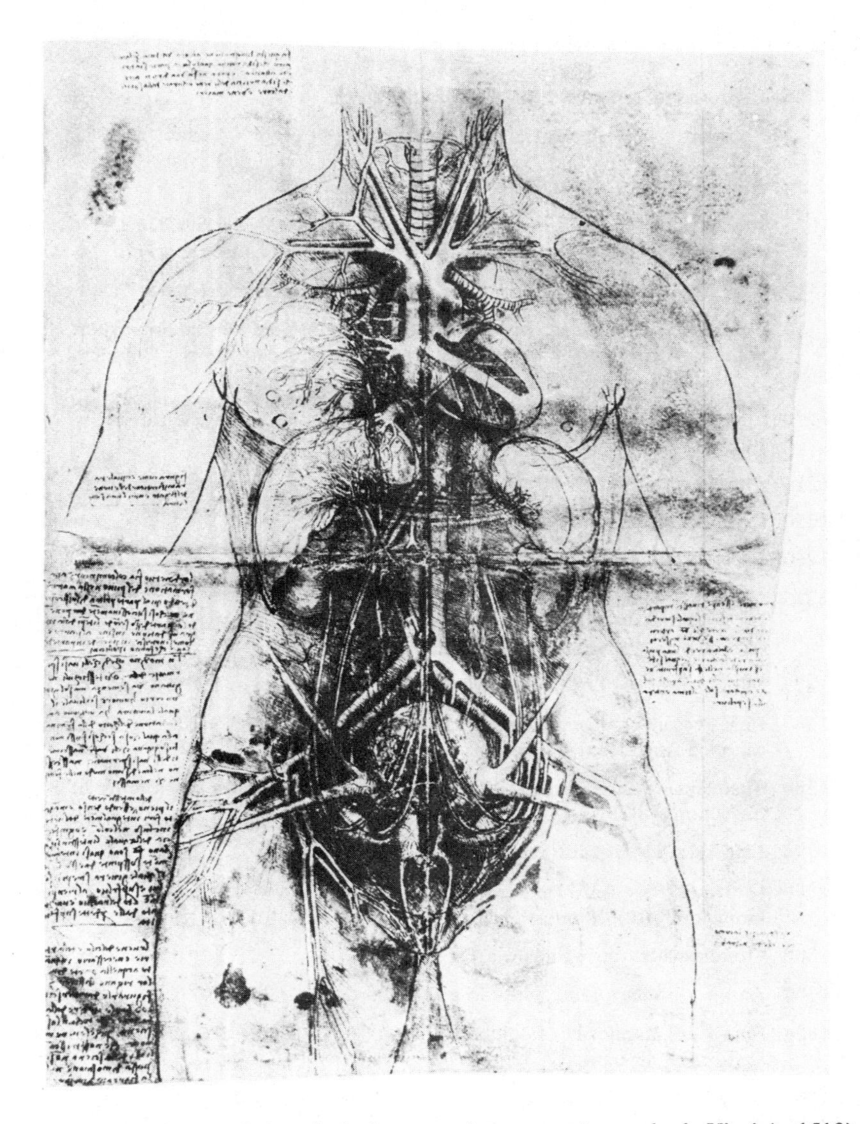

Figure 14.3. Dissection of the principal organs of a woman. Leonardo de Vinci (*c.* 1510).

stranger to anatomy. In the later part of his life his travelling kit normally included a bone saw and other anatomical equipment. His anatomical drawings, though in some places, where he sees with the eyes of Aristotle or Galen, incorrect, have never been surpassed. His approach, too, is in many respects thoroughly modern. He was not content merely to record what he saw, but sought always to go further and attempt to understand how the body

Table 14.1. The origins of anatomy in western Europe

– 985	Establishment of a medical school at Salerno[a]
1077	Constantine the African at Salerno and later at Monte Cassino (d. 1087); translates many important treatises from Arabic, including Haly Abas' medical encyclopaedia, *Liber Regalis* [a]
– 1085	Alphanus of Salerno translates, among other works, Nemesius' *On the Nature of Man*, and was responsible for *De Quatuor Humoribus* and *De Pulsibus*[a]
c. 1110	*Anatomia Porci:* two works on the anatomy of the pig translated at Salerno[b]
c. 1150	Gerard of Cremona (working at Toledo) translated several important Arabic compendia of medicine and also Arabic versions of Galen, Hippocrates and Aristotle
c. 1170	*Practica Chirurgia,* Roger of Salerno: This is the earliest European work on surgery[c]
1214	*Anatomia,* Maurus of Salerno[c]
– 1250	Organised medical teaching begins at Bologna[d]
1260	Greek Hippocrates translated by William of Moerbeke
1275	*Chirurgia,* William of Salicento[c] In this year William of Salicento is also recorded as opening a cadaver to ascertain the cause of death to settle a law suit[e]
1277	Various treatises of Galen translated by William of Moerbeke
1286	First record of a post-mortem: a body is opened at Cremona to ascertain the cause of death from a disease[e]
c. 1280	Thaddeus of Florence (1223 – 1303) probably inaugurates the practice of human dissection at Bologna[e]
1296	*Chirurgia Magna,* Lanfranchi of Milan[c]
c. 1300	Giotto (1266 – 1337) and his assistant Stefano, 'the Ape of Nature', institute a more naturalistic approach to painting
c. 1300	Mondino lecturing on anatomy at Bologna
1302	Autopsy by Bartolomeo da Varignana recorded at Bologna[d]
1304	Henri de Mondeville (Mondino's pupil) lecturing at Montpellier using large anatomical diagrams [b]
1308	Venetian republic publishes a statute allowing anatomies to be made each year
1316	*Anathomia,* Mondino of Bologna. This was the leading dissection manual of the middle ages. Its use persists into the sixteenth century. It is said to have been used by Leonardo at the turn of the fifteenth century[f] Although there is good evidence that Mondino dissected in person (unlike the majority of his successors until Vesalius), he nevertheless saw many anatomical structures through the eyes of the medieval Galen: the brain has three ventricles, the liver has five lobes, the uterus has seven 'cells', the stomach is spherical, and so on. His text, moreover, is full of poorly understood transliterations of Arabic anatomical terms.
1319	Proceedings recorded against four medical students at Bologna accused of body-snatching;[e] normally the few cadavers made available for anatomy would be those of executed criminals

1340 Statute initiating a biennial anatomy published at Montpellier[b]

1341 Public dissection by Gentile da Foligna at Padua[f]

1368 Initiation of human anatomy at Venice[b]

1388 Initiation of human anatomy at Florence[b]

1407 Initiation of human anatomy at Paris[b]

c. 1435 Donatello's bronze *David* the first naturalistic nude of the Renaissance

1442 Statute published at Bologna that two cadavers should be made available for anatomy each year

c. 1450 The development of printing, of very great importance for the dissemination of ungarbled anatomical information

c. 1465 Pollaiuolo (1432 – 1498), after apprenticeship first to Ucello and then to Donatello, concerns himself with at least surface anatomy to ensure the verisimilitude of his paintings: *Battle of the Nudes,* 1470[g]

1493 *Anatomice,* Benedetti: an important contribution towards the clarification of the prevailing morass of anatomical terminology[d]

1502 Berengaria da Carpi elected to the chair of anatomy at Bologna. He writes that he has dissected over two hundred bodies. He published his influential *Commentaria* (on Mondino's *Anathomia*) in 1514

c. 1500 Achillini (1463 – 1512), working at Bologna, shows considerable independence of judgement describing correctly among other structures the trochlear nerve and the seven-boned tarsus. His *Anatomiae Annotationes* was published posthumously in 1520.

c. 1500 Leonardo da Vinci at work on anatomy with the assistance of Antonio della Torre, who afterwards became professor of anatomy at Pavia (1506–1512)[h]

1528 *Die Kleine Chirurgie,* Paracelsus

1530 Charles Estienne at work on *De Dissectione Partium Corporis Humani* (published in 1545): this work is one of the first to incorporate contemporary illustrations, although these are for the most part poor

1543 *De Humani Corporis Fabrica,* Vesalius: the foundation work of modern anatomy

[a]P.O. Kristeller, *Bull. Hist. Med.,* **17** (1945), 138 – 94.

[b]C. Singer, *A Short History of Anatomy,* London (1957).

[c]R.L. Storey, *Chronology of the Medieval World,* London (1973).

[d]E.A. Underwood, *Ann. of Science,* **19** (1963), 1 – 26.

[e]C.D. O'Malley, *Andreas Vesalius of Brussels,* Berkeley (1964).

[f]H.P. Bayon, *Ann. of Science,* **3** (1938), 59–118, 435–56; **4**, 65–109, 329–89.

[g]E.C. Streeter, *Johns Hopkins Hosp. Bull.,* **27** (1916), 113 – 18.

[h]C.D. O'Malley and J.B. Saunders, *Leonardo da Vinci on the Human Body,* New York, (1962).

A minus sign in front of a date indicates that the date is the latest at which an event occurred or a development was complete.

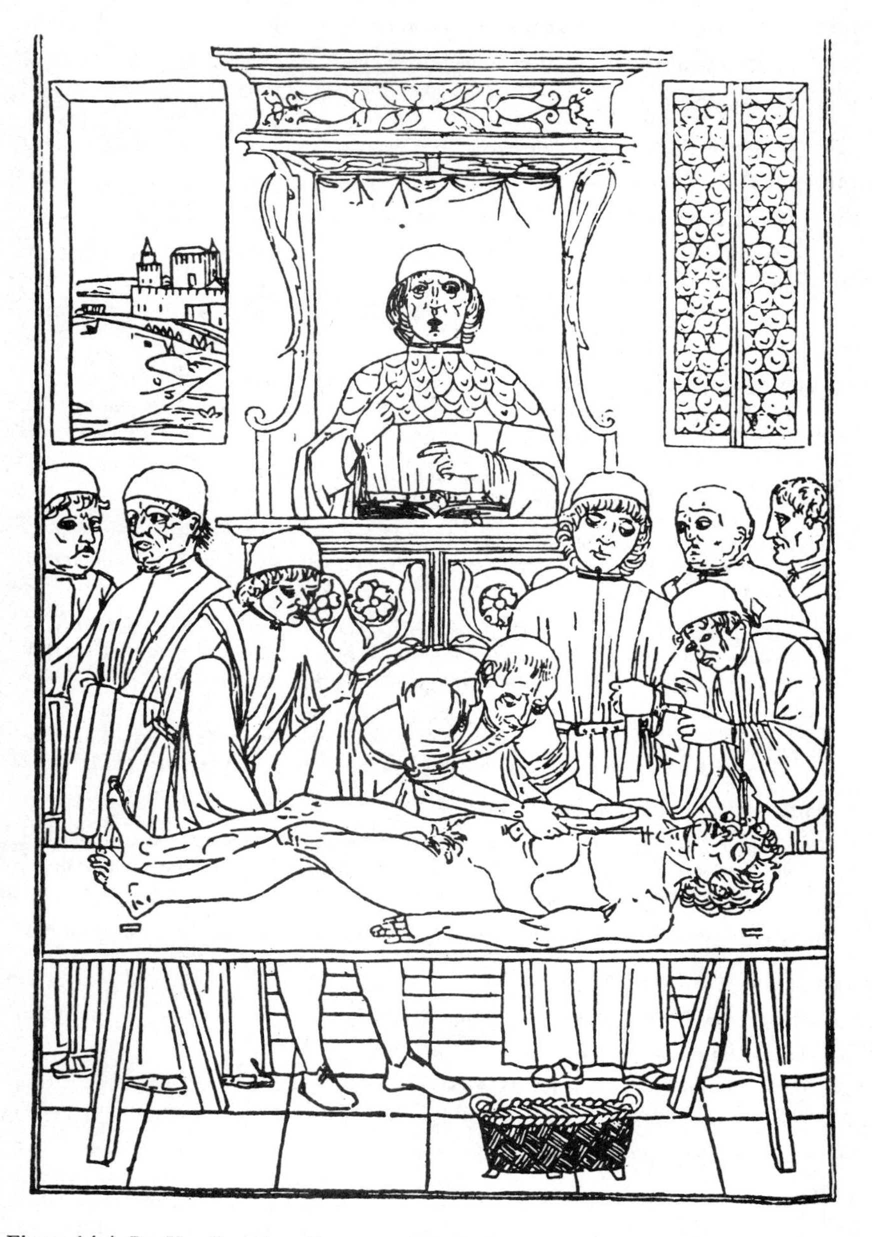

Figure 14.4. Pre-Vesalian dissection scene. The Barber-Surgeon opens the cadaver (normally a freshly executed criminal) with a mid-ventral incision. The Demonstrator (on the right) prepares to point out structures which the lecturer in his pulpit reads out from the traditional text, hardly glancing at the distasteful business below. The majority of the students chat amongst themselves.

functioned. Indeed, so enthusiastic an engineer as Leonardo could hardly escape treating the body as a divine machine. In drawing after drawing we see him endeavouring to show with his crayon how part acted upon part, how muscle acted on bone, how bone articulated with bone (16). His studies of the anatomies of birds were probably carried out in connection with his attempt to construct a flying machine. Leonardo, in addition to so much else, was an individual who understood with his hands.

Leonardo's work, however, was not published during his lifetime nor for centuries thereafter (17). Although it was known to exist, it seems to have had little or no influence on the subsequent history of anatomy. Partly, no doubt, this was due to the fact that Leonardo *was* an artist and not an anatomist. The output which was eventually bequeathed to Francesco Melzi was probably not organised into any very systematic form. Anatomists consequently followed their own more plodding paths. Leonardo, the eagle, remained largely unknown and the tradition from Mondinus passed through Berengaria da Carpi (*c.* 1460 – 1530) to the young Andreas Vesalius.

Vesalius (1514 – 1564) describes, in the preface to his great book, the state of anatomy as he found it in his youth. He points out that after the decline and fall of the Roman Empire all science fell into decay. In particular there developed that fatal disjunction between head and hands which was to prove so particularly disastrous for anatomy. It was this divorce of theory from practice which, he writes,

> ...introduced into the schools that detestable procedure by which usually some conduct the dissection of the human body and others present the account of its parts, the latter like jackdaws aloft in their high chair, with egregious arrogance croaking things they have never investigated but merely committed to memory from the books of others, or reading what has already been described. The former are so ignorant of languages that they are unable to explain their dissections to the spectators and muddle what ought to be displayed according to the instructions of the physician who, since he has never applied his hand to the dissection of the body, haughtily governs the ship from a manual. Thus everything is wrongly taught in the schools, and days are wasted in ridiculous questions so that in such confusion less is presented to the spectators than a butcher in his stall could teach a physician (18).

Vesalius changed all this. His revolutionary book, *De Humani Corporis*

(16) Leonardo's anatomical drawings are collected in C.D.O'Malley and J.B.Saunders, *Leonardo da Vinci on the human body*, Henry Schuman, New York (1962).

(17) Leonardo died at Amboise in 1519. His scientific manuscripts were bequeathed to Francesco Melzi, and after his death they passed to his nephew Orazio Melzi who, in turn, sold them to the sculptor Pompeo Leoni. Leoni carried the manuscripts to Spain, intending to present them to King Philip. Philip died, however, before the presentation could be made and the majority of the drawings were instead purchased by Thomas Howard, Earl of Arundel. They passed into the possession of the English Crown and, although seen by Huygens in 1690, were not 'rediscovered' until 1760. The Royal Librarian showed them to William Hunter, who set about seeing to their publication. He died, however, before this could be accomplished, and the first editions of Leonardo's masterpieces were not in fact published until the nineteenth century.

(18) A. Vesalius, *De Humani Corporis Fabrica Libri Septum,* Basiliae (1543): preface to Charles V; trans. C.D.O'Malley in *Andreas Vesalius of Brussels,* University of California Press (1964), p. 319.

Fabrica Libri Septum (seven books on the structure of the human body), was published in 1543 when its author was twenty-eight years old. It is a remarkable coincidence that this year also saw the publication of another revolutionary work: Copernicus' *De Revolutionibus Orbium Coelestium*. Just as one ushered in a new era in physical science, so the other initiated a new era in biological science.

The *Fabrica* was almost entirely based upon numerous first-hand dissections of the human body. It therefore contained many new insights and discoveries, and many contradictions of the received Galenical teaching. But in addition to the vastly important precept of always looking at the human body for authority rather than back to Galen or Aristotle, the *Fabrica* also broke fresh ground in another direction. Although a slow change was already at work in the production of anatomy texts, the *Fabrica* was the first important book to dispense with the old tradition of either no illustrations at all or merely the five ancient anatomical figures (19). Instead the book is graced with magnificent illustrations which are all carefully related to the text. Leonardo, as we have seen, had already attempted something similar, something indeed even more modern, but the *Fabrica* was the first printed book on anatomy to use this technique.

Figure 14.5. The anatomy theatre at Padua. The theatre was completed towards the end of 1594. The official photograph above gives some indication of the extreme steepness of the theatre; the spectators of an anatomy hung above the dissection like flies on the sides of a funnel. Underwood (1963) has estimated that the theatre held about two hundred on-lookers. The bottom galleries nearest the dissection table were reserved for professors of the university and other Paduan notables. The average student would have been about twenty feet away from the anatomy. Before 1594 anatomy theatres were temporary structures, as is the one shown in the frontispiece of Vesalius' *Fabrica.* Anatomies were also frequently carried out in private dwelling houses.

(19) Perhaps the most important of the *Fabrica's* forerunners in this respect was Charles Estiennes' *De dissectione partium corporis humani,* published in 1545 but in preparation some fifteen years earlier. The illustrations incorporated into this text, however, are crude and ugly when compared with those which adorn the *Fabrica.*

After the publication of his great work, Vesalius left the academic world and took up appointments as physician first to the Emperor Charles V and later to King Philip II. Although he seems at times to have hankered for the contemplative life of the Paduan faculty, he never returned to the university. His chair passed to Realdo Columbo, a former student, and eventually in 1551 to Gabrielle Fallopius. Fallopius's brilliant career (his *Observationes Anatomicae*, containing much new material, was published in 1561) was cut short by his death at the early age of 39 in 1563. Fallopius was succeeded by Fabricius ab Aquapendente, who quickly established a great reputation. It was during his tenure of the chair that the world's first permanent anatomy theatre was established at Padua (figure 14.5). Fabricius retired in 1613 after holding the chair for nearly fifty years. At the turn of the century, during the years 1600 – 1602, a young Englishman, William Harvey, had been one of his outstanding students.

William Harvey and the circulation theory

William Harvey (1578 – 1657) is usually regarded as the instaurator of the modern phase in physiology (20). We shall note, in the next chapter, that Descartes in some ways contends with him for this honour. Descartes' impulse was, however, ultimately metaphysical. Harvey was much more the specialist. Harvey's work, perhaps because of this feature, proved initially far more seminal. Yet Harvey, as we shall see, looked back to Aristotle for his general ideas, while Descartes looked forward to the mechanical philosophers of the enlightenment.

As is very well known, the crucial advance which secures for Harvey his place at the fountainhead of physiological thought lies in his demonstration of the circulation of the blood. Many previous workers had trembled on the brink of the discovery (21): but Harvey's demonstration proved ultimately indisputable. For Harvey's approach combined elegant and prolonged experimentation with a salient quantification. We read not only of the careful

(20) W. Harvey, *The Works of William Harvey translated from the Latin with a Life of the Author by Robert Willis M.D.*, Sydenham Society, London (1849). Some of the more important of Harvey's writings are conveniently collected in *Great Books of the Western World*, vol. 28 (ed. R.M. Hutchins), Encyclopedia Britannica Inc., Chicago, London, Toronto (1952), trans. Robert Willis: *An Anatomical Disquisition on the Motion of the Heart and Blood in Animals (Exercitatio Anatomica de Motu Cordis et Sanguinis in Animalibus)*, London (1628); *The First Disquisition to John Riolan on the Circulation of the Blood* (1649); *A Second Disquisition to John Riolan in which many objections to the Circulation of the Blood are refuted* (1649); *Anatomical Exercises on the Generation of Animals (Exercitationes de Generatione Animalium)* (1653).

(21) Accounts of Harvey's precursors and contemporaries may be found in H.P. Bayon, (1938 and 1939) and in W. Pagel (1966). Especially important were Michael Servetus and Realdo Columbo. Servetus' book, *Christianismi restitutio* (1553), containing an account of the pulmonary circulation, was condemned by the Inquisition and Calvin caused it to be burnt along with its author in the year of its publication. Realdo Columbo confirmed Servetus' belief in the pulmonary circulation by experiment and dissection and published his findings in *De re anatomica* in 1559.

vivisections of a multitude of different animals (Harvey mentions toads, frogs, serpents, small fish, crabs, shrimps, snails and shellfish in addition to the more usual warm-blooded animals such as hogs, dogs, or pigeons) but also of calculations of the quantity of blood expelled from the heart in unit time:

> ....I frequently and seriously bethought me and long revolved in my mind, what might be the quantity of blood which was transmitted, in how short a time its passage might be effected and the like...(22).

And answers to these ruminations were quickly obtained from experiment: '...in the course of half an hour...a larger quantity in every case than is contained in the whole body' (23). So large a quantity reduced to absurdity the ancient doctrine that the blood is continuously concocted from the food by the liver and 'sweats' across the interventricular septum into the heart's left ventricle.

Harvey's conception of the centrality of the cardiovascular system in the life of the body may be compared with that of his great predecessor Empedocles. Like Empedocles he conceived that a thorough understanding of this system went far towards answering the question 'What is life'? He is not, for example, enmeshed in theoretical discussions about spirits and souls. His position is that the vital principle is coeval with the blood. He is sardonic about the metaphysical implications of the word 'spirits' telling, by way of illustration, a tale of a remarkable stone brought home from the East Indies which possessed the most astonishing properties, 'filling the ambient air with beams scarcely bearable to any eyes', 'bursting forth if any attempt is made to cover it', 'of the most consummate beauty', and so on, and revealing after he has listed a vast number of poetic and fabulous properties that he is referring to nothing more unusual than 'flame'. Just so, he says, with blood: it can be described in similar fantastic words but, once outside the veins, it is but 'the unequivocal gore'.

Vesalius seems to have suspected something similar. But in sixteenth-century Italy the prudent man was cautious. Referring to the dissection of the heart, which, it will be recalled, was for the Galenist intimately concerned with the manufacture of the vital spirit, he writes that:

> ....you will find a great many censors of our holy and true religion. If they hear someone murmur anything about the opinions of Plato, Aristotle or his interpreters, or of Galen regarding the soul, even in the conduct of anatomy where these matters ought to be examined, immediately they judge him to be suspect in his faith and somewhat doubtful regarding the immortality of the soul (24).

Harvey, however, three quarters of a century later and living in the England of the Civil War, will have none of it. He writes (25):

(22) *The Motion of the Heart and Blood,* chapter 8.
(23) *Ibid.*, chapter 9.
(24) Vesalius, *Fabrica,* quoted by O'Malley, *ibid.,* p. 178.
(25) *A second disquisition to John Riolan,* p. 313.

> With regard...to spirits, there are many and opposing views as to which these are, and what is their state in the body, and their consistence, and whether they are separate and distinct from the blood and the solid parts, or mixed with these. So it is not surprising that these spirits with their nature thus left in doubt serve as a common subterfuge of ignorance. For smatterers not knowing what causes to assign to a happening promptly say that the spirits are responsible and introduce them as general factota. And like bad poets call this *deus ex machina* on to their stage to explain their plot and catastrophe.

A little further on he continues:

> ...I have, however, never found such in veins, nerves, arteries or parts of living subjects. Some make the spirits corporeal, others incorporeal. Sometimes they conceive of the spirits as contained in the blood (like flame in the aroma of cooking) and sustained by its continuous flow; sometimes of the spirits as distinct from the blood....

Clearly the old alchemical notions which we discussed in chapter 12 were still at work in the seventeenth century. William Harvey, however, is not concerned with essences or the striving after perfection, but only with the data which his senses provide. His is a thoroughly materialistic view (26). He concludes the discussion by comparing the physiological spirits to the bouquet of a glass of wine. They are no more, no less material than this:

> The spirits are no more separate from the blood than is a flame from its inflammable vapour. But in their different ways blood and spirit, like a generous wine and its bouquet, mean one and the same thing. For, as wine with all its bouquet gone is no longer wine but a flat vinegary fluid, so also is blood without spirit no longer blood but the unequivocal gore.

Harvey, then, will have nothing to do with the old pneumatic physiology. Indeed, he refutes it in great detail in his second letter to John Riolan. Moreover, it is clear that his great physiological discovery puts quite out of court the old tripartite division of the vital principle. In Harvey's consequent physiological monism it is, as we have seen, the blood which forms the substratum of animal life. In his final statement on the subject, published when he was in his seventy-third year, we read that 'the blood does not seem to differ in any respect from the soul or life itself; at all events, it is to be regarded as the substance whose act is the soul or the life' (27).

Harvey's Aristotelianism

Harvey has, by some, been called the Galileo of physiology. Certainly, like Galileo, he was not burdened with any general metaphysical theory which his specialist studies were undertaken to confirm. Like Galileo, too, he was not overmuch concerned with essences and spirits, and sought only to investigate and correlate properties which the scholastics would have dismissed as merely contingent. And like Galileo, finally, he retained a mediaeval aspect. Indeed, in his basic intellectual 'set' he was far less revolutionary a figure than the physicist. The first inkling of his great discovery came to him from a consideration of the *purpose* of the venous valves. This must have occurred

(26) Materialistic, however, in the Aristotelian sense where matter is endowed with a 'living' force.

(27) *On the Generation of Animals,* Exercise 71.

quite early in his career, for his Paduan teacher Fabricius, of whom he always speaks with veneration, had published a comprehensive treatise on the valves in 1603. But Fabricius had not understood their *raison d'être*. Perhaps it was the contemporary development of one-way valves in the technology of pumping which helped Harvey to his insight. Whatever the immediate psychological background, it was meditation on the meaning of the valves which, by his own account, opened the way to the circulation theory; and this attitude of mind is, of course, profoundly Aristotelian. It is a crucial example of an argument from final causes, the very type of argument which had exercised the sardonic wit of Harvey's older contemporary, the Lord Chancellor and philosopher of science, Francis Bacon.

Throughout Harvey's work the influence of Aristotle can be felt. No doubt this was partly due to his early training at Padua, which was in his time an epicentre of the Peripatetic philosophy, but it was also due to the fact that Harvey was through and through a biologist and not a physicist, astronomer or metaphysician. Other mediaeval elements also seem to condition his thought. He accepts the Aristotelian centrality of the heart to the body's physiology. He goes further. He compares the heart to the centre of a circle: an analogy which does not appear in the Stagirite's work. Indeed, in the eighth chapter of *De Motu*, where he introduces the circulation theory with the famous phrase: 'I began to think whether there might not be a motion, as it were in a circle', he has a great deal to say about circles. The ancient idea, which as we have seen was still held by Galileo, though soon to be discredited, that circular motion was somehow perfect motion, had sunk deep into Harvey's thinking. Harvey, moreover, was still inclined to feel the force of analogies drawn between the microcosm and the supralunary macrocosm. He proposes that just as the circular movement of the heavens holds the whole world together, so does the circular movement of the blood hold the body together. Eliminate this movement and the body dies. The microcosm/macrocosm analogy is stressed in many parts of his work. In *De Motu*, for example, he writes: 'The heart, consequently, is the beginning of life; the sun of the microcosm; even as the sun in his turn might well be designated the heart of the world' (28).

In the last analysis, then, it is Aristotle's world that Harvey still inhabits. He fully accepts the Stagirite's conclusion that everything in the living body develops for a purpose, that nature does nothing in vain. This is particularly evident in the embryological studies which occupied the later part of his life. '...the whole', he writes in *On the Generation of Animals*, 'seems to be referrable to one principle, *viz*: the perfection of Nature, who in her works does nothing in vain...' (29). And, as we have stressed, the impulse which

(28) *The Motion of the Heart and Blood*, chapter 8. This passage may be compared with a famous passage in *De Revolutionibus Orbium Coelestium* in which Copernicus praises the sun as 'the soul of the world – placed on a Royal Throne in the centre of the Universe, where it guides the family of stars circling around it'. (Quoted in H. Kesten, *Copernicus and his World* (trans. E. Ashton and N. Guterman), Roy Publishers, New York (1945)).

(29) *On the Generation of Animals*, Exercise 55.

directed him towards his greatest discovery was the recognition that, as he puts it, 'so provident a cause as Nature (had) not placed so many (venous) valves without design' (30).

On the Generation of Animals is Harvey's last great study. It is a book saturated in the Aristotelian tradition. Although Harvey's admiration of 'The Philosopher', as he still calls him, was not entirely uncritical, he nevertheless observes that 'the authority of Aristotle has always had such weight with me that I never think of differing from him inconsiderately' (31).

It is, of course, not surprising that Harvey should have seen the processes of embryology through Aristotelian eyes. For, as we noted in earlier chapters, Aristotle was not only an embryologist himself but may almost be said to have 'seen' the processes of the world as a case of 'macro-embryology'. His treatment of causation, for instance, makes most sense if we have in mind, as its referent, the development of an organism. Harvey evidently felt at home in this conceptual frame.

Now it will be recalled that Aristotle viewed the facts of embryology as one of the greatest obstacles in the way of an acceptance of the atomic theory. Harvey was of the same opinion. He was convinced, after a great deal of painstaking research, that the development of animals was something quite different from the coming together and mixture either of the four Empedoclean elements or of the Democritean atoms.

> Neither in the production of animals, nor in the generation of any 'similar' body (whether it were of animal parts, or of plants, stones, minerals etc.) have I ever been able to observe any congregation of such a kind, or any divers miscibles pre-existing for union in the work of reproduction (32).

The identification of the germ cells awaited, of course, the development of the compound microscope in the second half of the seventeenth century. Harvey continued his line of thought by emphasising that 'similar bodies', by which he means entities with a single unified function, exist before their elements 'being naturally more perfect than these'. This must remind us strongly of the Peripatetic position, and emphasises that students of embryology have long felt their subject to be the most difficult of all to harmonise with the mechanistic world picture. We shall return to this dilemma in chapter 21, 'for', as William Harvey points out (33), 'the nature of generation, and the order which prevails in it, are truly admirable and divine, beyond all that thought can conceive or understanding comprehend'.

William Harvey, then, in contrast to the great physicists of the seventeenth century, was still deeply immersed in the Peripatetic universe. Although the sixteenth- and seventeenth-century revolutions in anatomy and physiology, like those in the physical sciences, heralded a new era, they did not penetrate

(30) Robert Boyle reports this of Harvey in *A disquisition about the final causes of natural things*. . ., London (1688), p. 157.
(31) *On the Generation of Animals*, Exercise 11.
(32) *Ibid.*, Exercise 72.
(33) *Ibid.*, Exercise 71.

quite so deeply into the structure of science. Whereas Galilean physics revolutionised the whole logic of mediaeval science so that it could never be the same again, Vesalian and Harveyan biology was, in a sense, the retrieval of an old tradition: the tradition of going to the facts, a tradition which Aristotle and Galen worked within. The new biology of the seventeenth century emerged from a shaking off of the obfuscations of scholastic talk and a return to a concern with the things which the talk was about. In this sense it, too, was an escape from a labyrinth: the arachnid labyrinth of Scholastic disputation. The restructuring of biology was still in the future when Harvey died in 1657. The physical revolution had not yet had time to work its way through to the sciences of life. Only one thinker in the first half of the seventeenth century had fully grasped its consequences for biology. That thinker was René Descartes.

15 DESCARTES

Discretion the better part of valour

René Descartes was born in 1596 at La Haye, a small town not far from Poitiers. After a varied early career he published his first work, *A Discourse on Method*, in 1637. Although the first work he published this was not, in fact, the first work he completed. In 1633 he had finished the manuscript of a treatise entitled *Le Monde* which included a highly mechanistic account of man—*L'Homme*. This work, however, he held back on hearing of the trial and condemnation of Galileo by the Holy Office. His was not a fanatic temperament. Possibly he valued his private freedom to investigate for the truth more highly than any putative duty to display his discoveries to the public. However this may be, 'certain considerations', he says, 'prevented me from publishing' (1). Thus it was that, although written in 1632, *L'Homme* was only published, posthumously, in 1662.

Cogito ergo sum

Descartes' thought was both wide and deep (2). He made fundamental and lasting contributions to mathematics and metaphysics, and was unlucky not to have been similarly successful in physics and physiology. As with the great figures of classical antiquity, his metaphysics and his science were closely integrated. Thus to understand his physiology we must first examine his metaphysics.

The central phrase in Descartes' metaphysics, *cogito ergo sum*, has become almost a cliché. Descartes describes in the *Discourse on Method* how, having isolated himself in the incurious and commercially-minded Netherlands, he gave himself up entirely 'to the search after Truth'. He describes, in an autobiographical section at the beginning of the *Discourse*, how of all the subjects with which he had come into contact during his education, mathematics had seemed the most delightful: delightful because of its superior clarity and certainty. Thus it is not surprising to find that the

(1) R. Descartes, *Discourse on Method,* part V.
(2) There are several collected editions of Descartes' writings. The edition used in this chapter is R. Descartes, *Oeuvres complètes* (ed. V. Cousin), Paris (1824). *The Philosophical Works of Descartes* (ed. and trans. E.S.Haldane and G.R.T.Ross), 2 vols – Cambridge University Press (1911): most of the important works, but not *Traité de L'Homme.*

'search after Truth' with which he occupied his solitary (though comfortable) life in Holland consisted in a systematic probing of the evidence on which each of his opinions rested. This systematic doubt led him at first to believe that he 'knew' nothing, in the hard sense of that verb. All his so-called knowledge was no more than mere opinion and supposition. Even the demonstrations of geometry he doubted, for he had often observed men to whom certain geometrical propositions appeared clear and obviously true although he, Descartes, realised them to be false. And if other men could be so deceived, why not, says Descartes in all humility, why not myself unwittingly too? Even the possession of a body he believed to be far from indubitable. For do we not suffer strange illusions in our dream life, and how are we certain that we do not suffer similar illusions when awake? Thus Descartes believed himself to be on the verge of a universal scepticism. One thing, however, struck him before he crossed the threshold: it was impossible to doubt while he doubted that there was something which doubted. From this enlightening certainty sprang his famous phrase: 'I think, therefore I am'.

'Examining attentively', as he says, this discovery, Descartes concluded that the one indubitable thing in the world was that he was something, a substance,

> the whole essence, or nature of which is to think, and that for its existence there is no need of any place, nor does it depend on any material thing; so that this 'me', that is to say, the soul by which I am what I am, is entirely distinct from body and is even more easy to know than is the latter; and even if the body were not, the soul would not cease to be what it is (3).

This, then, is Descartes' major philosophical discovery and it forms the core of his thought. It is perhaps possible to see in it the philosophical expression of that striving for individuality which Burckhardt suggests motivated the Italian Renaissance; perhaps, too, we can detect the influence of the Protestant Lowlands, the home of a creed which required each man to develop his own individual relation with his creator, on the intellect of a Catholic of genius. But perhaps most of all we can detect in it a first answer to the problems which a universal mechanism posed to traditional ideas about man's place in nature. Whatever the causes, the consequences for western philosophy have been and still are profound. We are still plagued by a popular belief in what Gilbert Ryle has aptly termed 'the ghost in the machine'.

Res cogitans *and* res extensa

'Thought' in Descartes' system is evidently of paramount importance. It is the only thing of whose existence we have the right to be sure. Descartes is, moreover, quite explicit as to what he intends by the concept of thought. It seems to be coterminous with what we nowadays mean by the term

(3) *Discourse on Method,* Part IV.

'consciousness'. 'Not alone understanding', writes Descartes (4), 'but also willing, imagining and feeling are here the same thing as thought'.

Having to his own satisfaction established the epistemological superiority of mind, Descartes next applies himself to the task of discovering what certainty we have of the existence of matter. We have already seen that he is prepared to doubt even the tautological certainties of mathematics, suggesting that it is at least conceivable that we are systematically deceived by some supernatural power. It thus becomes the essential next step in his argument to establish the existence of such a power and to determine whether or not it might be inclined to play tricks upon the race of men. Descartes finds little difficulty in rehearsing the long-familiar ontological proof for the existence of a benign deity. He is able to satisfy himself, in other words, that because the idea of a perfect being can exist in our minds He must, necessarily, exist in reality to complete His perfection. Having thus established the existence of an omnipotent, omniscient, just and good God, it at once follows that our sensory information cannot be deceptive or illusory (5).

Thus Descartes believes himself to have saved from the chaos of universal dubiety two certainties: thought, or various modifications of thought such as imagination, feeling, will, and so on; and body, characterised by length, breadth, depth, in a word, by extension (6). The grounds for our certainty that our reasoning is valid and that our senses yield us truths about our situation are in both cases the same: the existence of an undeceitful God (7).

Mind and brain

In the last section we have seen how Descartes sets up the fateful distinction between *res cogitans* and *res extensa*. But, sharply distinct as the thinking and corporeal substances may be, it has to be admitted, of course, that in man they are somehow united. 'Pain and other of our sensations' he points out (8),

> occur without our foreseeing them; and the mind is conscious that these do not arise from itself alone or pertain to it in so far as it is a thinking thing, but only in so far as it is united to another thing, extended and mobile, which is called the human body.

Where, then, does this strange union take place? Descartes is quite clear. The mind, he says, is in the brain. Perhaps, though, this is too bald a statement of his rather sophisticated theory. In Meditation VI of his *Meditations on First Philosophy* we find him asserting that 'the mind does not receive impressions from all parts of the body at once, but only from the brain...'. He seems to imply that the brain, to use a much later analogy, acts the part of a radio receiver to which the 'mind' can tune in. That it is the

(4) *Principles of Philosophy*, Part 1, Principle IX.
(5) *Ibid.*, Principle XXX.
(6) *Ibid.*, Principle LII.
(7) *Meditations on First Philosophy*, Meditation VI.
(8) *Principles of Philosophy*, Part 2, Principle II.

brain which has this vital role is, says, Descartes, 'testified by innumerable experiments which it is unnecessary here to recount'.

Although in the *Meditations* Descartes finds it unnecessary to outline the experiments which establish for him the psychological primacy of the brain, we can discover what he probably had in mind by turning to another of his works, *The Passions of the Soul.* In this book he points out that the Aristotelian position can be demolished by the anatomical demonstration of a nerve connecting the heart to the brain. Cut this, he implies, and we should certainly feel nothing of the heart. This, of course, is merely a 'thought experiment', but Descartes shows himself to be well informed of the role of the more easily approached sensory nerves. Interfere with a nerve leading from the foot to the brain at some intermediate point, say at the loins or at the neck, and, says Descartes (9), the subject experiences a sensation in the foot. It is no more necessary, he goes on, for 'the soul to be in the heavens to see the stars' than for the soul to be in the heart 'in order to feel its passions'.

Descartes' anatomical knowledge was by no means all second-hand. He is known to have made a practice of visiting slaughterhouses while living in Holland, and quite probably carried out dissections himself. In the *Principles of Philosophy* he gives evidence of a more than merely hearsay acquaintance with the 'phantom limb' phenomenon experienced by amputees. It is likely that his 'New Method', which prevented him from following the dictates of any sect, would have impelled him in this, as in other matters, to see for himself.

It will have been noticed that the subject of discussion in the previous paragraphs has changed from 'mind' to 'soul'. In fact Descartes does not distinguish at all sharply between the two. 'We have reason to believe', he writes (10), 'that every kind of thought which exists in us belongs to the soul'. Remembering how wide the Cartesian notion of thought is, it is clear that soul and mind are interchangeable terms. Descartes does not, however, accept the classical belief that 'soul' is especially concerned with animal movement. He observes that many inanimate objects move in many and diverse ways, and these by definition lack 'soul'. His sharp dichotomy enables him to assert that passions pertain to the soul, actions to the body; movements pertain only to the body, every kind of thought belongs solely to the soul.

Cardiovascular physiology

If movement and action pertain to the body and do not necessarily require, as Descartes implies, the intervention of the soul, it follows that the body's activities can be treated as the activities of a piece of intricate machinery. And this Descartes proceeds to do. He explicitly compares the body to a

(9) *Passions of the Soul,* Articles XXX–XXXIV.
(10) *Ibid.,* Article IV.

finely designed watch or other mechanism. In *The Passions of the Soul* he briefly describes the organisation of this mechanism giving pride of place to Harvey's discovery of the blood's circulation. He ends by describing the movement of the muscles, and declares that 'all these movements of the muscles...depend on the nerves which resemble small filaments, or little tubes, which all proceed from the brain and thus contain like it a certain very subtle air or wind which is called the animal spirits' (11).

Thus we see that the ancient explanatory concept of the pneuma lived on well into the seventeenth century to confuse the would-be physiologist. Descartes, although trying to incorporate as much of the new science as possible into his system, remained in many respects very much a man of his age. Maynard Keynes called Newton the last of the alchemists, and Descartes, living some fifty years earlier, although clear in his mathematical thought, also remained in other matters very much influenced by the older ways.

Nowhere is this seen more clearly than in his cardiovascular physiology. Although, as we have already mentioned, he fully accepts Harvey's circulation theory, he nevertheless still believes that in the heart is to be found 'a species of fire...a continual heat...which the blood of the veins there maintains'. In the treatise on man—*L'Homme*—the theory is explained in more detail. After describing how the food is digested in the stomach and intestine, absorbed into the portal veins and carried to the liver where it is transformed into blood (which he compares to wine-making), Descartes asserts that the blood passes into the vena cava and then into the right ventricle of the heart, in the pores of whose wall lurks 'one of those dark fires [un de ces feux sans lumière] '(12). This classical notion is one of the cardinal points in the Cartesian physiology and responsible for one of its major mistakes.

Instead of the work of the heart being done at systole, as Harvey had shown, Descartes conceived it to be done at diastole. The blood entering the right and left ventricles from the vena cava and the pulmonary veins is, according to Descartes, volatilised by the central heat and, being volatilised, expands. This expansion is observed as the diastole of the heart. The volatilised blood quickly fumes off up the pulmonary artery and the aorta, and the heart, lacking blood, collapses back to its original size.

This view, which seems to us so obviously wrong, was far from seeming so incredible in the first half of the seventeenth century. Indeed, Descartes adduces much evidence in its favour. If the thorax of an animal is opened, he says, the heart can be felt by the hand to be warm. In order to assure ourselves that dilation is the active phase, Descartes advises the examination of the heart of a fairly large animal. He describes how the anatomy of the heart, its valves, and the vessels leading from and to it may be discovered by

(11) *Ibid.*, XII.
(12) *L'Homme*, p.338.

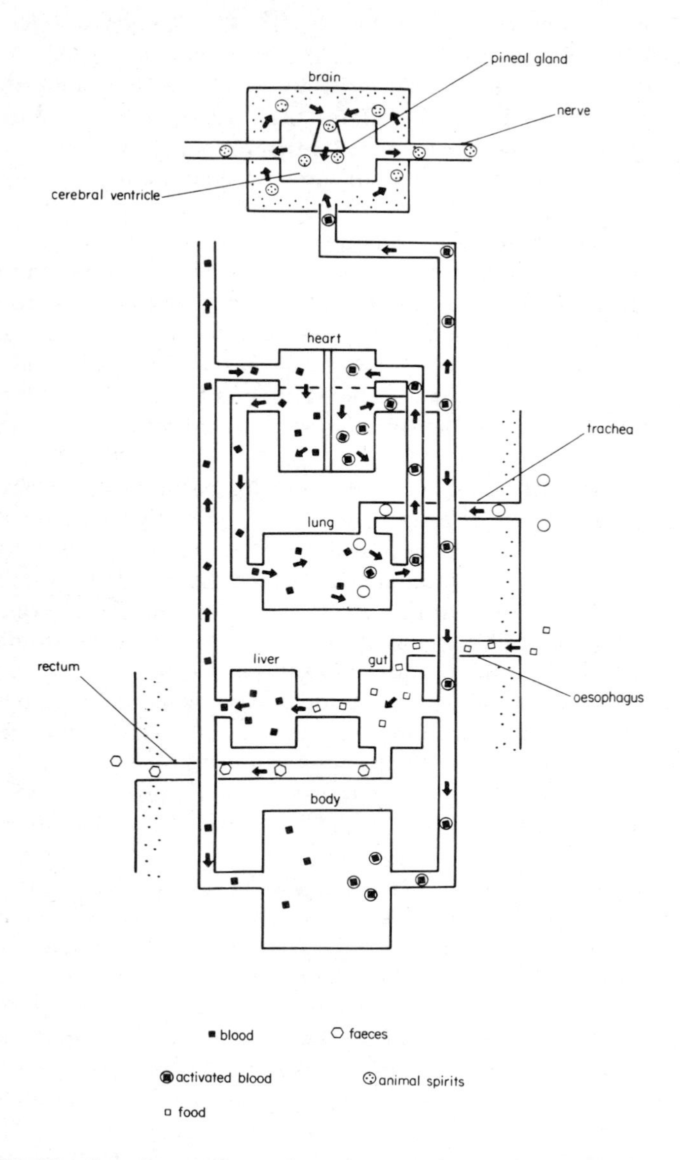

Figure 15.1. The Cartesian cardiovascular system. The chyle is transformed into blood by the liver very much, says Descartes, as wine is formed from grapes. The newly formed blood passes (sluggishly) to the right auricle and ventricle. Under the influence of the 'dark fire' in the ventricle it expands explosively and passes through the lungs. Here it reacts with the air and is 'activated'. The 'activated' blood fuels the furnace of the left ventricle. Finally the most active parts of the blood boil up the carotids towards the brain. As the blood passes through the substances of the brain, the animal spirits are filtered out. These jet out of the pineal into the cerebral ventricle. They escape from the ventricles through the nerves.

cutting across the apex of the ventricle and exploring the interior with the fingers (13). He goes on to consider in detail Harvey's theory of the heart. He argues vigorously, however, that Harvey was wrong to believe that the cavities of the heart shortened and hardened during what we now call systole. Descartes 'sees' the hardening and shortening as a response to the heating and consequent expansion of the blood within the ventricles. The rarefied blood beats against the ventricular walls, and it is this which causes the appearance of hardening and shortening. The 'expanding' blood soon overcomes the resistance of the ventricular walls and the whole heart enlarges. The ebullient blood, however, quickly escapes up the arteries and away from the heart which then collapses back to its original size.

Anyone who has watched a mammalian heart beating will recognise the difficulty in deciding between the Harveyan and the Cartesian theory. Harvey, however, in being the better comparative anatomist, had spent many hours observing the slower action of the hearts of cold-blooded vertebrates. It was largely on the basis of this experience that he had arrived at the correct interpretation. Descartes' theory, too, for all his protestations about being his own man in the fields of science and philosophy, was based at a crucial point on a received idea, the idea that on being heated blood undergoes an immediate and great expansion. 'The blood', he writes, 'is of such a nature that when it is heated only a little more than usual it instantly expands' (14).

This view of cardiac physiology is central to Descartes' endeavour which, as he states at the beginning of *L'Homme*, is to describe the human body as a machine similar to, though infinitely more complicated than, a watch, a windmill or a fountain. The idea of the blood volatilising and expanding is thus used to account for its distribution to the different organs of the body. The most direct route out of the left ventricle (15) is via the aorta and the carotid arteries to the brain. When the carotid artery reaches the brain it branches into innumerable minute vessels which ramify throughout its substance. Now the blood fuming up from the left ventricle will have its more rarefied and subtle parts in the van. This fraction alone, according to Descartes, will be capable of penetrating the labyrinthine channels of the cerebrum. The heavier and less rarefied moiety boiling up behind will be unable to penetrate the cerebral vessels and will consequently make its way through more patent channels to the other and less important organs of the body (16). For Descartes has, of course, a weighty purpose for the first moiety, consisting as it does of 'les plus vives, les plus fortes et les plus subtiles parties de ce sang' (17).

(13) *On the Formation of the Foetus,* Second Part, p. 437.
(14) *Ibid.,* Second Part, p. 440.
(15) Descartes perfectly understands the nature of the pulmonary circulation from the right ventricle: see *On the Formation of the Foetus,* p. 446.
(16) This conception of the circulation is strangely reminiscent of the account given of the circulation in *Rana* which was still to be found in zoology texts in the first half of the twentieth century.
(17) *L'Homme,* p. 344.

The lively subtle fluid which makes its way through the porous substance of the brain is destined to form the animal spirits. Thus we read in *L'Homme* that this moiety

> 'serves not only to nourish and sustain its substance [the brain's] but principally to produce a very subtle wind [un certain vent très subtil], or, better, 'une flamme très vive et très pure', which one calls the animal spirits (18).

We have already seen that Harvey and indeed Vesalius regarded the spirits with considerable scepticism. Descartes, too, as the above account shows, does not regard them as anything more than the most subtle fraction of the blood. In his eyes this distillate consists of the most minute particles which just because of their minuteness can move with extreme rapidity—'like the particles of flame issuing from a torch' (19). This most rarefied and subtilised portion of the blood, which we can now refer to as 'animal spirits', ultimately makes its way through the meshwork of vessels in the brain to the pineal gland. This gland hangs down into the third ventricle, and the spirits issuing from it are directed to all parts of the brain's cavities. Thus the ventricles of the brain, which we now believe to be filled with a clear watery cerebro-spinal fluid, were for Descartes filled with 'les esprits animaux'.

The above account of the Cartesian cardiovascular physiology bears out our suggestion that Descartes, like Newton, was still half-immersed in the old thought. Although accepting the Harveyan circulation theory, his overall view of the cardiovascular system remained thoroughly classical. This also connects with a remark we made in chapter 14: that Descartes, unlike Harvey, remained a generalist. In consequence his physiological insight was impeded by extraphysiological, indeed extrascientific, lumber. In the next section we shall see that Descartes has an important job for 'les esprits animaux' now that they are safely ensconced within the cerebral ventricles.

Myoneural physiology

The role which Descartes reserved for the animal spirits is the causation of bodily movement. It is here, in fact, that Descartes made his major contribution to animal physiology. Although his actual theories were soon discarded, his general approach provided later physiologists with a programme of research which has led to our present understanding of neuromuscular physiology. Towards the end of the nineteenth century we find T H Huxley writing: 'Descartes did for the physiology of motion and sensation that which Harvey had done for the circulation of the blood, and opened up a road to the mechanical theory of these processes, which has been followed by all his successors' (20).

Descartes recognises that bodily movements are brought about by muscles.

(18) *Ibid.*, pp. 345–6.
(19) *Passions of the Soul,* Article X.
(20) T.H. Huxley, *On the Hypothesis that Animals are Automata and its History* (address to the 1874 meeting of the British Association for the Advancement of Science), reprinted in *Collected Essays,* vol. 1, Macmillan, London (1898).

He recognises also that skeletal muscles are customarily arranged in antagonistic groups, so that when one muscle contracts another elongates, and *vice versa*. The crucial question for Descartes is: what causes a muscle to contract?—a question whose answer, we might note in parenthesis, is still not known. Descartes' answer is to postulate the presence of animal spirits in the muscles and the existence of passages between antagonistic muscles through which these spirits can flow. A flexor muscle, for example, might be caused to contract or, what in Descartes' eyes amounts to much the same, to balloon up, by spirits flowing into it from its paired extensor which becomes in consequence elongate and flaccid. Thus by distributing the spirits between the pair of antagonistic muscles the joint may be extended or flexed.

Clearly, however, skeletal movements do not just occur at random. They must in some way be controlled; and the control must, according to the Cartesian physiology, be exerted from the brain. Descartes is fully equal to this problem. Not only does he grasp, as we have already seen, the significance of the sensory nerves, he is also quite clear about the function of the motor nerves. In both cases the nerves are, for Descartes, fine tubes filled with animal spirits in the midst of which are suspended slender filaments (21). It is in this way that the animal spirits reach the muscles. Normally agonist and antagonist muscles contain carefully balanced quantities of spirits. As soon, however, as an extra packet of spirits arrives at one or the other muscle via the nerve, the balance is upset. One muscle now contains significantly more spirits than the other. A snowball or avalanche effect follows. The extra quantity of spirits in the one muscle is able to open the passages leading to it from the other and to close the passages leading in the opposite direction. Spirits in consequence cascade across into the innervated muscle which balloons up or contracts (22).

In this account of the control of muscular activity, the significance of the slender filaments suspended in the centre of the nerves has not been made clear. This is because their function is in sensation rather than in muscular control. According to Descartes they connect the sense organs and the general surface of the body to the brain. When a sense organ or any part of the body is stimulated, these intraneuronal filaments are pulled like strings. The cerebral ends of the filaments are attached to valves in the walls of the ventricles. When the string is pulled the valve opens and spirits flow out to the appropriate muscle. This is the Cartesian reflex. As an example he describes the blink reflex which ensues when a friend jestingly shakes his fist at one's eyes. The blink, he points out, is certainly not willed, indeed it is against the will. Descartes' explanation is that the raised fist affects the retina in such a way that the filaments within the optic nerve are pulled, and this leads to an opening of the cocks of the nerves leading to the muscles responsible for the blink (23).

(21) *Passions of the Soul,* Article XII.
(22) *Ibid.,* Article XI.
(23) *Ibid.,* Articles XII, XIII.

Figure 15.2. Reflex action. This diagram from *L'Homme* illustrates Descartes' theory of reflex action. The cord c – c – c to which Descartes refers may be seen marked at the base of the toes, in the thigh and in the shoulder.

Perhaps the most famous of Descartes' examples of reflex action is illustrated in figure 15.2.

In this figure, reproduced from *L'Homme*, the particles of fire beat upon the flesh of the foot and pull the slender cord c—c—c which opens the valve, d, in the brain at which the cord terminates. This allows the animal spirits in the cerebral ventricle, F, to escape and to flow 'partly into the muscles which serve to retract the foot from the fire, partly into those which serve to turn the eyes and the head to look at the fire, and partly into those which serve to advance the hands and bend the body...' (24).

(24) *L'Homme*, p. 359.

Descartes develops his mechanistic model further. He points out that nerves run not only to and from the brain, sense organs and body surfaces, but also to and from the brain and the viscera. Thus, he says, nerves leading from the stomach actuate reflexes based on hunger and nerves from the heart govern behaviour springing from feelings of sadness or joy (25).

Following this account of reflex mechanics, Descartes draws his reader's attention to what he believes to be an illuminating analogy. He suggests that we compare the reflex automaton to a cathedral organ. The cerebral ventricles he likens to the organ's reservoir of high pressure air. The heart and the arteries play the part of bellows which continuously top up the reservoir. External objects are to be compared with the organist's fingers which, pressing the keys on the keyboard, open passages from the reservoir into particular organ pipes. And, says Descartes, just as the harmony of an organ does not depend on the layout of the organ pipes, or on the shape of the reservoir or other parts, but solely on how air is pumped by the bellows, how the pipes are caused to vibrate, and on how the air is distributed among the pipes, so the action of the body does not depend on the external form of the parts distinguished by anatomists, or even on the shape of the cerebral ventricles but only on the spirits produced by the heart, the cerebral pores through which they pass and the distribution of the spirits in these pores (26).

Determinants of the personality

In the previous section we have seen how Descartes describes the behaviour of the body as the outcome of a reflex mechanism. His cathedral organ analogy emphasises, however, that the organist's fingers, or the external stimuli, do not entirely determine the automaton's response. This depends also on the machine's internal constitution, that is, on the nature of the spirits produced by the heart and their distribution in the cerebral pores.

The spirits, which are of course for Descartes a sort of monatomic gas, completely material, may be more or less abundant, more or less agitated, their particles more or less heavy and more or less equal among themselves. From these four sets of differences Descartes derives by various combinations the multifarious humours or inclinations which characterise the human personality—promptness, calm, liberality, perseverance, hardiness, timidity, anxiety, and so on. In the process of mixing the spirits to produce an individual's characteristic humour, not only external perceptions but also the lungs, the liver, the gall bladder and the spleen exert an influence. In these passages we see, once again, how strong a hold the ancient physiology still maintained over an enlightened seventeenth-century mind.

Turning to the distribution of the animal spirits in the cerebral 'pores', the second of the internal determinants of Cartesian behaviour theory, we come once again to Descartes' favourite organ: the pineal gland. It will be recalled

(25) *Ibid.*, pp. 384–5.
(26) *Ibid.*, p. 386.

that the animal spirits, having forced a tortuous passage through the labyrinth of the brain, finally emerge from the pineal into the cerebral ventricles. Now according to Descartes this gland is able to move to and fro in the ventricle and thus direct its jet of spirits from one position to the next. When the gland is at rest the spirits are directed forwards, and if they are unable to escape elsewhere the pressure may build up to such an extent that an explosive passage is forced out through the cribriform plate to the nostrils: an event commonly termed sneezing.

The pineal gland

The pineal gland, however, does not always remain at rest within the cavities of the brain. Far from it: Descartes' theory of voluntary action depends upon the mobility of this gland.

We have seen in an earlier section that the mind, for Descartes, is associated with the brain. And, in *The Passions of the Soul* and *L'Homme*, we read that it exerts its power principally through the pineal (27). Descartes maintains that he has arrived at this conclusion on strictly anatomical grounds. Whereas, he observes, all other parts of the brain are double, as are the organs of special sense, only the pineal is single. Introspection assures us that the doubting ego is unitary: it is impossible to fractionate our psychical selves. Furthermore our sense impressions, despite our two eyes, our two ears, are also unitary. In what appears to be one of the worst *non sequiturs* of his system, Descartes identifies the median unitary pineal as the anatomical substratum of the unitary, but we remember according to the philosopher himself, unextended, mind (28). It is with this gland that the soul both exerts its effects and perceives the external world. Descartes, in these passages, seems not to have understood the significance of his own theory—seems to be quite unworried by the consequences which his system has forced upon all subsequent philosophers and physiologists. For in so sharply distinguishing *res cogitans* from *res extensa*, mind from body, he posed the intolerable problem of their interaction. 'Physics tells me', writes C S Sherrington,

> that my arm cannot be bent without it disturbing the sun. Physics tells me that unless my mind is energy it cannot disturb the sun. My mind then does not bend my arm. If it does, the theoretically impossible happens. Let me prefer to think the theoretically impossible does happen. Despite the theoretical I take it my mind *does* bend my arm, and that it disturbs the sun (29).

If we turn from the rather superficial account given in *The Passions of the Soul* to the more detailed physiology of *L'Homme* we can discover exactly how, in Descartes' view, the all-important pineal performed its functions. We

(27) *Passions of the Soul,* XXXII.
(28) *Ibid.,* XXXIV, XXXV.
(29) C.S. Sherrington, *Man on his Nature,* Penguin, Harmondsworth (1955), p. 258.

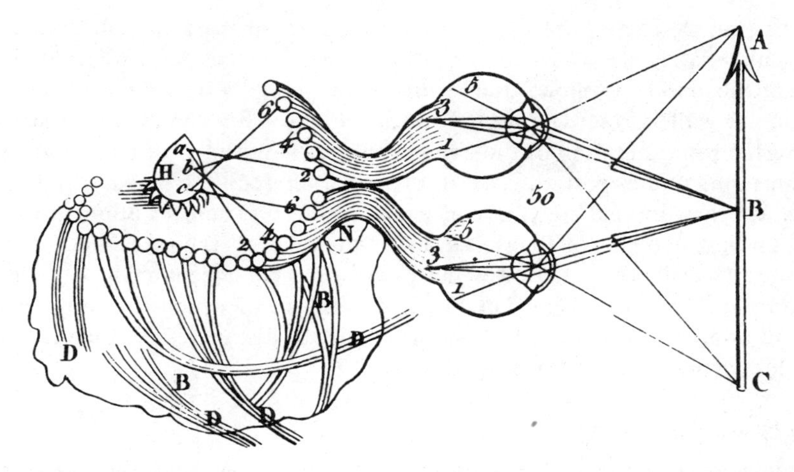

Figure 15.3. Pineal imagery. This diagram from *L'Homme* shows how the form of an object is traced upon the pineal gland, H. Note how the gland is represented as floating upon an efflux of animal spirits.

find first of all a masterly account of physiological optics (30). Figure 15.3, reproduced from *L'Homme*, shows how the form of an object is traced upon the gland. Rays of light from the object are focused on the retina and, as in the case of reflex withdrawal of the foot from a flame, open corresponding valves in the wall of the ventricle. Through these opened valves the compressed spirits of the ventricles escape.

We now come to a vital move in the Cartesian psychophysiological argument. It is a move which Descartes is able to make because of the nature of the physical ideas he sets out in the *Principles of Philosophy*. In this work we find that matter for Descartes, as for Aristotle, is a continuum: there are no interstices between things. Although, a continuum, it is, however, particulate. The particles are closely packed together, so closely, indeed, that there are no spaces left between them. In consequence movement can only occur by interchange of position. The analogy is to a close-packed heap of sand. Now, if we return to our concern with the escape of spirits through openings in the ventricular wall, we can draw an analogy with the escape of sand through, for example, the opening in the centre of an hour-glass. A depression immediately appears on the surface of the sand. In the Cartesian neurophysiology this cavity appears at the interface between the ventricular spirits and the surface of the pineal gland. In consequence the spiritual pressure on that part of the gland is reduced.

(30) Physiological optics is also treated at length in the *Dioptrics*. Here we read of experimental demonstration of part of the Cartesian scheme. The reader is advised to obtain the eye of a newly dead man or, failing that, of an ox or other large animal, and to cut a window in the back of the orbit. If a transparent screen is inserted in this window, says Descartes, an image of the objects in front of the eye may be observed displayed upon it (*Dioptrics*, Discourse V).

Thus, for Descartes, the escape of spirits through certain points in the ventricular wall leads to a reduction of the pressure exerted on corresponding parts of the gland. It follows that a mirror image of the two pictures on the ventricular wall is traced upon the pineal. Figure 15.3 shows that a single uninverted representation of the external object is outlined upon the gland. Furthermore, and importantly, this representation upon the pineal *is* the idea which appears 'immediately' in consciousness (31). Descartes appears to be propounding a version of the well-known 'dual aspect' theory of the mental and the physical. He generalises his psychophysical mechanism so that it accounts not only for vision but for all the things of which the mind may become aware: '...le mouvement, la grandeur, la distance, les couleurs, les sons, les odeurs, et autres telles qualités...'.

The physical basis of memory

The next step in the Cartesian argument is to show that the proposed mechanism also accounts for memory. Descartes adopts a theory which seems strikingly modern. The physical basis of memory, we read in *L'Homme* (32), is the substance of the brain which, for Descartes, is riddled with pores or tubes. These tubes vary in size and consequently in the ease with which they may be traversed by the animal spirits. The more often the spirits flow through these channels, the wider and easier they become. Substitute synapses for tubes and we have a very contemporary theory of memory. The neurophysiological bases of conditioned reflexes are often explained along these lines. Our first experience of a flickering, glittering object may lead to our reaching out to touch it. The subsequent pain ensures that we never make the same mistake again. The flickering flame is ever after associated with the sensation of burning. The tubes within the brain which connect the visual area with the muscles of withdrawal are ever afterwards enlarged and widened. The sight of a flame sends spirits coursing through them ensuring that never again do we attempt to touch it. Subjectively we 'remember' the pain.

The springs of human action

After these preliminaries, Descartes is now able to take up the general problem of developing a mechanical explanation of human behaviour. First he points out that the pineal is very delicately suspended within the cerebral ventricle. It is, he says, rather like an object borne up by the rising fumes and heated air above a fire. The fumes and hot gases are analogous to the animal spirits which incessantly pour from the gland. If the exodus of spirits is exactly balanced in all directions, the gland will remain stationary, suspended in the centre of the ventricular cavity. The slightest inequality in

(31) *L'Homme*, p. 398.
(32) *Ibid.*, p. 400ff.

the escape of spirits will, however, cause it to incline to one side or the other. But this, if it happens, affects the arteries carrying the spirits which run up the pineal stalk. They are twisted and bent this way and that. This has the consequence that the spirits leave preferentially from one side or the other of the gland. This, in turn, results in an increased pressure on the tubes in one side or other of the cerebrum. In many cases these tubes lead out to skeletal muscles and fluxes of spirits down them result, as we have already seen, in the movement of a limb. In this way the swaying of the gland among its funing spirits causes bodily movement. But, says Descartes, note one important feature. The first cause of bodily movement is 'la façon dont ces esprits sortent pour lors de cette glande' (33) and this flow of spirits away from the gland traces upon its surface a particular low-pressure contour. And, as we noticed above, such a contour is for Descartes the physical basis of an idea. Thus, he concludes, the true cause of bodily movement in his scheme is an idea: '...c'est son idée qui le cause'.

A second and, for Descartes, a more important cause of movement is provided by external objects. Here we remember Descartes' analogy of the forces of the external world to an organist's fingers playing out the harmony of movements which constitute behaviour. We also recall that Descartes believes that when an external force 'touches' an organ of sense its effect is to open corresponding pores in the wall of the cerebral ventricle. Spirits surging out from the pineal pour through these openings to the appropriate muscles. Now, says Descartes, the spirits on leaving the gland pull it to a small extent after them. A positive feedback effect occurs: being closer to the appropriate pores, more spirits than before pour through and this in turn pulls the gland yet further towards them, and so on. 'Ce qui', writes Descartes, 'rend l'idée que forment ces esprits d'autant plus parfaite; et c'est en quoi consiste le premier effet' (34). Once again we see Descartes' psychophysical theory deployed. The escape of spirits from the gland traces upon it a low-density contour and this is the physical basis of an idea. The idea $\rightleftharpoons$ contour is the cause of the movement.

Descartes continues his account of the body's machinery by describing in terms similar to those already outlined the mechanics of walking, of sleeping and waking, of dreams and reveries. He satisfies himself that none of these need escape the mechanistic schemata he has developed. He concludes by pointing out to his readers that he has described a machine which in every point, from 'la digestion des viandes' to 'les mouvements intérieurs des appétits et des passions; et, enfin, les mouvements extérieurs de tous les membres', is completely indistinguishable from a man. The activity, or life, of this artefact, he says, follows 'toutes naturellement...de la seule disposition de ces organes, ne plus ne moins que font les mouvements d'une horloge, ou autre automate, de celle de ses contre-poids et de ses roues...' (35). Thus it is

(33) *Ibid.*, p. 403.
(34) *Ibid.*, p. 408.
(35) *Ibid.*, pp. 427–8.

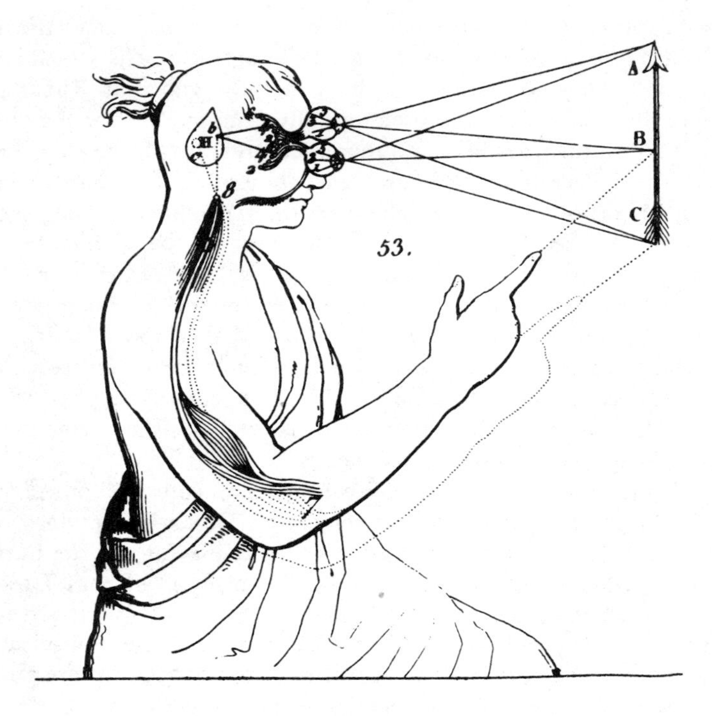

Figure 15.4. Automatic response to a visual image. The image of an object traced upon the pineal gland automatically leads to an appropriate bodily response. The animal spirits escaping from the pineal enter the nerve at 8 and inflate the biceps muscle, thus flexing the arm.

quite unnecessary to postulate any vegetable or animal soul, or any other principle of life 'que son sang et ses esprits agités par la chaleur du feu qui brûle continuellement dans son coeur, et qui n'est point d'autre nature que tous les feux qui sont dans les corps inanimés'. It is this insight, this programme, which assures Descartes of his place at the spring of modern physiology.

Descartes' programme for physiology

Bertrand Russell assents to the widely held view that Descartes stands at the origin of modern philosophy, saying that with him the sixteenth- and seventeenth-century scientific revolution at last makes its appearance in the world of philosophy. Something similar might be said about Descartes' position in the history of neurophysiology. He mapped out the programme which we, three centuries later, are still working within. J H Woodger puts it this way:

> Descartes' physiology of the nervous system has served as the foundation for all that has since been done in the interpretation of that system, and the modern view has *in principle* departed but little from the lead that Descartes gave it (36).

However, to map out a programme is not of course the same thing as fulfilling it. Descartes died in 1650, and it was not until half way through the twentieth century that a thorough understanding of the way in which nerves conduct impulses was finally achieved. Descartes would, however, have recognised twentieth-century neurophysiology as a variant of his own vision. Aristotle, we feel, would have been far more astonished; for with Descartes the world view of seventeenth-century physics begins to take over the old biological realm. This physics, as we have seen, is quite alien to the old Peripatetic science. With Descartes, then, the old organismic paradigm whose mutation we followed in earlier chapters finally begins to undergo a metamorphosis in its central fastness.

For in the precise mathematical mind of Descartes the tripartite soul of the ancients, a notion which we met first in Plato's *Timaeus*, is finally analysed away. The rational soul, pronounced an unextended thinking substance, is taken right out of the physiologist's province. It can only be examined by introspection. The sensitive soul remains in the physiological arena, but any immateriality it might be thought to have is exorcised. It consists of spirits which are in no important way different from methylated 'spirits'. The vegetative soul is similarly replaced, though not in such detail, by refined clockwork.

In fact Descartes does not dwell overmuch on the subject of nutrition, and what he does say (37) is very theoretical and not very enlightening. This is hardly surprising. The subsequent history of physiology shows, as we shall see, that the functions of the nervous system have been very largely worked out by biophysicists, while the phenomena of nutrition have mostly been the province of the biochemist. And in Descartes' day the subject of chemistry, far less of biochemistry, had hardly begun.

This brings us to a final point; for closely connected to the subject of nutrition is that of embryology. Harvey, certainly, saw epigenesis almost as an aspect of nutrition, and Harvey, of course, saw it through Aristotelian eyes. What does Descartes, the mechanist, have to say about this most Aristotelian of subjects? Again, as in the case of nutrition, very little. It is true that the title of one of his books is *De la Formation du Foetus*, but it is largely anatomical in content having, indeed, the alternative title *La Description du Corps humain*. Furthermore, in the parts of the book directly concerned with development Descartes is remarkably tentative. He begins his discussion with the qualification that he has not yet had time to do the various experiments necessary to sustain the views he is about to put forward.

(36) J.H. Woodger, *Biological Principles: a critical study,* Routledge and Kegan Paul, London (1967), p. 48.
(37) *De la Formation du Foetus,* Part III.

And in a letter written at the very end of his life (between November 1648 and January 1649) he admits to 'having lost almost all hope of finding the causes of the [self-] formation of animals...' (38). His account need not therefore detain us. Indeed, the mechanisation of the processes of embryological development has hardly yet been accomplished. We shall see in chapter 21 that a genuine 'entwicklungsmechanik' remains still, in 1976, a hope for the future.

(38) Quoted in N. Kemp Smith, *New Studies in the Philosophy of Descartes,* Macmillan, London (1963) p. 361.

16 ENLIGHTENMENT

The consequences of Descartes

Descartes died in 1650, allegedly of a chill brought on by the early morning lectures required by his patroness, Queen Christina of Sweden. His work outlined, as we have seen, a mechanistic physiology: a physiology which had no need of any concept which could not be equally well applied to the processes of the non-living world. As he points out in *L'Homme*, one has only to consider the ingenious machines worked by water to be found in royal gardens, some of which play musical instruments and others utter words, to be convinced that the activities of the human organism may be similarly caused. This casting forth of the ancient concept of a 'self-moving mover', active centre, or soul, climaxes the long withdrawal of vitality from nature which we have followed in the previous chapters.

But as with many pioneers, Descartes' effort was premature. The common wisdom does not nowadays look back to Descartes as the founder of modern physiology, but to Harvey—to Harvey who, although responsible for the hydraulic model of the vascular system, was in his deepest intuitions a follower of Aristotle. The Cartesian vision is rightly seen as essentially theoretical; the sciences of physics and chemistry which had begun to flow from the Galilean initiative were in 1650 still puny. Centuries of development were necessary before they were mature enough to sustain the Cartesian programme. In the late seventeenth and early eighteenth centuries, so little was accurately known of the sciences of the inorganic that the Cartesian vision could hardly be fulfilled.

This of course is not to say that no attempt was made. Several important 'iatromechanical' systems were in fact developed in the second half of the seventeenth century. Perhaps the most influential of these systems was that of Giovanni Borelli (1608 – 1679). Many of the ideas elaborated by Borelli are also to be found in the slightly earlier work of William Croone (1633 – 1684).

William Croone's iatrophysics

William Croone explicitly follows Descartes in his approach, writing 'we shall consider the living body to be nothing else but a kind of machine or automaton...' (1). In other respects, however, he differs radically from the

(1) W. Croone, *De ratione motus musculorum*, Amsterdam (1667), trans. L.G. Wilson in *Notes and Records of the Royal Society*, 16 (1961), 158–78.

Frenchman. He will only believe in what can be demonstrated to exist in an animal. And where, he asks, when considering the Cartesian theory of muscular contraction, '...are the cavities hollowed out to receive these breaths, or from what place does such a great quantity come, in order that they may so suddenly inflate the aforesaid muscles, unprovided as they are, with any cavity...'? (2) Croone is clearly very sceptical about the Cartesian neuromuscular physiology: he can find no cavity in the muscles, nor can he find any visible cavity in the nerves. He concludes that the Cartesian animal spirits are very largely fraudulent, writing '...as often, therefore, as I say the Animal Spirits, I mean that most subtle, active and highly volatile liquor of the nerves' (3).

In William Croone's eyes, therefore, the nerves did not convey a jet of high-pressure spirits to the muscles but instead small drops, merely, of the nerve liquor, or *succus nerveus*. There was virtually no flow of this juice; instead it was expressed from the nerve endings by a pressure wave initiated at the proximal endings of the nerves in the brain. Only in this way, he believes, is it possible to account for what he conceives to be the extremely rapid rate of impulse propagation in nerves. 'For', he writes, 'at the same time as the fibrils of the taut nerve...are struck in the brain, immediately these droplets of liquor exude from all its branchlets just as, when the piston of a syringe is pushed very lightly, liquid at once spurts out' (4). It is interesting to notice how similar this theory is to some of the concepts which Galen and the Stoics held in the early centuries of our era. Croone's theory is also rather similar to some of the ideas which Isaac Newton published at about the same time (in 1675) (5).

We have already noticed that Croone was highly sceptical of the Cartesian explanation of muscular contraction. He did, however, still adhere to the belief that the contraction of a muscle was in some way bound up with an increase in its volume. Indeed, he had performed an experiment before the Royal Society in 1661 which seemed to him to lend credence to this belief. He had shown how a weight attached to a bladder may be raised by inflating the bladder with air or water (figure 16.1).

He proposed that something rather similar occurred during muscular contraction. But instead of the Cartesian 'subtle wind' inflating the muscle he proposed that the droplets of succus nerveus expressed from the distal endings of the nerve interact with the blood present in the muscle resulting in

(2) *Ibid.*
(3) *Ibid.*
(4) *Ibid.*
(5) Letter of Isaac Newton to Henry Oldenburg in *The Correspondence of Isaac Newton*, Vol. 1, *1661–1675*, Cambridge University Press, London (1959), pp. 362–89. Also in *Opticks* (1704), Book 3, Q. 24: 'Is not animal motion performed by vibrations excited in the brain by the power of the will, and propagated from thence through the solid, pellucid and uniform capillamenta of the nerves into muscles for the contracting and dilating of them?'

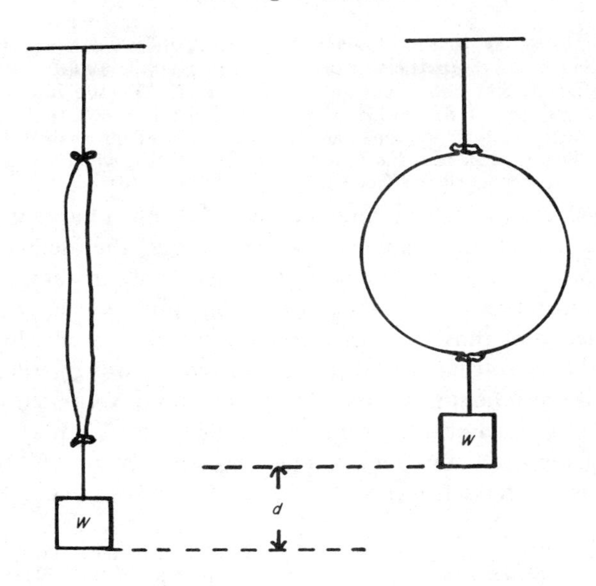

Figure 16.1. William Croone's analogy between bladders and muscles. The diagram illustrates the principle of an experiment performed by Croone before the members of the Royal Society in 1661. The weight *W* is lifted through distance *d* when the bladder is inflated.

an ebullition which in turn causes the observed expansion. The 'very active spirits' produced by the ebullition or effervescence rapidly escape through the boundary membrane of the muscle. Thus we observe that violent muscular exercise causes us to break out into a sweat.

However, at about the time that these theories were being discussed (1660s) the compound microscope was beginning to be used for anatomical investigation. In 1667 Steno published his *Elementorum myologiae specimen*, which contains evidence of microscopical investigation, and Leeuwenhoek was also concerning himself with the microscopical structure of muscle. In 1677 Leeuwenhoek, writing to the Royal Society from Delft, describes his observations as follows:

> I took the flesh of a cow; this I cut asunder with a sharp knife. [He removed the membrane and saw] . . . very clearly the Carneous threads, which in this piece of flesh were as thick as a hair on one's hand. Where they lay rather thick upon one another they appear'd red; but the thinner they were spread, the clearer they showed.
>
> I have used several methods of observing to see the particles of these Carneous filaments, and have always found, that they are composed of such parts, to which I can give no other figure than globular. Moreover, I have divided before my Eye, into many small parts very small pieces of these Carneous filaments, which pieces were several times smaller than a grain of Sand; and I have observed besides, that, when the flesh is fresh and moist and the globuls thereof are pressed or rubbed, they dissolve and run together, as if you saw an oily or thick waterish matter. These globuls, of which I say that the Carneous filaments do consist, are so small, that if I may judge by my

> sight, I must needs say, that ten hundred thousand of them would not make one grain of gravel-Sand . . . the particles which do constitute the flesh . . . (which I call globuls) are not perfect globuls but only come near such. I desire you to consider only, that a great number of Sheeps-bladders, filled with water, and held in the Air, and everywhere surrounded by the same, are round, but if you throw them together into a Tun, they will lose their roundness, and fall close together, whereby each bladder will come to have its peculiar figure, they being very flexible . . (6).

These observations soon became incorporated into physiological theories of muscular movement. Equally important was the demonstration by Swammerdam and Glisson that a muscle does not, in fact, swell during contraction but retains a constant volume. This finding clearly attacked the Croonian theory of muscle contraction at its roots. Accordingly Croone altered his theory so that, as he thought, it could still operate within the constraints of the Swammerdamian demonstration of volume invariance. He suggested that instead of being composed of one great bladder a muscle was composed of multitudinous minute bladders. In this way, he believed, the essence of his effervescence theory could still be maintained. He writes as follows:

> I am the more willing, first, because Mr Leeuwenhoek has since told us, That he finds by his Microscope the Texture of carneous Fibre to be of innumerable small Vesicles or Globules . . . and then because a sheet or two and two or three Schemes of that long expected work of Borelli, *De Motu Animalium,* having been sent to the *Royal Society,* I find there some Schemes for explicating Muscular Motion, the very same with those I make use of . . . (7).

Borelli's De Motu

What, then, were these 'Schemes'? Giovanni Borelli, far more than William Croone, was a mathematician and a physicist. He had early come under the influence of Galileo and was for ten of the most active years of his life a member of the Florentine Accademia del Cimento, a society founded by a group of Galileo's disciples. It is thus not surprising to find that his *De Motu Animalium* makes full use of all the mathematical and physical methods which had been developed by Galileo and his school. The full title of the book which he wrote about muscle was *Dissertationibus physico-mechanicis de motu musculorum et effervescentia et fermentatione*. The key words here are 'physico-mechanicis', and the schemes which Croone refers to are geometrical in nature.

The Borellian exposition is fully within the iatrophysical tradition stemming from Descartes. Indeed, it is interesting to notice that Borelli dedicates his major work, *De Motu Animalium*, of which the treatise on muscle mentioned above forms a part, to the same Queen Christina, now

(6) A. van Leeuwenhoek, *Phil. Trans. Roy. Soc.*, **12** (1677), 899–905 (895).

(7) W. Croone, *Phil. Coll. Roy. Soc.*, **2** (1681), 22–5. Croone is very tentative about the theory he proposes in this paper, writing that his account '. . . is only in the way of an *hypothesis,* not as if I did presume to believe I had found out the true cause of *Animal Motion* . . . but because I reckon such speculations among the best Entertainments of our Mind'.

exiled in Rome, whom Descartes had tutored in the colder climate of Stockholm. In the dedication to this treatise he sets out his approach to the topic of animal movement in the following terms: 'For as the bodies or functions of animals are at all times associated with movements they are subject to mathematics and therefore the mathematical consideration of them must be geometrical' (8).

Now a fellow member of the Accademia del Cimento with Borelli was the anatomist and microscopist Nicolaus Steno. We have already noticed that Steno had interested himself in the structure of muscle. Like Leeuwenhoek, he believed it to consist of innumerable minute globules or bladders. The mathematicising intellect of Borelli seized upon this interpretation, reshaping it so that the fine structure of muscle came to consist of a vast congeries of rhombs or parallelepipeds. This transformation is, of course, fully consonant with the Galilean approach to physical sciences which, as we saw in chapter 14, sought always to construct an ideal world behind the world of appearances, a world amenable to mathematical operations. Borelli then points out that, if a muscle is considered to consist of innumerable rhombs, then a theory somewhat similar to William Croone's can account for its contraction. First of all he points out that a weight may be lifted by pulling on the sides of a rhomb (figure 16.2).

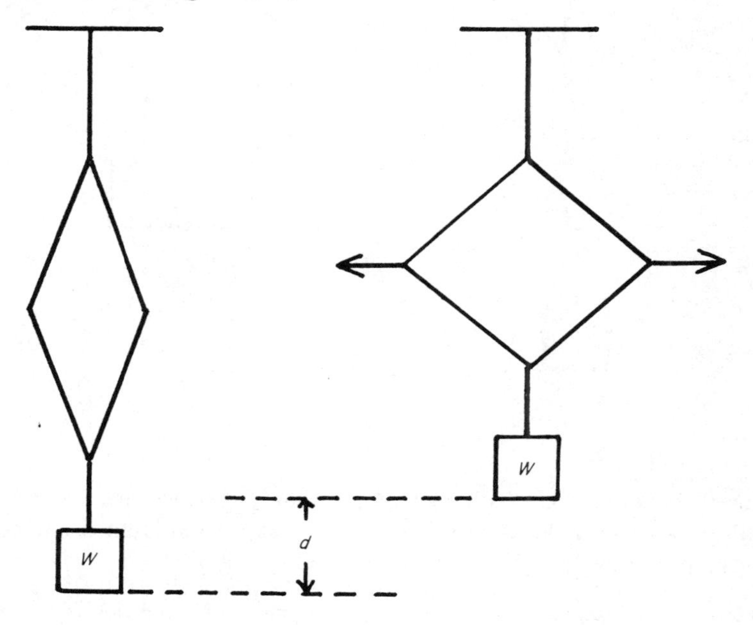

Figure 16.2.

(8) G.A. Borelli, *De Motu Animalium,* 2 vols., Roma (1681, 1682).

Now a square has the greatest area/side ratio of any rhombic or quadrilateral figure, just as a cube has the greatest volume/face ratio of any hexahedral body. Thus if a sudden effervescence or ebullition of the material within one of the muscle's 'unit cells' should occur, a square or cubical figure would inevitably result. And if this ebullition should occur in many of the muscle's 'unit cells' at one and the same time, then the whole muscle will contract (figure 16.3).

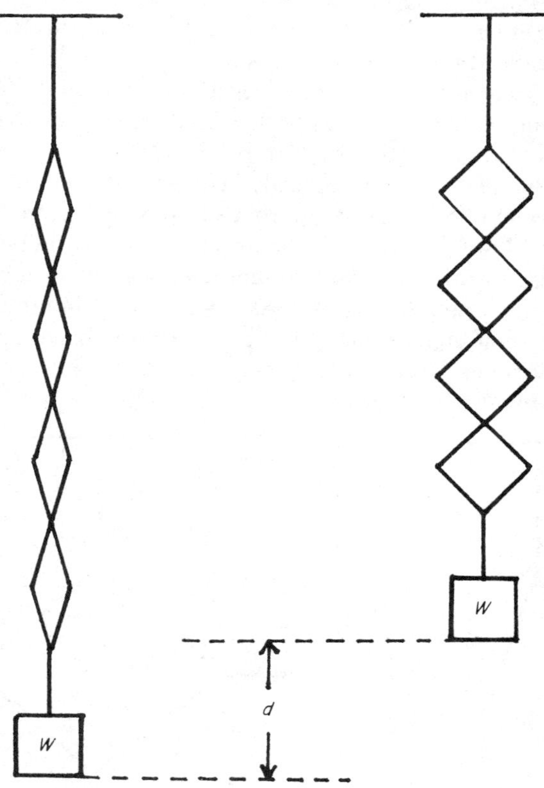

Figure 16.3.

Borelli believed, like William Croone, that the cause of such a sudden ebullition was the mixture of succus nerveus expressed from the nerve with material present within the rhombs. He writes as follows:

> . . . on the arrival of the influence transmitted by the nerves there takes place something like a fermentation or ebullience, by which the sudden inflation of the muscle is brought about. That such action is possible is rendered clear by innumerable experiments which are continually being made in chemical elaborations as when spirits of vitriol are poured on oil of tartar; indeed all acid spirits when mixed with fixed salts at once boil up with a sudden fermentation. In like manner, therefore, we may suppose

that there takes place in muscle a somewhat similar mixing from which a sudden fermentation and ebullition result, with the mass of which the porosities of the muscle are filled up and enlarged, thus bringing about turgescence and inflation (9).

For 'influence' read 'action potential and transmitter substance'; for 'sudden fermentation and ebullition' read 'disinhibition of S_1 ATP-ase and consequent activation of actinomyosin complexes'. But in 1681 iatrophysical and iatrochemical theories could perforce be no more than prophetic. The necessary energetics, biochemistry, pharmacology, biophysics and microscopy were simply not yet in existence. Without them the Borellian theory, like so many of its predecessors and contemporaries, was without foundation. The Borellian approach, which was far from unique at the end of the seventeenth century (John Mayow's *Tractatus quinque medico-physici* (1674) was also strongly influenced by the Cartesian tradition), does show, however, how strong a hold Galilean science had obtained over the scientific mind. Borelli himself regarded the animal organism as a machine, and where lack of instrumentation halted his research he hypostatised still further machines. For example, he believed that during respiration the animal organism takes in aerial corpuscles which are themselves minute oscillatory mechanisms. It is by the interaction of these vibrating particles with the particles of blood in an animal's body that all animal motion ultimately originates (10).

Iatromechanics reaches an impasse

The cul-de-sac into which the iatromechanists had worked themselves gradually became apparent during the first half of the succeeding century. By mid-century, mechanistic ideas in physiology were on the wane: living bodies were once again regarded as being in the possession of special powers not to be found in inorganic nature (11). In Germany, for example, Stahl (1660 – 1734) reverted to older ideas of an immaterial 'soul' animating and directing the activities of the material body. Later P J Barthez at Montpellier elaborated a triadic system in which a 'vital principle', distinct from both the conscious 'soul' and the material body, was responsible for all living activities. Clearly, before any further genuine progress could be made it was essential to advance the understanding of the sciences of physics and chemistry. It can, of course, be no part of the present book's endeavour to chart this advance. It is important, however, to notice certain salient developments. First amongst these must be an outline of the gradual transmutation of the 'subtle winds' which worked the Cartesian 'beast – machine' and the 'nitro-aerial' particles which performed a similar service in Mayow's physiology into the more prosaic notions common today. The 'subtle wind' and the 'nitro-aerial' particles became, in the course of time,

(9) *Ibid.*, vol. 2, chapter 3.
(10) See T.S. Hall, *Ideas of Life and Matter*, vol. 1, New York (1969), pp. 346–7.
(11) See R.E. Schofield, *Mechanism and Materialism*, Princeton University Press (1970), chapter 9.

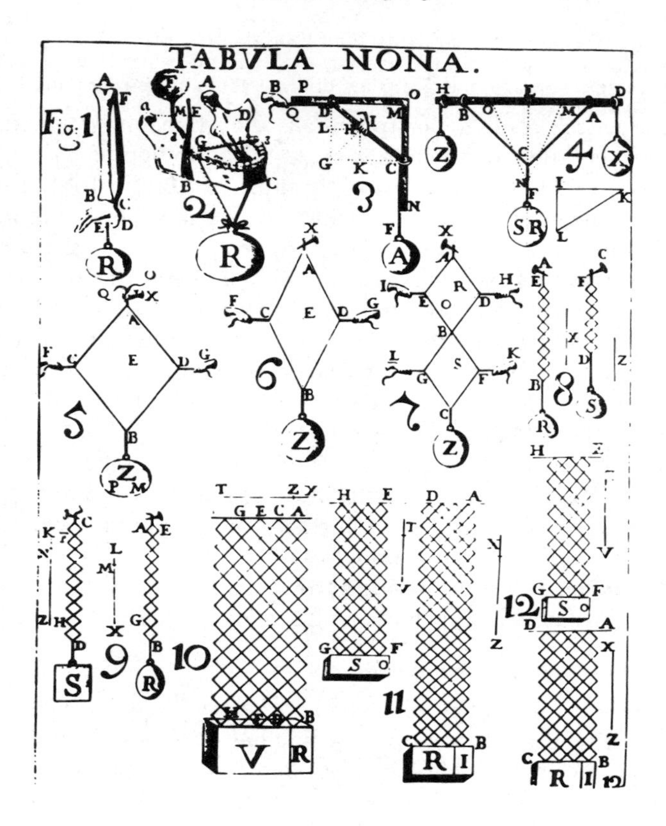

Figure 16.4. Plate IX from Borelli's *De Motu Animalium*. The plate illustrates various aspects of Borelli's 'physico-mechanical' treatment of muscle physiology.

subsumed under the general heading 'gas', and the vital gas whose existence physiologists had for so long suspected obtained, at the hands of Lavoisier, the name 'oxygen'.

Pneumatic chemistry

The term gas was invented by van Helmont and published in 1648. It was derived from the Greek *chaos* and denoted the destruction caused by the product of fermentation or combustion when an attempt was made to confine it in laboratory vessels: 'I call this spirit, hitherto unknown, by the new name of gas, which can neither be retained in vessels nor reduced to a visible form, unless the seed is first extinguished' (12).

(12) J. van Helmont, *Ortus Medicinae,* Amsterdam, 1648, trans. in J.R. Partington, *A History of Chemistry,* vol. 2, Macmillan, London (1961), p. 227.

To the exhalation from wood, the 'wild spirit', van Helmont gave the name 'gas sylvestre'. This gas, in its more modern guise as carbon dioxide, plays, of course, a considerable role in the history of the demise of the Greek physiological souls.

Van Helmont believed gases to be composed of atoms and, indeed, that intense cold could cause the condensation of these atoms into the form of liquid drops. It will be recalled from chapter 14 that Pierre Gassendi was also engaged in resurrecting the atomic theory at much the same time in the mid-seventeenth century.

Robert Boyle (1627 – 1691) also accepted an atomic, or more precisely a 'corpuscular', philosophy (13). To account for his famous observations on the 'spring' or compressibility of the air he supposed it to consist of large numbers of minute coiled springs, or fleeces of wool, an idea, as he says, that he borrows from Descartes:

> There is yet another way to explicate the spring of the air; namely by supposing with that most ingenious gentleman, Monsieur Des Cartes, that the air is nothing but a congeries or heap of small and (for the most part) of flexible particles . . . their elastical power is not made to depend upon their shape or structure, but upon their vehement agitation . . . . (14).

It is also well known that Robert Boyle prosecuted many researches into animal physiology. Included among these researches, some of which do not make very pleasant reading, are several which investigated the effects of subjecting living organisms to the 'vacuum Boyleanum'. One important conclusion emerges from this work. We find that, in 1674, he writes as follows:

> . . . the necessity of fresh air to the life of animals . . . suggests a great suspicion of some vital substance, if I may so call it, diffused through the air, whether it be volatile nitre or (rather) some anonimous substance, sydereal or subterraneal, but not improbably of kin to that, which I lately noted to be so necessary to the maintenance of life and of other flames (15).

'...noted to be so necessary to the maintenance of life and of other flames': clearly the analogy between life and a flame, perhaps only poetic in earlier centuries, was, in the seventeenth century, taking on a more concrete significance—a significance which has deepened as the centuries have passed. However, the 'vital substance' which Boyle suspected to be essential both to the continuance of life and of flames was not finally identified for over a century. Before it received its baptism at the hands of Lavoisier as the 'oxygenic principle', it was tracked through the phlogiston principle of Stahl and the experimental work of Black, Cavendish and Priestley. It was Priestley who made the discovery: a fruit of inspired serendipity:

(13) The difference was that Boyle was not concerned with the question of whether the the corpuscle could be split.

(14) *New Experiments Physico-mechanical Touching the Spring of the Air,* Oxford (1660), vol. 1, 8, quoted by Partington, *A History of Chemistry,* Macmillan, London (1961), vol.2, p. 523.

(15) *Suspicions about some Hidden Qualities in the Air,* Oxford (1674).

> Having acquired a considerable degree of readiness in making experiments of this kind [he writes (16)]a very slight and evanescent motive would be sufficient for me to do it. If, however, I had not happened, for some other purpose, to have had a lighted candle before me, I should probably never have made the trial; and the whole train of my future experiments relating to this kind of air might have been prevented.

The trial Priestley speaks of consisted in placing a candle in a bell jar enriched with the gas obtained by heating mercuric oxide with a powerful burning glass. The candle, he reports, burned with 'a vivid flame'. In the same series of experiments he showed that a mouse lived twice as long as usual in a jar filled with this 'uncommon air'.

But although Priestley may be credited with the discovery of oxygen and also with the recognition that animals kept in enclosed volumes quickly polluted their surroundings by the production of what he called 'fixed air', he never grasped the profounder significance of his work. This, as we noticed above, was achieved by his rival, Lavoisier, who, in so doing, revolutionised chemistry. Priestley, although outliving Lavoisier by a decade, retained until the end a firm belief in the phlogiston theory first propounded by Stahl in 1679. Thus to Lavoisier belongs the honour of recognising that:

> . . . the atmospheric air is composed of two gases, or aeriform fluids, one of which is capable, by respiration, of sustaining animal life, in which metals are calcinable and combustible bodies may burn; the other, on the contrary, has directly opposite properties . . . . We have given to the base of the former, or respirable portion of the air, the name *oxygen* . . . (17).

Lavoisier's revolution in chemistry had a significance far greater than the mere destruction of the phlogiston theory. His insistence on the discipline of the balance enabled him to show that oxygen is used up during animal respiration and is substituted by carbon dioxide in the expired air. Indeed Jacques Loeb, writing in 1912, is by no means alone in believing that 'the work of Lavoisier and Laplace (in the 1780s) marks the beginnings of a scientific biology...' (18). Lavoisier's insistence on the use of the balance also showed that heat is unweighable, but in this case instead of throwing out any material basis he called into being another of those subtle fluids which so plagued eighteenth-century science—*caloric*.

Energetics

Caloric is the starting concept of the nineteenth-century science of thermodynamics—a science which, although begun by a French *Polytechnicien* fascinated by the motive power of heat which energised the British industrial revolution, came as it developed to have a profound significance for the science of life. This need hardly surprise us, for the link between heat

(16) J. Priestley, *Experiments and Observations on Different Kinds of Air,* London (1774), quoted in H.M. Leicester and H.S. Klickstein (ed.), *A Source Book in Chemistry,* McGraw-Hill, New York (1952), p. 113.

(17) A. Lavoisier, *Traité élémentaire de Chimie,* tome 1, Paris (1789) p. 54, facsimile reprint by Culture et Civilisation, Bruxelles (1965).

(18) J. Loeb, *The mechanistic conception of life* (ed. D. Fleming), Harvard University Press (1964) (original edition, 1912), p. 4.

and life, cold and death, has always been commonplace. We noticed its prevalence in the physiological systems of antiquity in earlier chapters, and in the previous section of this chapter we saw that it was still prominent in the ingenious mind of Robert Boyle. Lavoisier and Laplace in their 1780 paper emphasised, moreover, that respiration, the source of animal warmth, was identical to the combustion of carbon. It is clear, therefore, that, as in the case of matter theory, any clarification of the phenomenon of heat in the inorganic realm is of considerable importance to the biologist. And, as it turns out, thermodynamics, especially the second law, has deep implications for the life sciences: not only in the biochemical study of metabolism but also for the structure of organisms and the complexification of that structure in organic evolution.

Again, as in the case of the development of pneumatic chemistry, no attempt can be made here to trace in detail the history of thermodynamics. It will be sufficient for our purposes merely to see how the major concepts of thermodynamics crystallised in the minds of nineteenth-century physicists.

The fact that the science of thermodynamics, or energetics, arose through contemplation of the engines of the Industrial Revolution has already been mentioned. W P D Wightman points out (19) that, until the late eighteenth century, physics was largely concerned with structures like bridges and fortifications, and with the flight of missiles; only when it became important for practical men like James Watt and Matthew Boulton to develop commercially viable steam-driven machinery did it become essential to find a relation between heat and energy. This technological interest also affects the world of letters. Blake's line 'energy is eternal delight' is the most famous of innumerable literary references to energy at the end of the eighteenth and the beginning of the nineteenth century.

But although 'energy' had become almost a cult word in the literary salon, its precise scientific connotation proved difficult to pin down. Its definition, in fact, awaited the prior demise of the Lavoisierian caloric (20).

Caloric, we have observed, is another of the 'subtle fluids' hypostatised by early modern science. It was believed that the phenomena of thermophysics could be accounted for by supposing that there existed 'a substance, real and material...a very subtle fluid which insinuates itself between the molecules of all bodies...' (21), a substance for which some bodies have a greater capacity than others and in consequence require more of to reach a similar temperature.

Sadi Carnot, the acknowledged founder of the science of thermodynamics,

(19) W.P.D. Wightman, *The Growth of Scientific Ideas,* Oliver and Boyd, Edinburgh (1951), p. 275.

(20) Although the term 'energy' was in constant colloquial usage at the beginning of the nineteenth century, it was not employed in the precisely defined sense to which physicists nowadays are accustomed. Although Thomas Young suggested in 1801 that the term 'energy' replace '*vis viva*' it was not until the 1850s and 1860s, after Rankine had popularised the term 'kinetic energy', that it became common in scientific discourse.

(21) Lavoisier, *ibid.,* tome 1, p. 4.

makes explicit use of this imponderable 'subtle fluid' in his seminal analysis of the motive power of heat. Indeed, so elegant is Carnot's formulation that students in the last quarter of the twentieth century are still introduced to the fundamentals of this science through his analysis. It has often been pointed out that whereas the British, James Watt, Matthew Boulton, Benjamin Thompson, were content to develop the technology of the heat engine, the French subjected the machinery to rigorous intellectual analysis. It has also been pointed out how similar this analysis is to that of Galilean mechanics. Once again mathematics is the key to the dark labyrinth, and once again mathematics requires that the concepts involved be stripped of their earthy connections, whether those of the Venetian dockyard or those of the satanic mills of Georgian England. Thus the science of thermodynamics makes use of abstractions such as frictionless pistons, infinitesimal movements, perfect insulations, and so on. For Sadi Carnot, the motion of, for example, a steam engine is caused by a flow of caloric from, as it were, a region of high potential (high temperature) to a region of low potential (low temperature). Caloric, once set free by the boiler, could not be destroyed, only equilibrated. Moreover, and very importantly, Carnot recognised that if it could not be destroyed it equally could not be created: 'Creation (of motive power) is entirely contrary to ideas now accepted, to the laws of mechanics and of sound physics. It is inadmissible' (22).

While French intellect formalised the phenomena on which the Industrial Revolution was based, the more empirical Anglo-Saxons continued to experiment. Benjamin Thompson, later Count Rumford, began the analysis which led to the eventual undermining and elimination of caloric. The story is well known. Boring cannon for the revolutionary wars, he concluded that heat could hardly be a 'subtle fluid' exuding from the pores of hot bodies but must, in some way, be related to friction. Caloric seemed to him to be a superfluous concept, and in the best Ockhamist tradition he believed that it should be excised from the body of science. The heat of a body should, in short, be looked upon merely as an expression of the motion of its constituent atoms and molecules.

This view, propounded in the first half of the nineteenth century, is still in essence the view which we hold today. Its firm establishment is due to the work of many men. Mayer, Joule, Helmholtz, Clausius and William Thomson are outstanding names in this history. Perhaps, however, it is with James Clerk Maxwell that we should end this short section. For he it was who finally showed how the macroscopic laws which Sadi Carnot had discerned arose from the random microscopic pandaemonium of the working substance. Heat flows from a hot to a cold gas, for example, simply because the molecules in the former are moving more rapidly than in the latter and

(22) S.N.L. Carnot, *Réflexions sur la puissance motrice du feu*, Dawson, London (1966): facsimile reprint of original edition, Paris (1824). English translation: *Reflections on the motive power of fire* (trans. R.H. Thurston), Macmillan, London (1890); Dover Books, New York (1960), p. 12.

hence invade the latter to a far greater extent than *vice versa*. Hence the second law. But the amount of molecular motion at the end (assuming the system is isolated from its surroundings) is exactly the same as at the beginning. Hence the first law.

Thus we see that the last of the 'subtle fluids'—caloric—is resolved into the movements of atoms in space. The interpretation of the phenomena of thermophysics is essentially statistical. The macroscopic properties noted by earlier workers, like the phenomena of pneumatic chemistry, received an explanation in terms of the mass action of innumerable microscopic particles. The laws of thermodynamics, like the gas laws, are system laws incapable of application to single Democritean or Newtonian point masses. The premature mechanists of the late seventeenth and early eighteenth centuries were at last vindicated against the exponents of mysterious subtle fluids. The world did indeed seem to be much as the corpuscular philosophers had suggested. The last great prerequisite for a genuine understanding of the physical bases of biology was thus an understanding of the nature of the corpuscles themselves.

The corpuscular hypothesis

The four ancient elements, earth, air, fire, water, had satisfied inquirers for nearly two millennia. It seems likely that Paracelsus was the first chemist whose alternative suggestions received wide publicity and acceptance.

Theophrastus Bombast von Hohenheim, self-styled Paracelsus, is one of the most mysterious and extraordinary personalities in the history of science. Although he is usually thought of as a chemist, he might with almost equal justification be designated alchemist, mystic, physician, charlatan or magician. Born in Switzerland in 1493 Paracelsus obtained an MD in Ferrara and was appointed medical officer of health in Basel in 1527. However, he very soon fell foul of the authorities and spent the rest of his life as a fugitive and itinerant teacher. He died in 1541.

Paracelsus seems to have accepted the ancient doctrine of the four elements but to have believed that they expressed themselves in the form of three principles—the *tria prima*. These principles were salt, sulphur and mercury. As an iatrochemist, Paracelsus made use of the *tria prima* in pathology, proposing a variant of the classical humoural doctrine in which the well-being of the body depends on a correct balance being struck between the three principles. Paracelsus, moreover, compares his three elements with the theological trinity of body, soul and spirit. Mercury, the principle of fusibility and ductility, was compared with spirit; sulphur, the principle of flammability, was compared with the soul; and salt, the principle of endurance and fixity, to the body.

The Paracelsian elements, the *tria prima*, entered the body of chemical and alchemical lore and were discussed and used by chemists well into the seventeenth century. However, throughout the seventeenth and especially the eighteenth centuries, a deepening stream of careful observation and analysis began to wash away the rather ill-defined concepts of the four elements of the ancients and the three principles of the Spagyrists.

Robert Boyle is conventionally accorded the honour of establishing the modern meaning of the term element, although in some parts of his voluminous writings he appears to back-track into obfuscation. In *The Sceptical Chymist*, however, he is quite clear:

> I mean by Elements, as those Chymists that speak plainest do by their principles, certain primitive and simple, or perfectly unmingled bodies; which not being made of any other bodies or of one another, are the Ingredients of which all those called perfectly mixt Bodies are immediately compounded, and into which they are ultimately resolved (23).

This has a modern ring, and although the more ancient ideas persisted long after Boyle's writings were published, the ideas of *The Sceptical Chymist* greatly influenced contemporary and subsequent chemistry. It was not, however, until the theory of phlogiston was falsified in the late eighteenth century that chemistry was finally free to develop into its modern form. The dephlogistication of chemistry was, as we have already seen, the great achievement of Antoine Lavoisier. Another aspect of the radical re-organisation which the subject received at the hands of Lavoisier was the invention and deployment of a new system of chemical nomenclature. 'That method', he writes (24), 'which it is so important to introduce into the study of chemistry is closely linked to the reform of its nomenclature...the logic of a science is related essentially to its language'.

So successful was this reform of the language of chemistry that, as Partington remarks (25), whereas earlier chemical works are nowadays unintelligible to all except the historian of science, the *Traité élémentaire de Chimie* can still be understood as a somewhat obsolete chemistry textbook. Lavoisier's definition of the term 'element' or 'principle' as used in chemistry is very similar to Boyle's: 'The connotation of the name "element" or "principle" is that it is the last point which analysis is capable of reaching' (26). This definition, he goes on to point out, is somewhat tentative, as it may be that our techniques are as yet incapable of showing that what is in fact a compound, or to use Boyle's term a 'mixt', is a compound. However, he says, the best rule is to regard all bodies as elements until experiment shows them to be otherwise. Accordingly he published in the *Traité* a list of 33 elements, including light and caloric. Of these 33 elements the majority still appear unchanged and bearing Lavoisier's names in modern chemistry texts.

(23) R. Boyle, *The Sceptical Chymist: or chymicophysical doubts and paradoxes touching the Spagyrist's principles commonly called hypostatical,* Cooke, London (1661): facsimile reprint, Dawson, London (1965), p. 350.

(24) Quoted in C.C. Gillispie, *The Edge of Objectivity,* Princeton University Press (1960), p. 233.

(25) Partington, J.R., *A History of Chemistry,* Macmillan, London (1962), vol. 3, p.484.

(26) Lavoisier, *ibid.,* p. 192.

Further progress in the understanding of the elements depended upon a further synthesis, a synthesis at two levels: intellectual and national. It depended firstly on a fusion of the chemist's and the physicist's vision. The corpuscular philosophy had gained wide popularity in the late seventeenth and early eighteenth centuries, but the ultimate particles of the 'mechanical philosophers' were generally regarded merely as physical principles. They were regarded as something altogether more abstract, more hypothetical, than the elements, or chemical principles, with which the practising chemists had to deal. Lavoisier, indeed, insisted that chemistry should concern itself with 'chemical atoms', or molecules, not with the speculative constructs of physicists. A synthesis between the physicist's and chemist's atoms was thus the first prerequisite.

Secondly, however, a synthesis between the systematising genius of French chemistry and the more workaday, less generalising, approach of the Anglo-Saxons was necessary. In the event the synthesis awaited the emergence of a Mancunian Unitarian to complement the most famous luminary of the Parisian *Académie des Sciences*.

John Dalton was born in Cockermouth in 1766 but spent most of his life in Manchester, where he died in 1844. His espousal of atomism seems to have arisen from an interest in meteorology and the solubility of gases, and also from a reading of Isaac Newton. He was able to explain very nicely by means of crude atomic models (apparently coloured blocks of wood) several of the laws which had been discovered about the behaviour of gases. Isaac Newton had already anticipated Boyle in showing how an atomic theory could explain the pressure – volume relations of a gas, and now Dalton was able to show how the theory could also account for the characteristics of gas solubility recently described by Henry (now known as Henry's law). He went further still and showed that the hypothesis would also account for his own law of partial pressures wherein, as Henry explained, 'every gas is a vacuum to every other gas'. It seems that it was only after these triumphs for the hypothesis that Dalton saw that it could also explain the observations made by Proust and several other chemists which came to be known as the law of 'multiple proportions'. Dalton recognised that the somewhat approximate ratios observed by the chemists could be explained if it were supposed that the chemical elements were in fact atomic in nature and that combinations consisted in atomic combinations. An entry in Dalton's notebook (27) of 1803 outlines the main characteristics of the chemical atomism from which he never subsequently departed:

1. Matter consists of small ultimate particles or atoms.
2. Atoms are indivisible and cannot be either created or destroyed [law of indestructibility of matter or of conservation of mass].
3. All atoms of a given element are identical and have the same invariant weight.
4. Atoms of different elements have different weights.
5. The particle of a compound is formed from a fixed number of atoms of its component

(27) Quoted in Partington, *ibid.*, vol. 3, pp. 784–6.

elements [law of fixed proportions].
6. The weight of a compound particle is the sum of the weights of its constituent atoms.
7. If more than one compound of two elements is known, the numbers of atoms of either element in the compound particles are in the ratio of whole numbers [law of multiple proportions].
8. The weight of an atom of an element is the same in all its compounds . . . [law of reciprocal proportions].
9. If only one compound of two kinds of atoms A and B is known, it is, unless there is some reason to the contrary, A + B. If there is more than one compound, one is A + B and the other 2A + B, or A + 2B and so on.
10. If the number of atoms m and n of two elements in a particle of a compound mA + nB are assumed, the relative weights of the atoms can be calculated from the ratio of the weights mA/nB of the elements found by analysis. [Dalton chose the atomic weight of hydrogen as unity and derived the weights of the other elements from this basis.]
11. Equal volumes of different gases at the same temperature and pressure cannot contain the same number of ultimate particles . . . .

These propositions mark the successful completion of the eighteenth-century chemical revolution. Dalton's notebook of the same year (1803) also contains the symbolic representation of these ideas: a forerunner of the elaborate structural formulae used by present-day chemists. Dalton's major work, *A New System of Chemical Philosophy,* was published in 1808 (28), and chemistry was set on the path from which is has never since deviated.

The 'chemical' atoms which Dalton envisaged differed from classical Democritean or Epicurean atoms in varying in both weight and quality. The atom of hydrogen, for example, differed from the atom of oxygen or sulphur not only in weight but also in its chemical characteristics. This feature is well brought out in Dalton's chemical symbolism (figure 16.5). The classical notion of an atom as an unsplittable ultimate unit of undifferentiated 'primary substance' seemed no longer appropriate. It only became meaningful once again when, right at the end of the century, physicists began to show that the Daltonian atom was itself composed of yet smaller particles. It then became clear that the classical notion of an atom fitted much more closely the subatomic particles, the electrons, neutrons and protons, from which, it emerged, atoms were built. The interesting properties of these particles, the conflation of continuity and discontinuity, of wave and point mass, will be touched on in a later chapter.

In addition to varying in weight and in chemical properties, the Daltonian atoms were also the centres of attractive force. In consequence of this, binary, ternary, quaternary and other compounds could be formed (see figure 16.6). Dalton believed, however, that such compounds were only formed between dissimilar atoms: like repelled like and attracted unlike. A similar idea informed the early 'dualistic' theories of valency held by Berzelius and others. According to these theories, the attraction between atoms was brought about by electrical forces. Oxygen, for example, was believed to possess a large negative charge of electricity, while potassium possessed a large positive

(28) J. Dalton, *A New System of Chemical Philosophy,* vol. 1, parts 1 and 2, Manchester (1808, 1810); vol. 2, Manchester (1827), facsimile reprint by Dawson, London (1965).

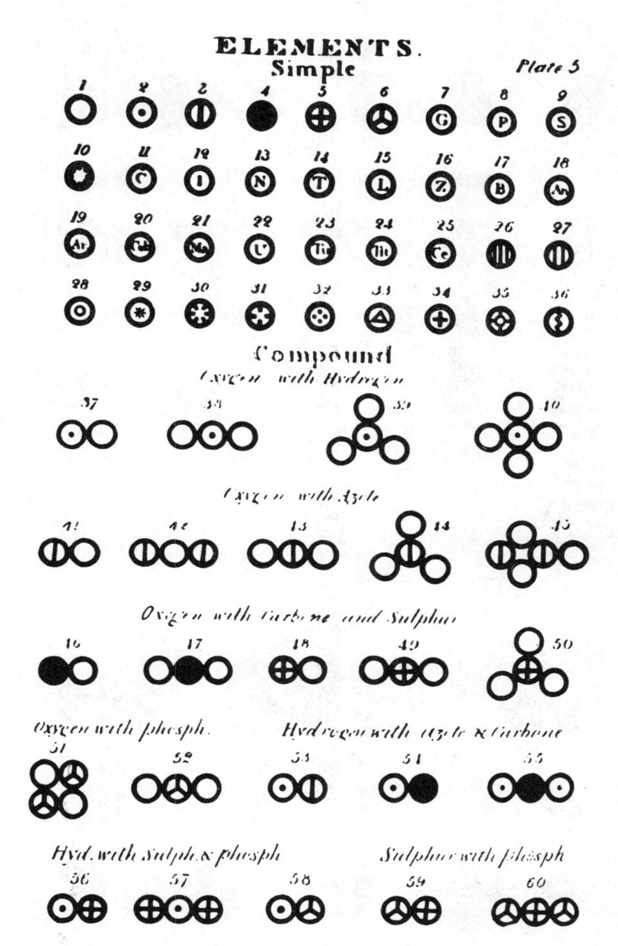

Figure 16.5. John Dalton's chemical symbolism.

Key: 1 Oxygen, 2 Hydrogen, 3 Azote, 4 Carbone, 5 Sulphur, 6 Phosphoros, 7 Gold, 8 Platina, 9 Silver, 10 Mercury, 11 Copper, 12 Iron, etc.
37 Water, 38 Fluoric acid, 39 Muriatic acid, etc.

Reproduced from Dalton's *New System of Chemical Philosophy,* vol. 1, part 2, 1827 (plate 5).

charge. Again like repelled like, unlikes alone associated.

This idea seems to have originated initially from Dalton's analysis of the ways in which gases dissolved in solvents: the law of partial pressures. Later workers were probably also influenced by ideas deriving from electrostatics (29). All these ideas had, however, one very serious drawback. It was impossible to account satisfactorily for certain well-known gas reactions. For

(29) Humphrey Davy from 1806 onward argued for an electrical basis to chemical affinity. His ideas received wide and favourable publicity.

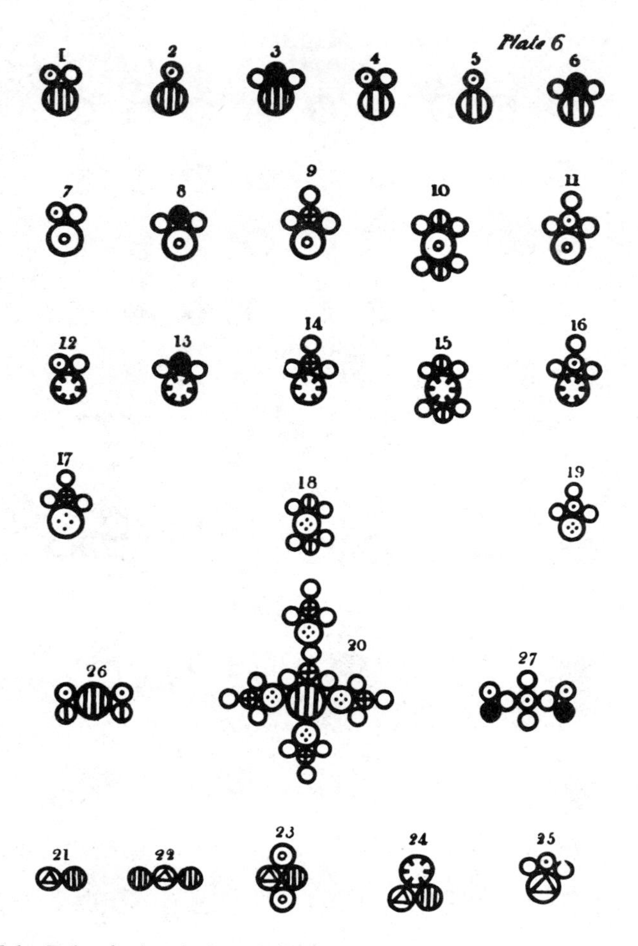

Figure 16.6 John Dalton's chemical symbolism (continued).

Key: 1 Hydrate of potash, 2 Hydruret of potash, 3 Carbonate of potash, 4 Hydrate of soda, etc.

Reproduced from Dalton's *New System of Chemical Philosophy,* vol. 1, part 2, 1827 (plate 6).

example, it was known that one volume of nitrogen reacts with one volume of oxygen to yield one volume of nitrous oxide. If the original gases were monatomic, it would follow from Dalton's rules that only half an atom of oxygen and half an atom of nitrogen would be involved in the reaction. And this notion cuts at the roots of the theory: atoms were indivisible, they were the final units of which elements were built. The difficulty was eventually resolved by Avogadro, who proposed that identical atoms could unite to form molecules. Although this idea was proposed in 1811, it was not until the

1850s that it became generally accepted. In this later period, modern conceptions of valency and chemical notation began to establish themselves. This development was largely due to the work of Frankland in England and Kekulé on the continent of Europe.

With these developments we must leave the evolution of the atomic theory of matter, now well on its way towards its eventual transmogrification at the hands of Bohr, de Broglie and Schroedinger. These developments will be touched on in chapter 21, where the ancient question of the relationship between atomism and organism will be raised once again and for the last time. It will be seen that the twentieth-century understanding of the atom is so different from that prevalent in earlier centuries that the apparent incompatibility between the biologist's and physicist's viewpoints very largely disappears.

17 THE MECHANISATION OF PHYSIOLOGY

A false principle?

The successful prosecution of the Cartesian physiological programme depended, as was stressed in chapter 16, on a prior understanding of the sciences of the inorganic. This, as we saw, was largely established by the eighteenth-century chemical revolution. Claude Bernard, who, as we shall see, is particularly well qualified to judge, writes as follows: '...it is only since the establishment of chemistry that the experimental method could be applied more thoroughly to physiology and pathology' (1).

This is a view from the heights of the century's end. In contrast, the beginning of the nineteenth century was full of doubts. Many prominent workers felt that it was almost a category mistake to attempt the explication of the organic in terms derived from the realm of the inorganic. The strongest reactions, inspired by the *Naturphilosophie* movement, came in the newly christened science of biology. This movement will be discussed in chapter 18. But even in the rather different milieu of anatomy/physiology, powerful voices of dissent were raised: '...to apply the science of natural philosophy to physiology would be to explain the phenomena of living bodies by the laws of an inert body. Here...is a false principle'.

So wrote one of the leading anatomists of the age, Marie François Xavier Bichat, in the book *Anatomie Générale* (Paris, 1801), which lays the foundations of histology. Bichat was not, however, the obscurantist vitalist which this quotation may, perhaps, suggest. He did not, like Stahl at the beginning of the previous century, suppose that an organism was best seen as an embodied soul. For Bichat life was, in a much-quoted phrase, 'the sum of the properties which resist dissolution, or disintegration'. He insisted that there was nothing particularly transcendent or mysterious about this 'sum of properties'; but he also insisted that they could only be studied in the *living* organism. In particular he believed that he could track these phenomena to a precise location in one or other of the tissues which he was the first to classify and analyse. This indeed formed the subject matter of the *Anatomie Générale.*

Bichat thus acts as a suitable *point de départ* for this chapter, for we shall

(1) *Pathologie expérimentale,* Paris (second edition, 1880), p. 412.

see that a major theme in the history of nineteenth-century physiology is that which considers how it is that organisms resist the forces of dissolution. A second major theme is the progress of minute anatomy, culminating in the enunciation of the cell theory by Schleiden and Schwann. Both these great issues find a consummation in the work of Claude Bernard. Beyond Claude Bernard at the beginning of the next century a further synthesis is achieved by C S Sherrington. The title of Sherrington's major work sums the matter up: *The Integrative Activity of the Nervous System.*

Search for the source of the body's warmth

From the earliest times it had been recognised that one of the forces which resisted dissolution as long as life remained was the force responsible for the body's warmth. The body's warmth, like the soul, departed at death; variation in body temperature accompanied (caused?) variation in psychical well-being. In the nineteenth century this ancient complex of ideas at last began to be unravelled. Again it was Claude Bernard, towards the end of the century, who was able to conceive the seminal idea, the idea of a *milieu intérieur*, which was to triumph over the last obfuscations of 'vital heat' and *vis viva*.

But this was in 1854, and at the beginning of the nineteenth century the old physiology was still strong. The eighteenth-century revolution in chemistry had replaced the ancient idea of the four elements and the Spagyrical idea of the three with a growing list—a list which continued to grow, in fact, throughout the century. The eighteenth century, too, had shown that certain essential processes were, seemingly, identical in the inanimate and the animate worlds. Mayow and Crawford, for instance, had indicated, and Lavoisier had proved, that respiration and combustion were cognate processes. But physiology at the beginning of the century still retained many ancient notions. The gulf between the inanimate and the animate, the gulf which Lavoisier so brilliantly bridged, seemed in many cases, and to many investigators, unbridgeable. Indeed, the sharpness of this dichotomy can itself be seen as one of the fruits of the chemical revolution of which Lavoisier was so preeminent an exemplar; for the establishment of a science of chemistry depended upon the prior development of a concept of the inorganic. In contrast to the striving, active material which the alchemist knew, the nineteenth-century chemist operated on something which was by definition inert and lumpish.

Bichat, we have already seen, considered any attempt to import chemical explanations into the biologic realm an example of faulty logic. And Bichat was far from alone; similar positions were taken by many of his eminent contemporaries. Chemists, in particular, were chary of seeming to extend their empires to include the phenomena of animate creation. Chaptal, in the influential *Elements of Chemistry* (2), sharply distinguished the uniformity

(2) J.A.C.Chaptal, *Elements of Chemistry* (trans. W. Nicholson), London (1791).

which seemed to reign in inanimate nature with the variability and flexibility of response which seemed the signature of a living body. Berzelius, in his book *A View of the Progress and Present State of Animal Chemistry* (1813), took a similar position: '...the cause of most of the phenomena within the Animal Body lies so deeply hidden from our view that it will certainly never be found' (3).

Berzelius, however, does not despair of animal chemistry in general so much as of neurochemistry in particular. But it is precisely the nervous system which, for Berzelius, hides the unfathomable secrets of the vital force. A chemistry of the brain seems to him almost inconceivable. 'But still more astonishing', he writes, 'are the operations of the brain. Is it probable that the human understanding, which is capable of so much cultivation...may one day explore itself and its nature? I am convinced that it will not' (4).

However, once the nineteenth century began to unfold, a spate of discoveries and fresh interpretations began to flow from the laboratories of chemists, physiologists and biologists. These discoveries were increasingly of a quantitative and mechanistic nature. Gradually the question of animal heat, of exactly how life did or did not resemble a flame, became clarified. Essential to this development was the perception of the living animal as a species of 'heat engine'. Lavoisier and Laplace, in their 1780 paper *Mémoire sur la chaleur*, had described a technique of measuring heat output by means of an ice calorimeter. Investigators of animal heat in the nineteenth century continued to make use of calorimetry and other physical techniques and of the conservation ideas being developed contemporaneously by the thermodynamicists. The animal body came by slow degrees to be seen not only as a heat engine but also as a chemical machine. There is no space to follow the development of these ideas in detail in this book. Lavoisier and Laplace, for instance, believed that the respiratory combustion occurred in the lungs from whence the heat was transported to all parts of the body by the blood. The search for the sites of the body's heat generation is in itself a fascinating and elaborate story (5). All that will be attempted here is to note a point of culmination and synthesis in the work of Justus Liebig. Liebig's *Animal Chemistry* avowedly sets out to bring together physiology and chemistry. As such it may perhaps be taken as the foundation work of the great science of biochemistry. Although a vitalist, Liebig may nevertheless be regarded as the man who finally established beyond all reasonable doubt that the source of the body's warmth was in all particulars no different from that of a fire's heat.

(3) J.J. Berzelius, *A View of the Progress and Present State of Animal Chemistry* (trans. G. Brunnmark), London (1813), p. 4; quoted in G.J. Goodfield, *The Growth of Scientific Physiology*, Hutchinson, London (1960).
(4) *Ibid.*, p. 8.
(5) The true sites (the tissues) were not in fact discovered until the researches of Edouard Pflüger in 1872 and 1875.

Liebig originates biochemistry

Liebig based his physiological chemistry on the painstaking analytical chemistry by which he had established the constitution of the foods consumed by animals and also the chemical constitution of many animal tissues, body fluids and excretory materials. In so doing Liebig established organic chemistry as an exact science and pointed the way to our present-day biochemistry and molecular biology. Meditating on the mass of raw data thus accumulated, Liebig brings out the fact that although a human being takes in oxygen throughout his life his body at the end is no heavier than it was at maturity. 'According to the experiments of Lavoisier', he writes (6),

> an adult man takes into his system, from the atmosphere, in one year, 746 lbs, according to Menzies, 837 lbs of oxygen; yet we find his weight, at the beginning and end of the year, either quite the same, or differing, one way or the other, by at most a few pounds.
>
> What, it may be asked, has become of the enormous weight of oxygen thus introduced, in the course of a year into the human system?
>
> This question may be answered satisfactorily: no part of this oxygen remains in the system; but it is given out again in the form of a compound of carbon or of hydrogen.
>
> The carbon and hydrogen of certain parts of the body have entered into combination with the oxygen introduced through the lungs and through the skin, and have been given out in the forms of carbonic acid gas and the vapour of water.

Liebig is not prepared to say exactly where the combination occurs, whether within the tissues, or in the blood. He is, however, prepared to say exactly where the necessary carbon and hydrogen for combination come from. They are, he says, 'furnished in the food'.

These biochemical insights also allowed Liebig to account for the phenomena of animal heat. 'In the animal body', he writes, 'the food is the fuel; with a proper supply of oxygen we obtain the heat given out during its oxidation or combustion.' For Liebig perfectly understands that all bodily activity is motivated by biochemical machinery. And he perfectly understands that 'in whatever way carbon (and hydrogen) may combine with oxygen, the act of combination cannot take place without the disengagement of heat'. There is, he says, just no other source of animal heat. Whether we examine the muscles or the nervous system, the heart or the liver, it is always 'the mutual action between the elements of the food and the oxygen conveyed by the circulation of the blood to every part of the animal body (that) is the source of animal heat'.

Thus, for Liebig, as for a long line of distinguished predecessors, respiration and the vascular system are closely allied. But whereas the ancients theorised about a connection between respiratory movements and the pulse, Liebig and his contemporaries were taking the first steps towards an analysis of this connection into its physicochemical elements. Although Liebig, with his relatively crude techniques, was still unable to gain insight into the processes we now call oxidative metabolism, he was nevertheless able

(6) J. Liebig, *Animal Chemistry* (trans. W. Gregory), London (1842), p. 12.

to understand in a general way the chemical and physiological bases of animal heat.

Yet, in spite of his considerable biochemical insight, Liebig remained, as we noticed above, a vitalist. He was not, however, a vitalist in the old, Stahlian, sense of one who considered the body to be motivated by, to exist for, an immaterial something: the soul. Quite the contrary: Liebig's vital force was explicitly conceived as analogous to other physical forces. Like Newton, he was not prepared to 'feign hypotheses' about the exact nature of this force: he preferred not to guess. But, just as the phenomena of gravitating bodies showed a force to exist between them, so, thought Liebig, the phenomenon of life showed a special force to exist within organisms (7). He believed that it was necessary to posit this force if we were to explain how it was that animal bodies resisted dissolution, how they retained their integrity of form in a world which otherwise tended to maximise disorder. This, he believed, was a fundamental and 'irreducible' characteristic of a living organism. A second and similar characteristic was embryological development and growth. Indeed, *Animal Chemistry* opens with a discussion of this vital issue:

> In the animal ovum as well as in the seed of a plant, we recognise a certain remarkable force, the source of growth, or increase in mass, and of reproduction, or of supply of the matter consumed; a force in a state of rest . . . ; entering into a state of motion or activity, it exhibits itself in the production of a series of forms, which although occasionally bounded by right lines, are as yet widely distinct from geometrical forms, such as we observe in crystallised minerals. This force is called the *vital force, vis vitae,* or *vitality.*

Liebig is clearly no mere 'test-tube' biochemist. He fully grasps the problem with which the would-be reductionist is faced. And Liebig's answer is probably the best that, in the 1840s, a chemist could provide. For his *vis vitae* was in reality no more, and no less, mechanistic than Newton's 'gravitation', or the chemical forces which were at that time exercising the minds of his colleagues. Indeed Liebig makes the point that 'chemical forces' are 'infinitely nearer' in their *modus operandi* than gravitational forces. 'The unequal capacity of chemical compounds', he writes (8), 'to offer resistance to external disturbing influences...in a word, the active force of a compound depends on a certain order and arrangement in which its elementary particles touch each other'. And so it is with the vital force. It is inseparable from 'a certain arrangement of elementary particles' (9), a certain form and organisation of matter. Thus we see that Liebig's concept of the vital force harks back, in a way, to the Aristotelian concept of soul as form but, far more importantly, it reaches forward to the biology of the twentieth century. It foreshadows the contemporary notion that matter begins to display lifelike properties when it assumes a sufficient complexity of form.

When, however, Liebig turns to the nervous system he is not so surefooted.

(7) *Ibid.*, p. 7.
(8) *Ibid.*, p. 205.
(9) *Ibid.*, p. 209.

Like Berzelius before him, he tends to regard the operations of the nervous system as lying somewhat beyond the purviews of his science. Notions derived from a more ancient physiology glimmer, for example, through the following passage:

> The nerves are the conductors and propagators of mechanical effects . . . by means of the nerves all parts of the body, all the limbs, receive the moving force . . . . The excess of force generated in one place is conducted to the other parts by the nerves . . . the [vital] force which one organ cannot produce in itself is conveyed to it from other quarters (10).

Even granting that the energy concept was in 1842 still unclear (Helmholtz's work was not published until 1847) and that, in consequence, Liebig used the term 'force' where we today might more appropriately use the term 'energy', the idea of a disembodied vital force, or energy, transported along the nerves from one organ to another has a distinctly unmodern ring. The nervous system in the 1840s and 50s remained *terra incognita* for the biochemist. It seems to belong to a different intellectual tradition: a tradition which, as we have seen in the last few chapters, takes its origins from physicists rather than chemists. In the first half of the nineteenth century it was a tradition in which the early 'electricians' felt far more at home than the pioneer biochemists. Even in the twentieth century, biochemists and neurophysiologists have tended to go their separate ways. It is one of the most encouraging signs for the future of brain science that, in the last third of the present century, there are at last indications of a confluence of the two traditions.

The origins of neurophysiology

It is thus to the development of neurophysiology that we must next turn our attention. For only when the nervous system had yielded up its spirits could a thorough-going mechanisation of physiology be accomplished.

It will be recalled from chapter 16 that during the seventeenth and early eighteenth centuries the flow of 'animal spirits' down the supposedly hollow tubes of the nerves had been metamorphosed into the viscous flow of a nervous fluid, the *succus nerveus*. Albrecht von Haller lent the weight of his great authority to this concept in the pages of the most famous and widely read physiology text of the eighteenth century: the *Elementa Physiologiae* (1757 – 1765). However, several workers at about this time (11) were pointing out that, as a matter of fact, no cavities could be found within freshly dissected nerves and, moreover, no drops of nerve juice could be detected when a nerve was sectioned. Also, at about this time in the mid-eighteenth century, much interest was accumulating around the topic of electricity, and many investigators began to consider whether it might be the case that electric or magnetic forces were involved in nervous activity.

It is this conjunction of physicists interested in the properties of the electric 'fluid' with biologists interested in the action of nerves and muscles which

(10) *Ibid.*, p. 219.

(11) For example, Alexander Monroe: *The Works of Alexander Monroe*, Charles Eliot, Edinburgh (1781).

continues and strengthens the iatrophysical tradition which we noticed originating in the seventeenth-century researches of Descartes and Borelli. This conjunction gives neurophysiology an orientation and a 'taste' markedly different from the vegetative physiology which we saw in the previous section rapidly growing into the great modern subject of biochemistry; for it was the physicist and not the chemist who concerned himself with the most important of the eighteenth-century 'subtle fluids'. And from the very beginning physicists recognised and investigated the effects of electricity in and on living organisms, including human organisms. One very well-known piece of pop-science was introduced by Stephen Gray in the mid-eighteenth century. This consisted in suspending a boy, on many occasions a pupil at Charterhouse School, by silk threads from the rafters. It could then be shown that if the boy's feet were touched by an electrically charged body his face immediately became capable of attracting feathers. As it had been previously established that only an electrically charged body possessed this attractive power, it followed that the human body was able to conduct the subtle electrical fluid. This experiment was repeated at the French court by Nollet and at Leipzig by Hausen.

Many scientists and socialites concerned themselves with the production of electrically charged bodies: numerous devices, principally based on friction, were developed. The first insights into methods of storing electrical charge were also obtained. Van Mussenbroek developed the Leyden jar (1746), and Bennet, the gold-leaf electrometer, a means of very accurately measuring the quantity of electrical charge possessed by a body. Cavendish distinguished in 1776 between the intensity and the quantity of electricity discharged from a capacitor such as a Leyden jar, and Benjamin Franklin, somewhat isolated on the other side of the Atlantic, developed the 'one-fluid' theory of electricity which found an immediate practical application in the lightning conductor. Electrical phenomena sparked off dozens of experiments, and dozens of theories circulated the learned academies of Europe as the eighteenth century drew to its close.

Then, in the 1780s, came the observations which set in motion two great sciences: neurophysiology and electromagnetism. The story has often been told. Here are Galvani's own words:

> I had dissected and prepared a frog and laid it on a table, on which, at some distance to the frog, was an electric machine. It happened by chance that one of my assistants touched the inner crural nerve of the frog, with the point of a scalpel; whereupon at once the muscles of the limbs were violently convulsed.
>
> Another of those who used to help me in electrical experiments thought he had noticed that at this instant a spark was drawn from the conductor of the machine. I myself was at the time occupied with a totally different matter; but when he drew my attention to this, I greatly desired to try it myself, and discover the hidden principle (12).

(12) L. Galvani, *De viribus electricitatus*, Bologna (1791), p. 4. A facsimile reproduction and English translation by M. G. Foley were published by Burndy Library Inc., Norwalk, Connecticut (1954).

That animal limbs would contract convulsively in response to electricity had in fact been known for several decades. Robert Whytt had, for example, used an 'influence machine' to deliver a severe shock to a hysterical female patient as early as 1757. Even Galvani's subsequent discovery, that frog muscles contracted when brought into contact with two dissimilar metals, had been anticipated by Swammerdam in the seventeenth century (13). Galvani, however, pursued the latter observations with a single-minded intensity which in the electrically-informed times of late eighteenth century gained widespread publicity. Again, let us allow him to speak for himself:

> Since from time to time I had seen frog preparations, hung from brass hooks in their spinal cords along an iron fence surrounding our home's sunken garden, exhibit the usual contractions, not only during electrical storms but also with the sky quietly serene, I believed that those contractions were the results of the changes which occur during the day in the atmosphere's electricity. Hence it was not without hope that I began a careful study of the effects of these changes on such muscular movements, attempting a variety of procedures. At different times over a period of many days I observed the animals suitably arranged for this purpose, but there was scarcely any movement in their muscles.

Galvani, disappointed, tried pressing and squeezing the brass hooks, upon which the frogs were hung, against the iron palings. These hooks passed through the spinal cords of the animals. He did, in these trials, observe frequent contractions of the muscles but, as he says, they seemed to bear no relation whatever to the state of the atmosphere. Persevering in his efforts (although it should be noted that the account was published in 1791 whereas the experiments were carried out in 1786 and hence may not have been quite so logical as hindsight suggested), Galvani brought his frogs indoors, away from the puzzling influence or lack of influence of the open sky. It was indoors that he realised the full import of his observations:

> But when I brought the animal into a closed room, placed it on an iron plate, and began to press the hook fixed in the spinal cord against the plate, behold, the same contractions and the same movements. I performed the same experiment over and over, using different metals, at different places, and at different hours and days, with the same result except that the contractions differed according to the metals used . . . (14).

Galvani, however, failed to recognise the crucial significance of the two dissimilar metals in his experiments. He held that the electricity resided in the animals themselves either diffused throughout the entire organism or, more likely, restricted to the muscles and nerves. His contemporary 'electrician' Volta, however, recognised the truth and from his physics evolved the voltaic pile and, eventually, what is sometimes called the second industrial revolution.

Volta published his interpretations of Galvani's experiments, which he had repeated, in 1792, and initiated a controversy which continued until the latter's death. In order to refute Volta's interpretation and to maintain the

(13) Swammerdam had shown in 1658 that frog's legs contracted when simultaneously touched with copper and silver wires.

(14) L. Galvani, *ibid.* Translated by E. Clarke and C. D. O'Malley in *The Human Brain and Spinal Cord,* University of California Press (1968), pp. 180–1.

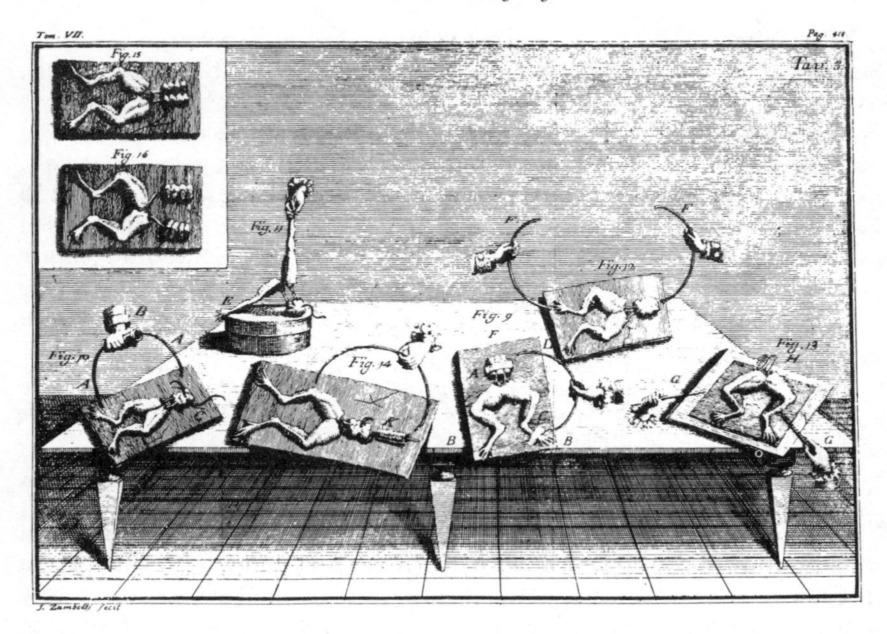

Figure 17.1. Plate 3 from Galvani's *De Viribus Electricitatus,* showing various experiments on frog 'nerve-muscle preparations'.

idea that the animal body possessed its own peculiar electricity, it was necessary to repeat the 1786 experiments without the aid of metals. Galvani accordingly devised an experiment which seemed to him finally to prove the existence of animal electricity. He laid a frog's nerve over the injured surface of the muscle to which it ran and, behold, once again obtained contractions. He devised a number of variations on this experiment, all of which, we now recognise, make use of the so-called injury potential of a wounded tissue to stimulate a nerve. Disliking public controversy, Galvani published his findings anonymously in 1794 in a treatise entitled *Dell'uso e dell'attivita dell'arco conduttore nelle contazioni dei muscoli.* Galvani died four years later. His nephew, Giovanni Aldini, did much to publicise his views on animal electricity in the early years of the nineteenth century.

Galvani conceived 'animal electricity' in very much the same way that his predecessors had conceived animal spirits:

> We believe, therefore, that the electric fluid is produced by the activity of the cerebrum, that it is extracted in all probability from the blood, and that it enters the nerves and circulates within them in the event that they are hollow and empty, or, as seems more likely, they are carriers for a very fine lymph or other similarly subtle fluid which is secreted from the cortical substance of the brain, as many believe. If this be the case, perhaps at last the nature of animal spirits, which has been hidden and vainly sought after for so long will be brought to light with clarity (15).

(15) *Ibid.*, p. 42 (trans. M. G. Foley).

In this last sentence we can perhaps glimpse the motive which drove an otherwise quiet and retiring man, a man of delicate and somewhat refined sensibility, to enter the rough and tumble of public controversy and to spend his days in the none too salubrious task of examining the convulsions of decapitated frogs. We may also note how tenaciously the ancient pneumatic physiology lingered on to bias the thoughts of physiologists. We have now entered the nineteenth century AD and still the ideas promulgated in the fifth century BC have not been fully exorcised. Animal electricity is the last of the guises in which the Greek physiological soul haunts the mind of the animal physiologist.

And, in fact, very little further progress was achieved in the understanding of the physiology of muscles and nerves for another quarter of a century. Chemistry, as we saw in the previous section, advanced apace; it seemed, however, as we noticed in our account of Liebig's work, that it had rather little bearing on the subject matter of neurophysiology. The techniques which were to finally initiate the modern science of nerves and muscles came, once again, from physics and not from chemistry. They came, moreover, from that branch of physics initiated by Galvani's opponent: Volta. For in order to develop a better understanding of 'animal electricity' it was necessary to find some way of measuring it. Just as Liebig's animal chemistry was made possible by his great tables of organic and biochemical analysis, so neurophysiology only became possible when techniques for analysing and quantifying electricity became available.

A start was made in the development of this essential prerequisite by the invention of the galvanometer, especially the astatic galvanometer in 1825. With this instrument it was possible to detect and measure the so-called 'frog current'. This current, first demonstrated by Nobili in 1827, seemed to pass up from the leg muscles of a flayed frog towards the spinal cord. Its nature and cause was for many years a source of confusion and debate: in the 1820s, when Oersted and Ampère were only just beginning the disentanglement of current electricity from the older concepts of static, or frictional, electricity, its significance, though completely obscure, seemed considerable. In the 1840s Carlo Matteuci, the harbinger of du Bois Reymond, published many experiments in which he attempted to analyse this current. His results, as is only to be expected at this early date in the history of the science of electricity, were confused and muddled. He did, however, show that electrical phenomena analogous to those found by Galvani and Aldini in the frog could be detected in the muscles and nerves of many animals. Indeed, believing that the nerves were strictly analogous to metallic conductors, he attempted to measure currents in the horse sciatic nerve. His slowly reacting instruments were, however, quite incapable of picking up the action potentials which must have been present. It is, nevertheless, proper to ascribe to this industrious worker the first experimental detection of an action potential. He showed that if the nerve to one muscle (muscle A) is laid over the surface of a second muscle (muscle B), then if muscle B is caused to contract muscle A is also convulsed. Matteuci, however, although distinguishing this electrical phenomenon from the frog current, was confused when it came to explaining its nature. Indeed, later, after he had published these observations, he denied that the induction of contraction in the second muscle (muscle B) was electrical at all.

We have already hinted that Matteuci and his fellow workers in the 1840s were, in a sense, preparing the ground for a more incisive mind. In 1841 Emil du Bois Reymond, then twenty-two years old, was given by his professor, Johannes Müller, a copy of Matteuci's book *Essai sur les phénomènes électro-physiologiques des animaux*. Müller suggested that here was a topic worthy of intensive investigation. Du Bois Reymond agreed and pursuing the subject for some forty years established the modern discipline of electrophysiology. In particular he was able to detect and measure the action potential (16), a feat which demonstrated that the nervous system could be treated as a physicochemical mechanism like any other part of the body. In a famous sentence he writes: 'If I have not completely deluded myself, I have succeeded in restoring to life in full reality that three-hundred-year-old dream of the physicist and the physiologist, the identity of the nerve principle with electricity' (17).

Du Bois Reymond, in showing that a negative potential change could be detected in working muscle and nerve, took the first major step towards the modern understanding of the physical basis of the nerve impulse. His work marks the end of the beginning in neurophysiology. He showed that there was no need to invoke mysterious spirits to account for the operation of the neuromuscular system. The only requirements were more accurate, more precise and more sensitive physical instruments. Du Bois Reymond's vision of a biophysics of the nervous system was soon confirmed by a remarkable experiment carried out by von Helmholtz in 1850. Helmholtz, whose polymathic genius we have already noticed in the context of thermodynamics, succeeded in what many of his contemporaries regarded as a futile and impossible task: the measurement of the velocity of impulse propagation in a nerve. Estimates (18) in the eighteenth and early nineteenth centuries had varied from 9000 feet/minute up to 57 600 million feet/second, and most workers simply believed, often on metaphysical grounds, that the actual velocity could never be ascertained. Helmholtz, however, by means of a brilliantly simple experiment, succeeded in showing that frog nerve conducted an impulse at a rate of about 30 metres/second (ranging between

(16) The true dimensions of the action potential were not, of course, determined until the twentieth century, when it became possible by means of intracellular microelectrodes to measure the potential change not only from the surface of the membrane, but also across the membrane. It then became clear that the physical basis of the action potential was not merely a depolarisation of the neuronal membrane but a reversal of its polarity.

(17) E. du Bois Reymond, *Untersuchungen über thierische Elektricität*, vol. 1, Introduction, Berlin (1848); trans. Clarke and O'Malley, *ibid.*

(18) See E.G.T. Liddell, *The Discovery of Reflexes*, Clarendon Press, Oxford (1960), p. 46.

24.6 and 38.4 m/s). This result, as we remarked above, confirmed physiologists in their physicalist approach to the nervous system; on the other hand, it emphasised that nerve conduction could not be regarded as identical to the transmission of electricity in metallic conductors.

The work of du Bois Reymond and Helmholtz set neurophysiologists on the path from which they have never since strayed. Advance along this path has continued to depend, as it had from its inception, on the development and use of instruments for the recording of minute electrical events. In the twentieth century, the invention and application of the nearly inertia-free cathode ray oscilloscope and associated electronics has been the *sine qua non* of our present fairly complete understanding of neuronal physiology. Here, if anywhere, we have a clear instance of our physiological understanding depending upon our technological competence. The case is interesting, also, in the dialectical form which the advance shows: electrotechnology was largely initiated by Volta's appreciation of Galvani's work on the frog neuromuscular system, and then reacted back on physiology by providing the instruments necessary to continue and advance the investigation.

Before leaving the subject of neurophysiology, a subject which we shall look at again in chapter 22 with especial reference to the brain and spinal cord, it is important to note that du Bois Reymond's influence spread far outside its comparatively narrow realm. The success of his physicist's approach to the nervous system, a system which, it will be remembered, Berzelius and others regarded as the seat of the mysterious vital force, convinced him that any vitalist position was untenable. In consequence he conceived the destiny of physiology to be a closer and closer union with the sciences of physics and chemistry. Ultimately, indeed, he sees the science of physiology as a special subsection of the physical sciences. This viewpoint was not, of course, original with du Bois Reymond: we have already met it in the work of Descartes and his followers. It was a viewpoint, however, which was still in the nineteenth century, especially as it applied to the nervous sytem, impossible to substantiate. Du Bois Reymond set out his general position with great force and clarity in a famous lecture which he gave in Berlin in 1872. In this lecture, 'Uber die Grenzen des Naturkennes' (On the boundaries of science), he distinguished between certain questions the answers to which we do not yet know but which are, in principle, answerable, and certain other questions the answers to which again we do not know but which are, he thought, in principle incapable of solution. This distinction of questions to which we could reply *ignoramus* from those which called for *ignorabimus* formed something of a *cause célèbre* at the end of the nineteenth century. We shall return to it again in the context of the evolution theory (chapter 19), for the origin of life was, for du Bois Reymond, one of the 'world enigmas' to which the correct response was *ignoramus*, while the origin of sensation and consciousness (chapter 22) was a problem to which the only proper response was *ignorabimus*.

The units of life

But we are running ahead of our story. So far in this chapter we have seen how chemistry, especially in the hands of Liebig, had begun the long task of unravelling the complexities of intermediate metabolism or vegetative physiology, formerly the dominion of the lowest of the three Greek physiological souls. Similarly we have followed the progress of physics in interpreting the electrical phenomena of nerve and muscular activity, the one-time province of the middle physiological soul. One further development, however, needs to be chronicled before the Bernardian synthesis can be set in its true perspective. This is the great nineteenth-century advance in minute anatomy.

Cells had, of course, been seen and christened as long ago as the seventeenth century. The story is well known. Robert Hooke, in 1665, gazing through one of the earliest microscopes at a thin slice of cork, saw a multitude of minute closely packed cavities. He reported (19) his observations to the newly formed Royal Society of London in the following words:

> . . . I could exceedingly plainly perceive it [the cork] to be all perforated and porous, much like a honey comb but that the pores of it were not regular; yet it was not unlike a honey comb in these particulars.
>
> First, in that it had a very little solid substance, in comparison to the empty cavity that was contained between . . . for the Interstitia, or walls (as I may so call them) or partitions of those pores were near as thin in proportion to their pores, as those thin films of wax in a Honey Comb (which enclose and constitute the sexangular cells) are to theirs.
>
> Next, in that these pores or *cells,* were not very deep, but consisted of a great many little Boxes, separated out of one continued long pore, by certain Diaphragms.

Hooke's findings were confirmed by many other investigators during the late seventeenth century and throughout the eighteenth century. Similar compartments were not, however, so obvious in animal tissues and the notion that both plants and animals consisted of living units—the cells—was slow in coming. This tardiness was partly, perhaps mainly, due to imperfections in the early microscopes. Chief among these were defects arising from spherical and chromatic aberration. Microscopists using the early instruments saw objects shimmering in multicoloured haloes, and progress in the analysis of biological material was accordingly retarded throughout the whole of the eighteenth century. Only at the beginning of the nineteenth century did microscopes with achromatic objectives first make their appearance. Once this technical advance had been made, however, progress was rapid, and Schwann was able to publish the book which is nowadays regarded as inaugurating the 'cell theory' in 1839 (20).

Speculation that the body, like the macrocosmic world outside, is

(19) R. Hooke, *Micrographia,* London (1665); facsimile reprint by Culture et Civilisation, Bruxelles (1966), p. 113.

(20) *Mikroskopische Untersuchungen uber die Ubereinstimmung in der Struktur und dem Wachsthum der Tiere und Pflanzen,* Berlin (1839). The English translation (H. Smith) was published in 1847 with the title *Microscopical researches into the accordance in structure and growth of animals and plants,* by the Sydenham Society, London.

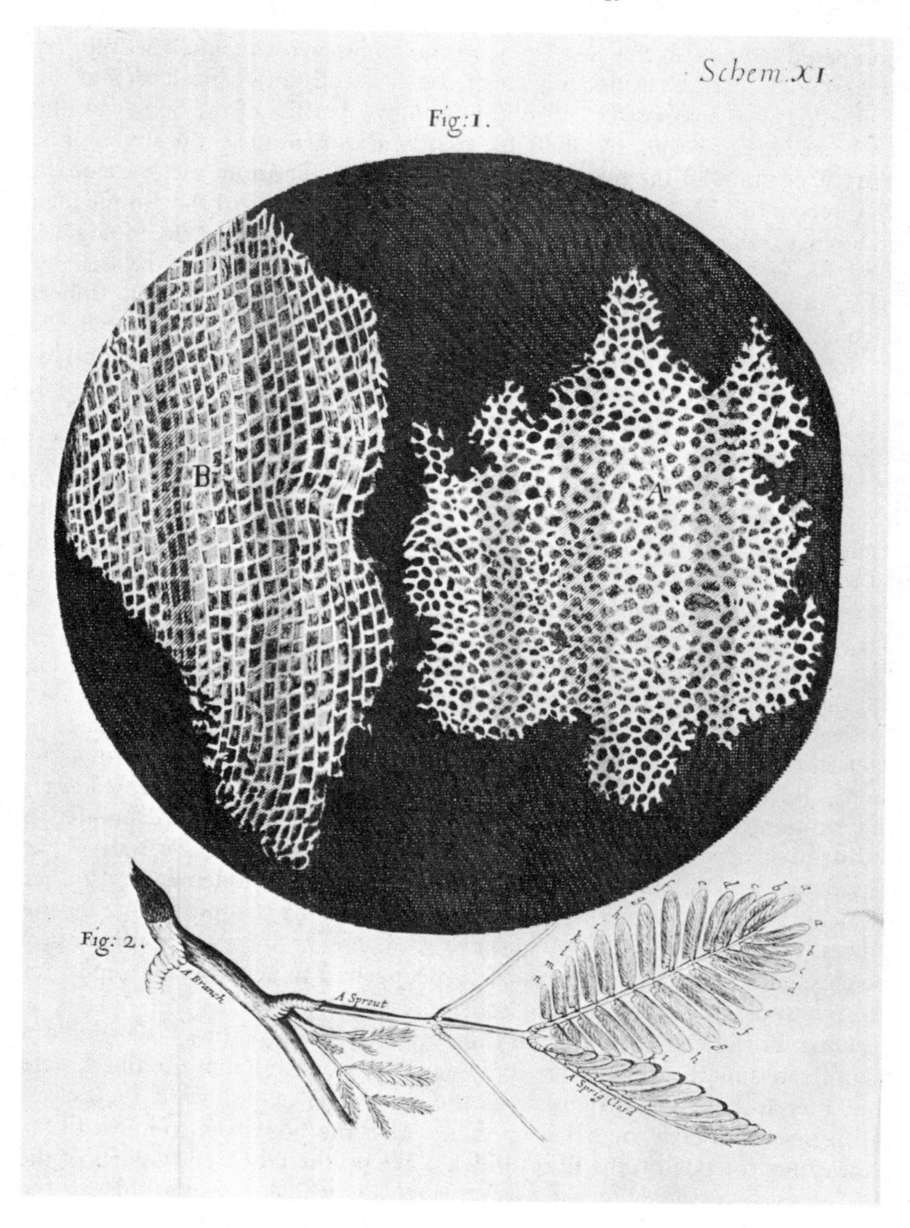

Figure 17.2. Drawing of cork cells published by Hooke in his *Micrographia* in 1665.

composed of units, in this case living units, is, however, very ancient. We saw in earlier chapters that the atomists of classical antiquity conceived that the living body was composed of unsplittable atoms. The life of the body was due to the activity of peculiarly small and swiftly moving atoms—the soul atoms. We traced some of the history of the latter idea's demise in the subsequent chapters of this book. We also noticed that the application of the atomic idea to biology was from the first strongly resisted, indeed derided, by biologists. And, in fact, the living units posited by post-seventeenth-century science bore, in general, rather little resemblance to the physicist's atom. Robert Hooke, for example, suggested in 1665 that the vital activities might be performed 'by small machines of nature...which a greater perfection of opticks may make discernible' (21). Rather different from Newton's 'solid, massy, hard, impenetrable, moveable particles...'! Several eighteenth-century thinkers, however, came rather closer to the physicist's concept. Buffon (1707–1788) and Maupertuis (1698–1759), for example, both believed organisms to consist ultimately of living particles or molecules (22). In the eighteenth century it was, however, anachronistic to pioneer the development of a molecular biology. Just as the biophysics of the seventeenth century was largely premature, so the concept of 'living molecules' of the eighteenth was largely imaginary and fruitless.

Somewhat closer to the realities of the time were the theoreticians working in another of the traditions which had originated in classical times. We noticed in chapter 11 that Erasistratus and also Galen conceived the body to consist of an interwoven fabric of artery, vein and nerve. This belief that the body is ultimately built of fibres rather than corpuscles emerges intermittently throughout the history of anatomy and found, for example, an influential exponent in Boerhaave during the eighteenth century. For Boerhaave and others, however, the fibre theory was complementary, rather than antagonistic, to the corpuscularian notions of the iatrophysicists; for the fibres, even the nerves, were, we have seen, hollow for eighteenth-century science, and they were for the most part conceived to act as mere 'pipes' for the transmission of the vitalising 'soul atoms'.

Later in the eighteenth century, however, Diderot (1713–1784) made a significant modification to the fibre theory by proposing that the fibres were not merely hollow transmitters of the vital effluvia but were themselves alive—were, in fact, the living units of which the body was composed (23). According to Diderot, the fibres which make up the fabric of all parts of the body exist in a state of tension: it is this tension which is responsible for the phenomenon of life. Death he interpreted as the loss of this vital tension.

This sketch of the ideas prevailing before the introduction of the

(21) Hooke, *ibid.*, preface, g.
(22) See account in T.S. Hall, *Ideas of Life and Matter*, vol. 2, Chicago University Press, London (1969).
(23) *Eléments de Physiologie*, Paris (1774–1780).

achromatic microscope objective serves to emphasise yet again how closely interwoven, acting and reacting, is the progress of science and technology. In this case the interaction is not at the 'philosophical' level but is a more 'practical' affair. Until the instrument or technique is available to control the situation, the human imagination freewheels fruitlessly. The date which is conventionally given for the production of the first satisfactory achromatic compound microscope is thus of considerable significance in the history of biology. That date is 1827, for in that year Amici demonstrated his first achromatic objective lens and from that time on satisfactory instruments began to diffuse through the great European research laboratories.

But even in the 1830s, with the increasing availability of these greatly improved instruments, it was not immediately clear that the bodies of all macroscopic organisms were constructed of basically similar units; or that, even if they were, these units were cells in the sense which is now commonly given to that term. As far back as the beginning of the nineteenth century, Lamarck in the *Philosophie zoologique* (1809) was insisting that the solid parts of animal bodies consisted of cellular material. However, Lamarck was not especially interested in the individual cells themselves but only in the type of tissue architecture they represented. Contemporary histologists would probably call Lamarck's *tissu cellulaire* areolar tissue. A closer approximation to a genuine cell theory was made by several workers in the first half of the nineteenth century who saw the body as composed of, constructed from, microscopic globules. Dutrochet was probably the most eminent of these nineteenth-century globulists. He writes as follows: 'Everywhere, in effect, one finds in animals only globular corpuscles, now united in linear and longitudinal series, now in a confused agglomeration' (24).

Clearly this concept is close to the modern cell concept. It fails, however, on two counts. First, the globule was never demonstrated to be anything more than merely an architectural unit of living tissue (25). Second, the term globule was applied to a very wide range of objects. Although Dutrochet and other globulists may indeed have seen what we nowadays regard as cells, they also undoubtedly saw all sorts of diffraction haloes, oil droplets, and so on, which they rather indiscriminately lumped together under the general heading: globule.

The globulists were thus in pursuit of an illusion, a Jack-o'-Lantern which was to lead them away from the true path into a morass of semantic confusion. The first steps on the road leading to the modern understanding were taken by the botanist Robert Brown when, in 1833, he established that plant tissues were composed of *nucleated* cells. The recognition that plant cells contained nuclei was crucial. In the succeeding five years, histologists also recognised the presence of nuclei in several animal tissues. Most

(24) R.J.H. Dutrochet, *Recherches anatomiques et physiologiques sur la structure intime des animaux et sur la motilité*, Paris (1824); quoted in Hall, *ibid.*, vol. 2, p. 185.

(25) Dutrochet, it is true, believed the globules to be units of motility in addition to being units of structure, but this remained (as in the time of Croone and Borelli) a speculation.

importantly, nuclei were detected in developing skeletal tissue, for it was clear that cartilage and bone possessed a cellular organisation. In particular, Theodore Schwann observed the presence of nucleated cells in the developing notochord of embryonic fish and amphibia. In 1839 he described the methods he used in the study of this tissue, and remarked: 'The interior (of the chorda dorsalis (notochord)) exactly resembles parenchymatous cellular tissue in plants' (26).

It required, however, the coming together of a botanist and a zoologist to generate the cell theory. The zoologist was Schwann, the botanist Schleiden. Schleiden seems to have been a somewhat unbalanced personality. He was liable to fits of extreme depression, in one case, at least, so powerful as to induce an unsuccessful attempt at suicide. His place in the history of science is, however, assured through his importation of a new significance to Brown's demonstration that plant cells possessed nuclei. Schleiden saw the nucleus as the cytoblast (*sic*) or generative centre of the cell. Indeed he went further. He proposed that the nucleolus, which he had discovered during his researches, first formed the nucleus and that this in turn formed the cell. Furthermore, Schleiden also believed that cells could be viewed from two different aspects: firstly in themselves alone, as independent units of life and development, and secondly as helping to constitute a multicellular community—the plant. He is thus one of the first to see a plant as a community of cells or, as he calls it, a *polypstock*.

In 1838 the botanist and the zoologist met for dinner. As they dined, Schleiden expounded his views (mistaken views, as we nowadays see) on the function of the nucleus in the development of plant cells. Immediately Schwann began to see the possibility of a wide-ranging synthesis.

> I at once recalled having seen a similar organ in the cells of the notochord, and at the same instant I grasped the extreme importance my discovery would have if I succeeded in showing that this nucleus plays the same role in the cells of the notochord as does the nucleus of plants in the development of plant cells (27).

In fact, as we noticed above, Schleiden's views on the functions of plant cell nuclei were somewhat eccentric. Schwann, under the influence of Schleiden, accepted these views and consequently strayed from the path which has led to the modern understanding. However, with the publication of the *Mikroskopische Untersuchungen* in 1839, Schwann established the point, which afterwards became quickly accepted, that animals and plants shared a common architectural principle: both are composed of cells.

But not only are plant and animal bodies composed of similar subunits, they are also composed of similar functional subunits. We have already noticed that this idea was in Schleiden's mind:

(26) Schwann, *ibid.*, p. 10.
(27) Quoted by J.R. Baker, The cell theory: a restatement, history and critique, *Q. J. Microsc. Sci.,* **90** (1949), 103.

> Each cell leads a double life: an independent one, pertaining to its own development alone; and another incidental, in so far as it has become an integral part of a plant. It is, however, easy to perceive that the vital process of the individual cells must form the very first, absolutely indispensable fundamental basis, both as regards vegetable physiology and comparative physiology in general (28).

This idea found powerful adherents. Perhaps the most influential voice among those who spoke of the cell as 'the primitive organism' or the 'unit of life' was that of the famous editor of the *Archiv für pathologische Anatomie und Physiologie und Klinische Medicin* (after the editor's death renamed simply *Virchows Archiv*): Rudolph Virchow. Virchow's major work *Die Cellular-pathologie...* was published in 1858 (29). In it he insists that no cells originate *de novo*; all are offspring of preexistent cells. This important insight is encapsulated in Virchow's famous and pregnant phrase: *omnis cellula e cellula*. This insight sets right the misplaced views of Schleiden and Schwann who believed cells to originate from acellular blastemae and the still murkier speculations of the Nature-philosophers who, again, conceived cells to originate *de novo* from a formless matrix (30).

Virchow brought out the meaning of the cell theory by pointing to analogies between biological and social 'organisms'. He explicitly compared cells to the individuals living in a nation-state. He frequently refers to organisms as 'cell-republics' or 'democratic cell states'. Disease—and he was, of course, first and foremost a pathologist—was caused not by some mysterious indwelling entity but by the malfunctioning of certain cells. The cell was thus both the unit of life and the unit of disease. The body was 'seen', as Hobbes saw the State, as the sum total of its individually striving cells.

Virchow's concept of what was essentially 'living' about cells varied, or, better, evolved throughout his life; nutrition, however, remained always fundamental. Some cells might be specialised to perform particular functions—a muscle cell to contract, a gland cell to secrete—but all alike possessed the power to transform food molecules into their own substance and the cognate power to reproduce themselves.

With the enunciation of the cell theory by Schleiden and Schwann and its development by Virchow we may leave the development of minute anatomy in the nineteenth century. Through much debate, much argument and much opposition, the cell theory in much the form envisaged by Virchow gradually overcame its critics during the last half of the nineteenth century. It gradually became clear that cells were, at least at one order of magnification, the body's atoms. It gradually became clear that a knowledge of their properties

(28) M.J. Schleiden, Beitrage zur Phytogenesis, *Arch. für Anat. und Physiol.* (1838), trans. H. Smith, Sydenham Society, London (1847).
(29) *Die Cellular-pathologie in ihrer Begrundung auf physiologische und pathologische Gewebelehre,* Berlin (1858); a translation into English was made by F. Chance, *Cellular Pathology,* London (1860).
(30) See Cellular-Pathologie, *Arch. für Anat. und Physiol.*, 8 (1855), I. Translations of several of Virchow's papers are to be found in L. J. Rather, *Disease, Life and Man: selected essays by Rudolph Virchow,* Stanford University Press, Stanford (1958).

explained much about the properties of an organism. So that Verworn, in 1894, was able to write:

> We have traced all the phenomena of change in matter, form and force back to the point where they disappear into the cell. But of what takes place in the muscle cell, the sense cell and so forth we have not the slightest conception (31).

The situation in mid-century

At this point it may be worth while to recapitulate briefly the moves made so far in this chapter. We have been concerned to trace the process which led to a mechanisation of physiology during the nineteenth century. We noticed that at the beginning of the century, for biologists such as Bichat, there seemed to be a fairly sharp distinction between the organic and the inorganic realms. Whereas living things appeared to react to external forces in an active manner, a manner which tended to ensure that they remained more or less unchanged, non-living things reacted in a completely passive fashion, a fashion which resulted, if the external forces were at all harsh, in their dissolution or disintegration.

In order to pick up the nineteenth century's response to this challenge to the mechanistic world picture which had, as we saw in earlier chapters, swept through first physics and then chemistry, we looked first at the origins of biochemistry. We saw how the mysterious property which had for millennia seemed almost synonymous with life itself, and certainly a property which strongly resisted outside attack, *animal warmth*, was shown to be a matter of chemistry: that there was, in fact, no difference between the oxidative reactions proceeding outside the body and those proceeding within and which were responsible for the body's warmth. However, we saw in our account of Liebig's *Animal Chemistry* that he, like other chemists at that time, still tended to believe that the nervous system somehow escaped the writ of physicochemical science. Berzelius, as we saw, certainly believed that the mysterious *élan vital* was hidden within, or was an unanalysable property of, the central nervous system.

Thus we next looked at the development of ideas in neurophysiology and traced some parts of the progressive interpretation of this system in terms of physics and, in particular, of the electrical branch of that subject. With du Bois Reymond and Helmholtz, the ancient notion of spirits or subtle fluids coursing through the nerves found its *terminus ad quem* and from the 1850s onward neurophysiology progressed along the road leading to modern times.

At this point in history it was clear that the parts of which an organism is composed were rapidly yielding to mechanistic interpretation. There seemed to be no aspect of physiology for which this was not the correct and appropriate method. Yet an organism seems to be something more than the mere sum of its parts. This was the criticism raised by biologists against the ancient atomic theory, and it still seemed valid. As Claude Bernard was to point out, the parts of an organism do not act independently of each other

(31) M. Verworn, Modern physiology, *Monist*, 4 (1893), 361.

but in 'reciprocal harmony' (32). The physicochemical approach, moreover, had not by mid-century answered Bichat's initial challenge: how can a mere mechanism, a congeries of parts, resist the forces of dissolution?

Thus, before Bichat's challenge could be effectively answered, a deeper knowledge of the body's parts, the units of life, or cells, had to develop. This development we traced in the last section and we saw that, again by the 1850s, the insight that cells could be regarded as the living units of which the body was composed had been achieved. The time was thus ripe for the advance in physiological insight which is always associated with the name of Claude Bernard.

Claude Bernard and le milieu intérieur

Claude Bernard was born near Villefranche in 1813. He studied under Magendie in Paris from 1839 but soon surpassed his teacher, producing a stream of publications on practically all aspects of physiology and pharmacology. From about 1854 Bernard began to include in his lectures a discussion of the *milieu intérieur* within animals and the mechanisms which tended to ensure the constancy of its properties.

This was an original and crucial insight. Bernard says: 'I think I was the first to urge the belief that animals have really two environments: a *milieu extérieur* in which the organism is situated, and a *milieu intérieur* in which the tissue elements live' (33).

Note that he speaks of 'tissue elements'. He is thinking of cells and thinking of them in much the same way as the founders of the cell theory: 'A complex organism', he writes (34), 'should be looked upon as an assemblage of simple organisms which are the anatomical elements that live in the liquid *milieu intérieur*'. And elsewhere he is even more explicit: 'In the organic environments of higher animals the histological units are like veritable infusoria, that is to say, they are provided with an environment proper to themselves...' (35).

Thus we can glimpse Bernard's insight. The organism consists of a myriad centres of life bathed in the fluid *milieu intérieur*—the plasma, lymph and tissue fluid. The stability of this internal environment is the *sine qua non* of an independent or 'free' existence. So long as the *milieu intérieur* remains invariant, the external conditions, the 'cosmic environment' as he sometimes calls it, may vary very widely. It is thus the mechanisms which control the *milieu intérieur* which are responsible for the 'resilience' of living things, for

(32) C. Bernard, *Introduction à l'étude de la médécine expérimentale,* Paris (1865); facsimile reprint by Culture et Civilisation, Bruxelles (1965); trans. H.C. Green, *Introduction to the Study of Experimental Medicine,* Dover, New York (1957), p. 225.
(33) C. Bernard, *Lecons sur les phenomènes de la vie communs aux animaux et aux végétaux* (ed. A. Dastre), page 112, vol. 1, Paris (1878); trans J. Fulton in Selected Readings in the History of Physiology, compiled by J. Fulton and L. Wilson, C. C. Thomas, Springfield, Illinois (1966).
(34) *Ibid.,* p. 113.
(35) *Introduction à l'étude de la médécine expérimentale,* p. 120.

their resistance to dissolution, for their flexible response to external stimuli. And, for Claude Bernard, it is the nervous system, more than anything else, which provides this mechanism.

> The necessary conditions for the life of the elements which must be brought together and kept constantly in the *milieu intérieur* if freedom and independence of existence are to be maintained are already known to us: water, oxygen, heat and reserve chemical substances.
>
> These are the same conditions necessary for life in simple organisms; but in the perfected animal, whose existence is independent, the nervous system is called upon to regulate the harmony which exists between all these conditions (36).

Bernard was himself responsible for several of the pioneering investigations which established the importance of the nervous system in controlling the parameters of the internal environment. Those investigations have been widened in subsequent years by other workers to penetrate and illumine broad areas of physiology. In no area of physiology, however, has Bernard's generalisation been more significant than in the study of animal heat: a topic with which, it will be recalled, we began this chapter. Bernard himself made important advances in this central subject, showing that two sets of nerves are involved in the regulation of body temperature. His experiments demonstrated, as he says, '...that the calorific function proper to warm-blooded animals is due to a perfecting of the nervous mechanism, which, by incessant compensation, maintains a practically fixed temperature in the internal environment' (37).

Since Bernard's time (he died in 1878, eulogised as the greatest physiologist of the age), research into the mechanisms responsible for thermoregulation has continued in full spate and is still very active today. The nervous mechanisms, the effector and sensory organs, are all nowadays known in much greater detail. With the clarification of the concept of nerve reflexes towards the end of the nineteenth century it was finally possible to understand how 'organismic', 'holistic' behaviour could be accounted for in mechanistic terms. It became clear that receptor organs located especially in the hypothalamus of the brain are able to detect variation in the temperature of the blood flowing in their immediate vicinity and automatically signal to effector organs, especially dermal sudorific glands, muscles attached to the roots of the hairs, and sphincter muscles in the walls of the arterioles leading into the sub-papillary plexus. The consequent loss or conservation of body heat brings the blood temperature back to its 'set' value (38).

It is clear, therefore, that Claude Bernard's work marks a major step towards the mechanisation of physiology. It showed how the Cartesian programme promulgated in the 1650s could finally, in the late 1850s, be established in realistic detail. It showed, in other words, that there was

(36) *Leçons,* p. 113–14.
(37) *Ibid.,* p. 118.
(38) Homeothermic mechanisms in mammals are, of course, far more complex than the above paragraph suggests. Fuller accounts may be found in any good modern textbook of physiology.

nothing mysterious, nothing which in principle escapes a mechanistic explanation, in the apparently purposive behaviour of organisms. This insight only became widely accepted when the technological advances of the late nineteenth and twentieth centuries surrounded mankind with apparently purposive artefacts. It became common knowledge that the flexible response of automated mechanisms depended merely on the monitoring of the output so that any variations could be 'fed back' to influence the production. It appeared, on the surface, that the automated machinery purposively altered its activity so that the product remained optimal. The close analogy between this and the regulatory mechanisms known to physiologists escaped neither engineer nor biologist. In 1947 the mathematician Norbert Wiener coined the term 'cybernetics' to describe the subject matter of this area of common interest. The behaviours of organisms could be accounted for as activities tending to maintain the major parameters of the internal environment constant. Physiologists consequently no longer needed to invoke a doctrine of 'final causes', to hypostatise 'entelechies' or to import a teleology; the concepts of the systems engineer seemed perfectly adequate. In the twentieth century these concepts have been carried beyond the entire organism to account also for the happenings in the 'elementary organism', the cell. We may suspect that the biochemistry of the future will become ever more concerned with feedback and control at the molecular level.

18 THE ROMANTIC REACTION

Is man a machine?

Man a Machine: a famous title, a famous book. La Mettrie, who published his work in 1748 (1), was in many ways a typical thinker of the European enlightenment. Like Voltaire, he attracted the wrath of churchmen. Like Voltaire, he was forced to flee from his orthodox enemies to a secular protector. Like Voltaire, he created enemies among his conservative countrymen by pointing out the seeming illogicalities of their thought, by poking fun at the absurdities in their intellectual position. And the conservatives, not as deeply schooled as they might have been, attempting to defend a position whose origins and structure had been almost forgotten, or perhaps never been made explicit, reacted with anger to hide their bafflement: reacted with anger at what seemed to be the easy superficiality of la Mettrie's philosophy, and yet were mostly unable to show exactly where it was wrong and why it was not a profound alternative to their own thought. For la Mettrie seemed merely to have taken the Cartesian system to its logical terminus. Whereas Descartes conceived the living world to be composed of two categories of beings—animals which were (mere) mechanisms, and humans which had superadded to the mechanism a soul— la Mettrie saw animals and humans as one creation: machines all.

Indeed, la Mettrie explicitly says that he suspects that had Descartes lived a century later and worked in the eighteenth rather than the seventeenth century, then he, too, would have taken his, la Mettrie's, position (2). He believes, in other words, that theological power in the seventeenth century caused Descartes to make discretion the better part of valour; that he endowed man with a soul to escape the fate of Galileo and Bruno. In the eighteenth century, however, the position is easier. *Autos da fé* are no longer popular, at any rate north of the Pyrenees.

La Mettrie is thus prepared to say that not only are animals machines, but so are we. We differ from the animals only as a superior mechanism does from an inferior, as, he says, the famous mechanical flute player of Vaucanson differs from the same roboteer's simpler mechanical duck. If Descartes distinguished *res cogitans*, mind/soul/consciousness, from *res*

(1) J.O. de la Mettrie, *L'Homme Machine,* Leyden (1748); original text and translation by G.C.Bussey and M.W.Calkins, Open Court Publishing Co., Chicago (1912).
(2) *Ibid.*, p. 143.

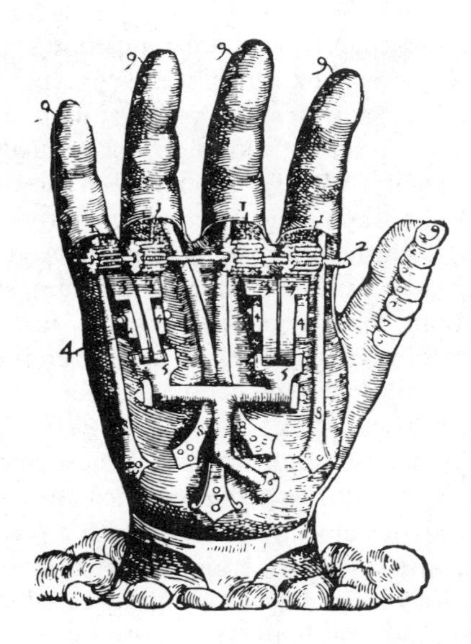

Figure 18.1. A sixteenth-century mechanical hand developed for an amputee. This picture was published by the famous French military surgeon Ambroise Paré in his *Dix Livres de la Chirurgie,* Paris (1564). The *Ten Books of Surgery* contain several other examples of such mechanical prostheses. La Mettrie was probably familiar with such crude artefacts in addition to the more delicately constructed puppets popular in the eighteenth century.

extensa, matter/energy, la Mettrie was prepared to lump everything, all that was known and knowable, on the *res extensa* side of the abyss. One can see that this is a bold move, one can see that it is a provocative move, but one can also see that it is not a profound move.

La Mettrie supported his position by bringing forward the clear dependence of our states of consciousness on the states of our brains. He points out how enfeeblement of the brain during an illness leads to an alteration of our state of consciousness, an alteration, as he puts it, of the soul. Opium, wine and coffee also affect the soul, as does excitement, as does diet. The meat-eating English are, says la Mettrie, for this reason fierce and savage souls. The soul and the brain are indissolubly bound together: indeed they are identical. There is but one thing in la Mettrie's world: extended substance. When matter becomes organised in particular ways properties such as life and mentality emerge. We shall return to a consideration of some of the vicissitudes of this proposition in chapter 22.

Here, however, la Mettrie has been introduced not so much for his own sake as to serve as a rather extreme example of the type of thought which thinkers in the latter part of the century were to react against. For although it was easy for a philosopher-scientist like la Mettrie to mock the guardians of the old order, it was nevertheless quickly realised, more quickly felt, perhaps, than realised, that the mechanico-rationalist position somehow lacked depth. Several times already in this book it has been suggested that insights such as those of a Democritus or a Descartes were brilliant, ultimately valid, but premature. La Mettrie's position, although not argued at the same depth, falls into the same category. We have already seen, at a more technical level, that the mechanical philosophy had already run into serious difficulties by the middle of the eighteenth century (chapter 16). And in both chapters 16 and 17 we have seen how much patient and brilliant work had to be done before, no earlier than the second half of the nineteenth century, the Cartesian position in physiology could be said to make detailed sense. La Mettrie's works were published a full century earlier. At a scientific level they could be no more than speculation.

And the consequences, the implications, of these speculations touched a tender nerve. Today, two centuries and more later, the nerve is still active; for it seems that to call humans machines brings forth a profound dissent. It is a pejorative move, an affront to something in our conception of ourselves. Today machines are, of course, far more sophisticated than they were in the eighteenth century. And, perhaps, to compare a human being to a chess-playing computer begins to approach acceptability. But to compare a man with Vaucanson's flautist, or with the Cartesian organ, even if only in principle, seemed to many in the eighteenth century to omit the essence of what it is to be human.

Now what this essence might be has filled the pages of countless books for millennia. It would be foolish to renew the attempt here. What did, on the face of it, seem obvious, however, was that humans were far more flexible in their behaviour, far less deterministic, than the robot flautist, that they were, in a word, *free*. Something of this perception about the nature, not only of humans, but of living things in general was discussed in the last chapter. It will be recalled that Claude Bernard's famous dictum emphasised this concept: 'La condition *de la vie libre* est le constance du milieu intérieur' (my italics). Humans were in some important sense free, their horizon of action contained limitless possibilities. In contrast, machines were: just machines. Their behaviour could be predicted 'like clockwork'. When they behaved correctly they were completely determined; only when they broke down, when, in other words, they were not acting as genuine machines, did the unexpected happen.

The revolt against this premature mechanisation of life and, in particular, by implication, human life gathered force throughout the second half of the eighteenth century. In the event perhaps the most influential voice raised in defence of human liberty was that of Jean-Jacques Rousseau. In 1755, in the *Discourse on the origin of inequality* (3), Rousseau defined the difference between men and animals. Although, he says, animals may be no more than

(3) J-J. Rousseau, *Discours sur l'Origine de L'Inégalité,* Paris (1755); Classiques Larousse, Paris (1939), pp. 35–6.

'ingenious machines' which are driven by instinct to behave in predictable ways, men have superadded to this mechanism, in some way, the attributes of a free agent. Men can *decide* to act or not to act in a given situation, and this ability, he says, cannot be accounted for by 'les lois de la mécanique'.

On the other side of the Channel other voices also struggled to stem the flood tide of the mechanical philosophy, struggled to maintain the sense of possibilities which Newtonian science seemed progressively to diminish. William Blake, for example, took the geometrising spirit to be Satan, to be behind all those forces which inhumanly concentrated men, women and children into the mills and factories of Albion. He felt profoundly, and was not alone in feeling, that the world of the mathematical physicist was deeply inhospitable to the human soul. He felt, moreover, that if it could in any sense be proved to be true it could only be partially true. The human world was full of colours, scents, music, good and evil, in a word, of qualities, with which the geometrician simply did not concern himself:

> May God us keep from single vision,
> And Newton's sleep.

And, finally, among these examples of a rising reaction to the apparent shallowness of 'enlightened' rationalism, we may mention William Wordsworth. Wordsworth, as a young man, had witnessed and enthused over the Revolution which the ideas of the enlightenment combined with those of Rousseau had done so much to bring about and in which, perhaps symbolically, that paragon of the enlightenment, Lavoisier, lost his head. But he, too, in his most inspired moments felt that there was something much more to human life and to the world in which we find ourselves than was dreamt of in a philosophy such as la Mettrie's:

> And now with gleams of half-extinguished thought
> With many recognitions dim and faint...
> I have felt
> A presence that disturbs me with the joy
> Of something far more deeply interfused
> Whose dwelling is the light of setting suns
> And the round ocean and the living air,
> And the blue sky, and in the mind of man:
> A motion and a spirit, that impels
> All thinking things, all objects of all thought
> And rolls through all things (4).

The Critical Philosophy

The essence of the Romantic reaction to the rationalism of the enlightenment thus seems to have lain in a refusal to accept that man could, in any important sense, be compared to a machine. The poets also wished to say

(4) W.Wordsworth, *Lines Composed a Few Miles Above Tintern Abbey* (1798). It is interesting to note that the passage quoted also appears on the title page of Haeckel's *The History of Creation* (1876). This indicates, as it is hoped the later part of this chapter and the next chapter will also indicate, the intellectual milieu in which so much of nineteenth-century evolution theory originated.

that nature was in some sense 'alive' as well, and not a dead mechanism to be weighed and measured and operated upon. To trace the sociological roots of this reaction would take us too far afield. However, it was clear to these reactionaries that man, although he might be everywhere enslaved to economic forces, to habit and tradition, was nevertheless born 'free': he was by 'nature' a free moral agent. Indeed, it seemed that without the possession of this essential faculty the very possibility of his performing good or bad acts, of being responsible or irresponsible, of being praised or blamed, was illusory. This, anyway, is how it appeared to Immanuel Kant.

Immanuel Kant (1724–1804) is a significant figure in our history, as he gives a certain intellectual respectability to some of the otherwise murky biological theories of the early nineteenth century (5). As is well known, the great German philosopher began his career deeply interested in natural science, and this interest remained throughout his life. In addition to his interest in astronomical matters, an interest which resulted in the first book which he published, *A General History of Nature and Theory of the Heavens* (1755), containing an original contribution to our knowledge of the origin of the solar system, Kant was also deeply concerned with biology and anthropology (6). In the *Critique of Judgment* (1790), the third and last of the great critiques, he includes several interesting and penetrating sections on the nature of organisms.

Although the *Critique of Pure Reason* (1781) was published first, it is possible to maintain that it is the *Critique of Practical Reason* which is most fundamental to Kant's philosophical position. Kant, as was mentioned above, was convinced that what was fundamental to moral action was human freedom; that acts performed instinctively, or under strong external pressure, were, simply, amoral. At the end of the *Critique of Practical Reason* he summarises his most profound perception: 'Two things fill the mind with ever new and increasing admiration, and awe, the oftener and the more steadily we reflect upon them: *the starry heavens above and the moral law within*'. These words are also inscribed upon his tomb.

But therein lay the apparent dilemma; the apparent contradiction which we outlined in the previous section—for the mechanical philosophy, especially Newtonian science, seemed to eliminate human freedom. If, *pace* la Mettrie, humans were machines, and if, *pace* Newton, machines were predictable, then it seemed to follow that men were not, could not be, free. This was the impasse from which Kant sought to escape in his greatest work, the *Critique of Pure Reason*, and the solution which he proposed altered the

(5) Kant's three important critiques – the *Critique of Pure Reason* (1781) (trans. J.M. D. Meiklejohn), the *Critique of Practical Reason* (1788) (trans. T. Abbott) and the *Critique of Judgment* (1790) (trans. J.C. Meredith) – are conveniently collected in Encyclopedia Britannica, *Great Books of the Western World*, vol. 42 (ed R.M. Hutchins), Chicago, London, Toronto (1952).

(6) Kant's first publication was a short treatise entitled 'Thoughts on the true estimation of Living Forces' (1747), in *Kant's inaugural dissertation and other writings on space* (trans. J. Handyside), Chicago (1929).

course of all subsequent Western philosophy. It is, moreover, the solution popularised by the English Romantic poets. The theory of the imagination proposed by Coleridge and Wordsworth (chapter 1) is derived from Königsberg. In essence, Kant proposed that the mind was not the passive *tabula rasa* of the empiricist philosophers but was active, was indeed, in a sense, creative. This perception constituted his self-styled 'Copernican revolution' (7).

Kant points out that his discovery was partially anticipated by David Hume (who, indeed, he admits awoke him from his 'dogmatic slumbers') in his demonstration that causality was not a connection which we observe between the world's events but a connection which we impose upon our observations. Similarly, says Kant, space, time, substance, causality, and so on, are not observables but the very conditions which make meaningful observations possible. They are ordering relations imposed on our 'sense data' by ourselves so that we can make sense of our observations. Thus, says Kant, just as Copernicus revolutionised astronomy by showing that the observer moved and not the firmament, so the Critical Philosophy revolutionises metaphysics by showing that the fundamental attributes of our knowledge about the world are properties of the observer and not of the observables.

The details of Kant's philosophy need not, of course, detain us. It is clear that he stands within the great tradition of Western thought, a tradition which we have seen included William of Occam, Galileo and Isaac Newton, a tradition which sees science as a 'calculus', a means of 'saving the appearances', a technique for connecting observations into a network of relations; but a technique which allows us to say nothing about what might or might not lie behind the phenomena, what might or might not be the nature of the substance in which the attributes which we measure inhere, what might or might not be the nature of, to use the Kantian term, the ***Ding an sich,*** the ***thing-in-itself.***

Thus whether the world really *is* a mechanism, however much it might *seem* to be one, remains for Kant ultimately unprovable. Indeed, says Kant, we may speculate or, to use his own terminology, we may use the speculative reason to suggest, that the world, the noumenal world, is quite other than what it seems to our limited sensory powers. And this is the gap through which he believes he can bring in the concept of human freedom. Kant believed that he possessed, that all human beings possessed, one direct line to noumenal reality: we are all conscious beings, and this consciousness, which was for Kant, as for Descartes, unimpugnable, was simply that we are free. This fact we know in a far more secure manner than that a sequence of phenomenal events related to each other in, as we say, a 'causal chain' have about them (*pace* Hume) any binding necessity ensuring that one event invariably follows another.

(7) *Critique of Pure Reason,* Preface to Second Edition (1787).

This establishment of the possibility of human freedom in a seemingly mechanical universe Kant promulgated in the *Critique of Practical Reason*. However, it is in the third of the critiques, the *Critique of Judgment*, that the force of Kant's reasoning reaches into the field of biological science. In this work he attacks the analogous problem in the biological realm: can mechanistic explanations, explanations using only efficient and material causes, completely account for the apparently teleological phenomena of biology? Is purposive activity possible in a mechanical universe? Kant is sceptical. He is sceptical, for example, that there could ever 'arise a second Newton who would make intelligible the production of a single blade of grass in accordance with laws of nature which were not arranged by some intention' (8).

His argument for teleology is akin to that with which he established the possibility of human freedom in a mechanistic universe. The mechanical philosophy is still the only base from which a scientific biology can be developed. But whether the world is really like this is quite unknowable, and in the life sciences, mechanism is awkward to apply and seems foreign to the subject matter. To minds like ours, organisms are far more easily described by teleological schemata. 'This principle', he writes (9),

> the statement of which serves to define what is meant by organisms is as follows: *an organised natural product is one in which every part is reciprocally both end and means.* In such a product nothing is in vain, without an end, or to be ascribed to a blind mechanism of nature.

This definition will stir a memory of what the great Stagirite taught two millennia and more before. The philosopher of Königsberg goes on to give an explanation of what he means by the above principle. Suppose, he writes (10), that

> in a seemingly uninhabited country a man perceived a geometrical figure, say a regular hexagon, inscribed on the sand, his reflection busied with such a concept would **attribute, although obscurely,** the unity in the principle of its genesis to Reason, and hence would not regard as a ground of the possibility of such a shape the sand, or the neighbouring sea...etc.

The explanation of the figure glimpsed in the sand would thus be, according to Kant, in terms of purpose. The castaway would not ask 'how?' but 'why?' The hexagon, Kant concludes, 'would be regarded as a purpose, but as the product of *art*, not as a natural purpose'.

But, Kant asserts, it is possible to observe 'natural purposes'. Indeed, we are surrounded by them! 'A thing exists as a natural purpose', says Kant (11), 'if it is...both cause and effect of itself'. And these things are for Kant, as they were for Aristotle, organisms, 'self-organising beings'. They differ from mere mechanisms, such as watches, in that in addition to possessing a moving power they also possess a formative power: 'a formative

(8) *Critique of Judgment,* § 75
(9) *Ibid.,* § 66.
(10 *Ibid.,* § 64.
(11) *Ibid.,* § 65.

power of a self-propagating kind which it communicates to its materials though they have it not themselves; it organises them, in fact, and this cannot be explained by the mere mechanical faculty of motion' (12). Kant cannot see how these entities can be described in mechanistic terms; he feels that although a 'natural purpose' is a most obscure idea (indeed, in the absence of an artificer he cannot see how it makes any sense at all), it is nevertheless 'given' in our experience of the phenomenal world, and hence we must make what shift we can with teleological reasoning. 'Organisms', he concludes, are 'unthinkable and inexplicable on any analogy to any known physical, or natural agency...' (13).

This perception of the nature of organism and the apparent impossibility of accounting for it solely in terms of efficient and material causes is, of course, no new discovery with Kant. We have noticed throughout this book that biologists have consistently resisted the premature mechanisation of their subject. Kant, however, seems to have been the first great philosopher since Aristotle to have considered the problem in its full depth. The advances of physical sciences since the seventeenth century had, perhaps, sharpened the dilemma. We have already noticed that by the end of the eighteenth century spirits had finally been driven from matter. The philosopher was unable to see how passive, inert matter could be the cause of the active bustle of the living process: 'The possibility of living matter cannot even be thought; its concept involves a contradiction because lifelessness, *inertia*, constitutes the essential characteristic of matter' (14). The reader will recognise here a perplexity identical to that with which we allowed Bichat to initiate chapter 17.

Kant, however, as befits a philosopher, draws certain wider conclusions from this paradox. Believing, as we have seen, that the only effective method for investigating the phenomenal world is the Newtonian method, the method which presupposes mechanism and efficient causation, he nevertheless holds, as we have also noticed, that this method tells us nothing of the noumenal world, of what it is really like 'out there'. He is consequently able to assert that 'it is left undecided whether or not in the unknown inner ground of nature, physico-mechanical and purposive combination may be united in the same things in one principle' (15).

He implies that organisms may in some sense represent a concentration of characteristics which are in fact immanent in all parts of nature. He suggests that just as the behaviour of certain isolated parts of organisms may be treated mechanistically and yet, when discussed *in vivo*, are better treated teleologically, so, he says, perhaps the success of mechanistic science in the inanimate realm is merely due to the fact that we are always only discussing

(12) *Ibid.*, § 65.
(13) *Ibid.*, § 65.
(14) *Ibid.*, § 73.
(15) *Ibid.*, § 71.

parts and never the entire system. 'By the example', he writes (16),

> that nature gives us in its organic products we are justified, nay called upon, to expect of it and of its laws nothing that is not purposive on the whole ... the principle must be regarded as valid in this way not merely for certain species of natural beings, but for the whole of nature as a system.

Thus we see that the great and ancient analogy of nature to an organism still arouses sympathy in the mind of the greatest of eighteenth-century philosophers. Although the true method for a scientific biology must remain based in physicochemical mechanism, yet, for mortals at any rate, the total view necessarily escapes the mechanistic schematum. Indeed, Kant's insight into the possibility of a genuinely scientific biology goes quite deep. He comes close, in fact, to some of the notions popularised by nineteenth-century evolutionists. He sees, for example, the dim outlines of a systematic explanation for the findings of comparative anatomy. Towards the end of the *Critique of Judgment*, he writes:

> When we consider the agreement of so many genera of animals in a certain common schema, which apparently underlies not only the structure of their bones, but also the disposition of their remaining parts, and when we find here the wonderful simplicity of the original plan, which has been able to produce such an immense variety of species by the shortening of one member and the lengthening of another, by the involution of this part or the evolution of that, there gleams upon the mind a ray of hope, however faint, that the principle of the mechanism of nature, apart from which there can be no natural science at all, may yet enable us to arrive at some explanation in the case of organic life. This analogy of forms, which in all their differences seem to be produced in accordance with a common type, strengthens the suspicion that they have an actual kinship due to descent from a common parent (17)..

Thus Immanuel Kant. The old and the new struggle together within his philosophy. His mind, moreover, is wide open to all the ideas struggling to be born in the eighteenth century (18). All are subjected to the same critical scrutiny: and among the ideas struggling to be born is the idea of a scientific biology. It glimmers dimly through the homologies and analogies of comparative anatomy. These beckoning similarities also affected the far less critical mind of Johann Wolfgang von Goethe (1749 – 1832).

A poet's biology

If the major part of Goethe's science is nowadays regarded as a pastiche, even as somewhat farcical (19), it is perhaps because we look at it through an alien tradition. A defeated science, like a defeated nation, suffers from the

(16) *Ibid.*, § 67.
(17) *Ibid.*, § 80.
(18) The poet Herder, who attended Kant's lectures in 1762–1764, describes his teacher's open-mindedness in the following terms: 'With the same genius with which he criticised Leibniz, Wolf, Crusius, Hume and expounded the laws of Newton and Kepler, he would also take up the writings of Rousseau, or any recent discovery of nature .... Natural history, natural philosophy, the history of nations and human nature, mathematics, and experience – these were all sources from which he enlivened his lecture and his conversation. Nothing worth knowing was indifferent to him...' Quoted in T.K.Abbott, *Kant's Theory of Ethics*, Longmans, London (1927), pp. xxx–xxxi.
(19) See, for example, C.S.Sherrington, *Goethe on Nature and on Science*, Cambridge University Press (1949).

distortion of history imposed by the victor's ideology. A similar fate for long befell the efforts of mediaeval scholastics: the malevolent propaganda put about by post-seventeenth-century scientists inhibited for many years a proper appreciation of mediaeval science (20). And Goethe's approach to biology is clean contrary to the mechanistic approach favoured by moderns. Looking back from our twentieth-century vantage point, now that the dust of the battle has settled, much of what Goethe attempted seems absurd. But in the closing years of the eighteenth and the opening years of the nineteenth centuries, the absurdity was by no means so obvious, the decisive evidence by no means so clear-cut.

Goethe, of course, saw nature with a poet's eye. For him she is alive, a creative being continually 'fathering forth' the phenomena we experience. Thus for Goethe, as for the Romantics in general, empirical research, the analysis of the world into a multitude of fragments, is a false method. It is as if we should isolate, and examine in isolation, each fragment of paint from a work of art. We should end up with a congeries of meaningless numbers. In exactly the same way, it seems to Goethe, the natural scientist takes the giant organism of nature to pieces:

> He who would study organic existence
> First drives out the soul with rigid persistence;
> Then the parts in his hand he may hold and class,
> But the spiritual link is lost, alas! (21).

In contrast, Goethe recommended a radically different method. He sought to grasp by contemplation the inner law, the inner coherence or meaning, of natural phenomena. Goethe, like Aristotle before him, saw in nature the formative power of an artist. And just as the artist creates his work to embody an idea, so, thought Goethe, nature also creates according to underlying patterns, or ideas.

This approach to nature is, of course, more appropriate to animate than to inanimate objects. Indeed, his optical researches, which he regarded as the crowning achievement of his scientific endeavours (though he seems hardly to have actually tried any experiments himself), never made any impression on physicists raised in the Newtonian tradition. However, his investigations of both animal and plant anatomy were surprisingly productive for an amateur. In animal anatomy he is credited with the rediscovery of the intermaxillary bone in the human skull (22); his botanical researches were more successful still. Agnes Arber refers to *Die Metamorphose der Pflanzen* (1790) as 'one of the minor classics of botany' (23).

(20) L. Thorndike, A History of Magic and Experimental Science, (1929–1956), vol. 5, chapter 1.

(21) J.W. von Goethe, *Faust* (trans. B. Taylor), *World's Classics,* Oxford University Press (1932), lines 1935–9.

(22) The intermaxillary bone had in fact been described by Vesalius in the sixteenth century, forgotten and rediscovered independently of Goethe by the French anatomist Vicq d'Azyr in 1784.

(23) A. Arber, *Goethe's Botany,* Waltham, Massachusetts (1964).

Both these sets of biological researches were inspired by the conviction that nature creates according to some underlying, unifying, 'ground plan'. Individual organisms were to be seen as variations played upon the abiding archetypal themes of the *urpflanze* and the *urtiere*. Indeed, so sure was Goethe of the actual existence of at least the *urpflanze* that he set forth into the gardens of Palermo convinced that he would find it growing among its fellows. Schiller, to whom Goethe had confided the distinctness of his quasi-eidetic image, responded in a sceptical manner: 'Das ist keine Erfahrung: das ist eine Idee' (24). Goethe was amazed, and insisted that his archetype was far more than mere idea. Whatever we may nowadays feel about Goethe's apparent conflation of idea and fact, his enthusiasm did lead him to make, as we have seen, some genuine contributions to biology.

He was, for example, convinced that the skeletons of all known vertebrates were but variations on a single common theme—an *ur*-skeleton. Hence, knowing that the intermaxillary bone was developed in many mammalian skulls, he was convinced that it should also be demonstrable in man's. Goethe was able to prove his point by examining foetal and juvenile skulls in which fusion with the maxillary bones had not had time to occur. This striking success for his approach encouraged him to proceed further, and it was while contemplating a sheep's skull during his sojourn in Italy in 1790 that the conviction seized him that the skull was nothing but a modified vertebra. Just as in *Die Metamorphose der Pflanzen* he believed himself to have proved that stem-leaves, sepals, petals and stamens could all be 'seen' as variations on one underlying archetype, so in vertebrate animals he conceived that all the different vertebrae and finally the skull itself were but variations on a fundamental ideal vertebra.

From what has been said, it can be seen that Goethe was groping towards a 'theoretical morphology', a set of generalisations which would tie together and give 'meaning' to the multitudinous different structures and forms which zoologists and comparative anatomists were listing. In this respect we can recognise his relation to Kant. And, indeed, Kant was the only philosopher whom Goethe acknowledged. To read a page of his philosophy, he said, was like entering a brightly lighted room. It was, however, the *Critique of Practical Reason*, with its emphasis on human freedom, which claimed his allegiance—the *Critique of Pure Reason* seemed to him too forbidding even to begin. He refers, however, to the comfort he felt on finding that the great philosopher had set out in precise and ordered detail the vaguer thoughts and feelings he had, himself, long entertained (25). We noticed in chapter 1 that Goethe believed that *Die Metamorphose der Pflanzen*, although written in complete ignorance of the Kantian philosophy, was 'entirely in the spirit of

(24) 'That is not an observational fact: that is an idea.' An account of this famous episode may be read in P. Hume-Brown, *The Life of Goethe*, vol. 2, John Murray, London (1920), p. 441.

(25) J.W. von Goethe, *Einwirkung der neueren Philosophie;* quoted in E. Cassirer, *ibid.*, (1963).

his (Kant's) ideas'.

The Romantic impulse which Goethe expressed, and which seems to have been particularly strongly marked among the immediate followers of Kant's philosophy, emerged powerfully in the biological science of the early part of the nineteenth century. It conflicted strongly with the more cool-headed rationalist tradition. The conflict came to a head in 1830. In August, news of France's 'three glorious days' reached Weimar. On the afternoon of 2 August, Soret visited the now aging Goethe. He was met with excited talk about the 'great event': 'the volcano has burst forth, all is in flames, and there are no more negotiations behind closed doors'. Soret agreed and began to discuss the political situation. Goethe would have none of it: 'I am not speaking of those people at all', he said.

> I am interested in something quite different. I mean the dispute between Cuvier and Geoffroy de St Hilaire, which has broken out in the Academy, and which is of such great importance to science .... In Geoffroy de St Hilaire we have a mighty ally for a long time to come ... the synthetic treatment of nature, introduced into France by Geoffroy, can now no longer be stopped. From the present time mind will rule over matter in the physical investigations of the French. There will be glimpses of the great maxims of creation, of the mysterious workshop of God! Besides, what is all intercourse with nature, if, by the analytical method, we merely occupy ourselves with individual material parts, and do not feel the breath of the spirit, which prescribes to every part its direction, and orders, or sanctions, every deviation, by means of an inherent law.
>
> I have exerted myself in this great affair for fifty years... at first I was alone... But then I gained kindred spirits in Sommering, Oken, Dalton, Carus and other equally excellent men. And now Geoffroy de St Hilaire... This occurrence is of incredible value to me; and I justly rejoice that I have at last witnessed the universal victory of a subject to which I have devoted my life, and which, moreover, is my own *par excellence* (26).

Naturphilosophie

In spite of this immense enthusiasm Goethe was, of course, wrong. The 'synthetic treatment of nature' although retaining powerful adherents was even before Darwin well outside the mainstream of biological advance. It was from its inception too closely tied to the excesses of German Romantic philosophy to receive full-hearted consent from the more analytically minded. Indeed to some extent its effect was negative. Its seeming inanities actively discouraged speculation in biology for nearly fifty years. Goethe, furthermore, was wrong about the facts: Geoffroy de St Hilaire had rather emphatically not won the argument with Cuvier.

What, then, was the argument all about? And who were Cuvier and Geoffroy de St Hilaire? The argument was about comparative anatomy, and Cuvier and Geoffroy de St Hilaire were two of the most prominent workers in the field.

The origins of comparative anatomy may be traced far back into classical antiquity. For two thousand years, no treatise on the subject approached, let alone surpassed, the Aristotelian compendium: *Historia Animalium*. With equal justice Aristotle, it will be remembered, can also be called the father of

(26) *Conversations of Goethe with Eckermann and Soret,* 2 August 1830.

embryology. The subsequent histories of the two subjects are, as we shall see, closely connected.

After Aristotle, few significant advances in *comparative* anatomy were made until the eighteenth century. Human anatomy, it need not be said, was reborn in the sixteenth century in the work of Vesalius and his contemporaries. The interests of the great Italians were, however, predominantly medical, and their anatomies were made with these interests principally in mind. Thus, apart from a few rather isolated investigations, notably those of Leonardo, of Belon (27) and of Perrault (28), the restoration of comparative anatomy awaited the work of John Hunter in the mid-eighteenth century. Hunter (1728–1793) was a tireless anatomist. He spent all he could spare both of money and time in collecting and arranging a museum of comparative anatomy. He is said to have personally dissected a wide range of animal forms from holothurians to whales. Hunter's theoretical concern, however, was rather with physiology than anatomy, and it fell to Vicq d'Azyr and, more importantly, to Georges Cuvier to establish comparative anatomy, once again, as a subject in its own right.

Georges Cuvier was born in 1769 at Montbéliard in Switzerland. After a thorough schooling in anatomy he found work as a tutor to a family living on the coast of Normandy. Here, like his great predecessor, Aristotle, he became deeply interested in the biology of the sea shore. He dissected all the animals he could find in the inter-tidal zone (here he was luckier than Aristotle who, at Mytilene, could only study the non-tidal Aegean) and compared their structure. Some of the drawings he made of these dissections came to the notice of Geoffroy de St Hilaire who was at that time professor of anatomy at Paris. St Hilaire immediately sent for the young Cuvier, and in a very short time Cuvier found himself appointed to a chair in anatomy.

Cuvier's subsequent career was brilliant both in science and in administration. He died in 1832, a peer of France and minister for ecclesiastical affairs. Cuvier's early career coincided with that of Napoleon Bonaparte, and it has been pointed out that both shared the same organisational genius. Cuvier's major works, *Leçons sur l'anatomie comparée* (1799–1805) and *Le Règne Animale* (1817), are great works of organisation and classification. Although the systematic zoology which we find in these books is not the same as that presented in more modern texts—the animal kingdom, for example, is divided into only four major groupings, Vertebrata, Mollusca, Articulata and Radiata—it marks a considerable advance on the preexisting Linnean system. More important, however, was Cuvier's thoroughly comparative approach. With his much greater insight into the anatomies of invertebrate animals, Cuvier was inclined to believe that whereas meaningful comparisons could be made

(27) Pierre Belon published some interesting comparisons of the skeletons of birds and humans in 1555.

(28) Claude Perrault took advantage of the opportunities afforded by the menagerie at the Jardin des Plantes established in Paris in 1699.

between the members of one of his groups, comparisons between the members of different groups were almost valueless. He conceived, in other words, that each group consisted of animals whose forms and structures varied only as variations on a single underlying theme. Something of the same concept still informs modern zoology, although the number of major homogeneous groups, or phyla, is nowadays much larger.

The idea that particular organisms were variations on an underlying ground plan is, of course, by no means original with Cuvier. Our previous discussions of Kant and Goethe show that it was 'in the air' at the end of the eighteenth century and it may, indeed, be traced back, in a sense, at least as far as Plato. Cuvier, however, had used the idea far more sparingly and, as we nowadays see it, far more realistically than many of his, particularly German, contemporaries. He had learnt from one side of Kant's philosphy the lesson that we only deal with phenomena and that the scientist's job is to classify these in the best possible way; all else, thought Cuvier, was merely 'metaphysics'—for him a pejorative term. The predominantly German school of *Naturphilosophie* was, in contrast, far more impressed by the more speculative and transcendental implications of Kant's system (29).

The titles of the books which the 'nature-philosophers' published are symptomatic of the intellectual climate of the movement: *Philosophie Anatomique, Lehrbuch der Naturphilosophie, Précis d'Anatomie Transcendente,* and so on. They searched always for the plan behind anatomical phenomena; in their desire to establish 'inner' laws for morphology and anatomy, they tended to regard those who toiled at the detailed gathering of empirical facts with a certain condescension. As late as 1866 Agassiz was writing that the facts of comparative anatomy 'show the immediate working of mind in the construction of the animal kingdom...(they are) evidence of the existence of a Creator, constantly and thoughtfully working among the complicated structures that He has made' (30). Earlier nature-philosophers looked back beyond Kant to Spinoza. Comparative anatomy seemed to them to display the working out in matter of the thoughts of the deity (31). Others, like Oken (1779–1851), the author of the *Lehrbuch der Naturphilosophie* and perhaps the most influential of its early devotees, compared the universe in an obscure way to a living organism. Each animal class he considered to represent one of the functions of the highest organism: man. 'Animals', he writes (32), 'are only the persistent foetal stages or conditions of man.' Serres,

(29) Schopenhauer, for instance, regards the findings of comparative anatomy as an explicit confirmation of Kant's metaphysics: 'for here artist, work and materials are one and the same. Each organism is a consummate masterpiece of exceeding perfection.' *On the Will in Nature* (1836), in *Schopenhauer Selections* (ed. D.H. Parker), Scribner, London (1928), p. 404.

(30) L. Agassiz, *The Structure of Animal Life,* Scribner, New York (1866), p.122.

(31) See E. Perrier, *La Philosophie Zoologique avant Darwin*(ed. F. Alcan), Ballière, Paris (1884), preface, 10.

(32) L. Oken, *Lehrbuch der Naturphilosophie*, Berlin (1809–1811);trans A. Tulk, *Elements of Physiophilosophy*, Ray Society, London 1847), p. 492.

in the *Précis d'Anatomie Transcendente*, puts forward much the same idea. The entire animal kingdom, in his opinion, is like a single organism which 'has been arrested in its development, here earlier, there later...'. 'Human development is a transitory comparative anatomy just as, in its turn, comparative anatomy shows fixed and permanent stages in the development of man' (33). It is from murky ideas like these that arose one of the best known generalisations of nineteenth-century biology: the recapitulation theory.

The widespread recognition, especially in German zoology, that the embryonic stages of 'higher' organisms resembled the adult stages of 'lower' organisms occasioned a great deal of research into embryology. It seemed that the process of development, like the range of bodies of adult organisms, also displayed the working-out and modification of an underlying thematic idea. If in place of idea we insert the term *archetype*, we should find ourselves speaking much the same language as Richard Owen, the doyen of British comparative anatomists (34). Indeed, the concept of an archetype and the somewhat Romantic associations conjured up by the term lingered on into the comparative anatomy of the twentieth century. Perhaps the most convincing indication of this persistence lies in the nomenclature: the *arch*enteron, the *arch*inephros, the *arch*ipterygium, and so on.

Returning, however, to 1830 and the controversy which so excited Goethe, we can now see what the argument was all about: for Geoffroy de St Hilaire and Cuvier, who, it will be remembered, had started as St Hilaire's protégé, had drifted apart. While Cuvier liked to remain close to the facts, St Hilaire's mind retained a more speculative bent. While Cuvier's great knowledge of anatomy had led him to recognise four great groups of animals, St Hilaire conceived that the anatomies of the entire animal kingdom were but variations on a single structural theme. There was, in his zoological philosophy, no vast variety of dissimilar individual animals but rather a single 'being', *animalité*, which showed itself through a multitude of forms. Believing this, St Hilaire proposed in a communication to the Académie des Sciences that a cuttle-fish could be compared to a vertebrate, point for point, if it was supposed that the vertebrate was bent like a hairpin half way down its abdomen. Cuvier found little difficulty in demonstrating the absurdity of this notion; his great and detailed knowledge of comparative anatomy enabled him to pour scorn on St Hilaire and his followers.

In spite of Goethe's sanguinity, the great majority of younger workers swung behind Cuvier and the *Naturphilosophie* movement was largely discredited. Years later, Weismann recorded that none of his teachers in the 1850s so much as mentioned evolution because 'the overspeculation of

(33) E. Serres, *Précis d'Anatomie Transcendente*, Paris (1842), p. 91.
(34) R. Owen, *On the Archetype and Homologies of the Vertebrate Skeleton*, London (1848). For the continuation of the tradition into the twentieth century, see E.S.Goodrich, *Studies in the Structure and Development of Vertebrates*, Macmillan, London (1930).

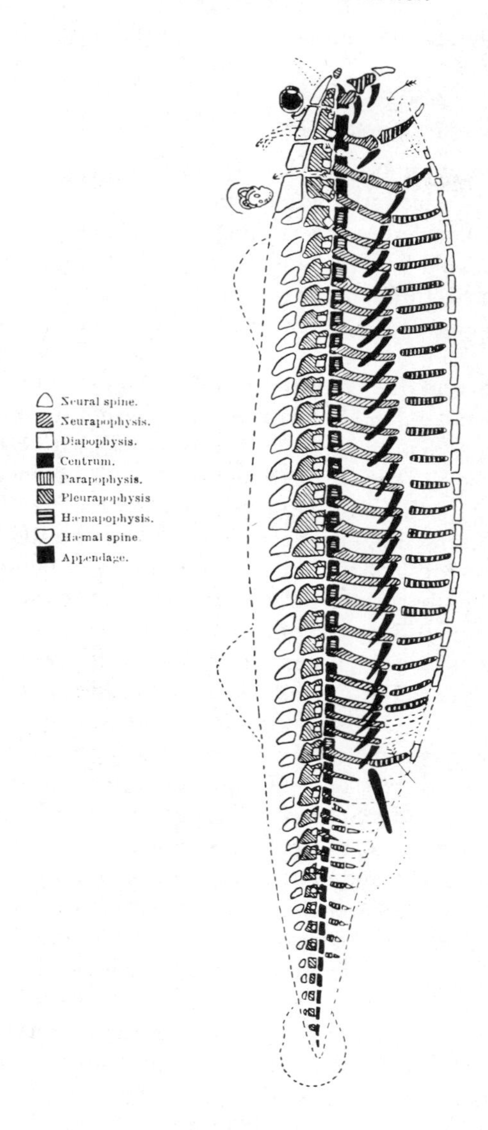

Figure 18.2. Richard Owen's concept of the archetypal vertebrate skeleton. Each segment of the body (including the four cranial segments) is composed of the same skeletal elements or 'vertebrae'. All vertebrate skeletons are, according to Owen, variations on this common theme. From *On the Archetype and Homologies of the Vertebrate Skeleton,* London (1848).

Naturphilosophie had left in their minds a deep antipathy to all far-reaching deductions...' (35).

Embryology and the recapitulation theory

Although the major impulse behind *Naturphilosophie* quickly died away after 1830, a residue of ideas was left behind which served the needs of zoologists in their task of ordering and making sense of the empirical data of comparative anatomy. This task was aided by the dawning realisation that the homologies detected throughout the animal kingdom became more understandable when anatomy's fourth dimension, embryology, was taken into account.

A celebrated instance of the appeal to embryology is provided by the continued attempts to understand the nature of the vertebrate skull. We have already noticed how the idea that the skull was a modified vertebra seized Goethe (although he was not the first to whom the idea had occurred) while he sat contemplating a specimen in the Jewish cemetery at Venice. Goethe's interpretation became something of a *cause célèbre* during the first half of the nineteenth century. Most of the great anatomists and embryologists of the period contributed to the argument. Richard Owen conceived the skull to be constructed of four modified vertebrae. Von Baer believed the jaws and the limbs to be variations on a single common theme. Carus went further and proposed that the whole skeleton was but a series of modified vertebrae.

Eventually, however, a more precise knowledge of the skull's development became available. It then became clear that the homologies detectable in the vertebrate skeleton were to some extent due to its embryological development. T H Huxley (1858) forcibly argued the case for an approach through embryology. Instead of comparing the adult forms of a series of vertebrate skulls (says Huxley) and extracting from them a cloudy archetype, the zoologist should investigate and compare the embryology of the different skulls. He would then find that the archetype, the underlying theme, is to be identified with the primordium from which all such skulls are found to develop (36).

This approach was supported by the work of many embryologists (in particular von Baer) who had shown that embryogenesis seems always to proceed from the general to the particular. The early embryos of a series of vertebrates are always very much more alike than the adults into which they develop. Thus the great development of comparative embryology in the first half of the nineteenth century provided a material basis for the 'metaphysical' archetype of the nature-philosophers.

(35) A. Weismann, *The Evolution Theory*, vol. 1 (trans. J.A. and M.R. Thomson), Edward Arnold, London (1904), p. 28.

(36) T.H. Huxley, On the theory of the vertebrate skull, *Proc. Roy. Soc.*, 9 (1858), 381–457; Reprinted in *Scientific Memoirs of T.H. Huxley* (ed. M. Foster and E.R. Lankester), vol. 1, Macmillan, London (1898), esp. pp. 542–585.

It did not, however, fully satisfy the enquiring mind. For, after all, why should animals develop in this way from the generalised to the specialised? Why should the embryos of all vertebrates display at some stage the gill clefts which are retained in the adult and have functional value only in fish? Why should animals be classifiable into natural groups? We have seen that the earlier exponents of *Naturphilosophie* had an answer. But we have also seen that such speculation, after 1830, was frowned upon. It was, indeed, not until 1859 that the thaw occurred. The publication of *The Origin of Species* and the acceptance of the doctrine of organic evolution unfroze the speculative intellect. It quickly became apparent that the homologies which had so forcibly impressed earlier thinkers, and which had caused them to dream of underlying divine ideas, could equally well, if not better, be accounted for in terms of evolutionary theory. Organs were homologous, organisms were comparable, not because the deity possessed only a certain number of master plans but because the organisms concerned shared a real common ancestor in the remote past. 'On my theory', wrote Charles Darwin, 'unity of type is explained by unity of descent' (37).

We may thus conclude this chapter with the insight that while zoologists believed in special creation and the fixity of species, the concepts donated by the Romantic nature-philosophers remained the best available. The non-mechanistic ideas of the archetype, of theme and variation, of progressive individuation, ordered and made sense of a great number of anatomical findings. Among the many ghosts exorcised by the Darwinian revolution, the Goethean *typus*, the archetype, was not the least important.

(37) C. Darwin, *The Origin of Species,* John Murray, London (sixth edition, 1872), p. 260.

19 THE ORIGIN OF SPECIES

The selection of natural selection

The movement of thought associated with the name of Charles Darwin was crucial in many ways and at many different levels. Organic evolution is an idea with a very lengthy history. Certainly, as we saw in the earlier chapters of this book, it was current in the classical civilisations of the ancient world. It was not, however, until the mid-nineteenth century that the conditions of the scientific and socio-economic environment were favourable to its selection and rapid spread. Even then, at the outset, it was, as Darwin himself points out (1), an unnatural idea to most educated, even biologically educated, minds. It was repugnant because it broke through a complex of deeply entrenched and deeply intertwined beliefs.

Systematics and the fixity of species

From its inception, western thought has wrestled with the problem which some have termed process and reality, others the many and the one, yet others opinion and knowledge. We noticed something of this debate in the earlier chapters of this book. In particular we saw how the unchanging *a priori* truths of geometry so impressed the mind of Plato that he was inclined to regard the flux of sense data as mere shadows of an abiding reality. Knowledge could only be of *being*, not of *becoming*. Perhaps, therefore, at some level in men's minds, this profoundly embedded precept formed a barrier to the notion that 'all is flux', that 'we never step into the same river twice'.

Certainly in the eighteenth and the beginning of the nineteenth centuries the idea that species were fixed and immutable was widely accepted. Indeed, without the vast duration of past time which we nowadays accept this is but the common sense of everyman. Organisms, in general, *do* breed true to type. We should be most sceptical of the nurseryman who sold us seeds purporting to be one plant when in the spring a totally different plant germinated! And, probably, without this belief in the fixity of species it would have been difficult for systematists, and one immediately thinks of Linnaeus, to bring

(1) C. Darwin, *Autobiography and Selected Letters* (ed. Francis Darwin), Appleton, London (1892), p. 45; a new edition of this work was published by Dover, New York (1958).

order into the great congeries of organic forms brought to light by the European diaspora. We find, for example, that Linnaeus, whose attitude toward conventional religion was by no means straightforward, writes in the *Philosophia Botanica* that the number of species to be discovered is identical to the number originally created: 'Species tot numeramus, quot diversae formae in primitione sunt creatae'. The *Philosophia Botanica* was published in 1751; the same thought, however, recurs in the earlier editions of the *Systema Naturae* (2). In this great and seminal work he reviews the plant and animal kingdoms as if they were exhibits in the gallery of a natural history museum. The exhibits were, of course, alive and the founder of the gallery was conceived to be their creator, God. He writes that he believes himself to have been privileged beyond other men in the talent 'to peep into his (God's) secret cabinet' (3), and in so doing he has been able to recognise, or at least to cross the threshold towards the recognition of, the nature of the Creator.

Contrivances without a contriver

Another of the entangled issues which the Darwinian theory cut through was the issue which we may sum up under the term hylozoism. Throughout the latter part of this book an attempt has been made to chronicle some of the stages on the road of retreat of this notion. We have watched a progressive replacement of the organismic analogy (represented by the Aristotelian system) by a mechanistic analogy. We saw in chapter 15 that Descartes pioneered a mechanistic physiology. We saw how he concluded that the mechanisms he described, or rather outlined, could not be distinguished from a living body. In chapter 17 we saw how the progress of physiologists working in the Cartesian tradition had largely vindicated Descartes' vision. But one great question clearly remained. It was the question which excited the Romantics, discussed in chapter 18. It was the perplexity which Isaac Newton succinctly expressed in the 28th Query of the *Opticks*: 'How came the bodies of animals to be contrived with so much art...?' It was the question which sustained Paley and Hume in their religious belief: 'consider, anatomise the eye; survey its structure and contrivance; and tell me from your own feeling, if the idea of a contriver does not immediately flow in upon you with a force like that of a sensation' (4). It was this ultimate redoubt of hylozoism which the Darwinian theory breached. The Newtonian query was answered in terms which Democritus and Descartes would have recognised.

(2) The first edition of the *Systema Naturae* was published in 1735. In subsequent editions Linnaeus expressed doubts about this undynamic approach to nature. This dubiety about the basis of his work reached a climax in the final (1766) edition. The tenth edition (1758) is taken as the foundation work of modern systematics.

(3) Quoted in K. Hagberg, *Carl Linnaeus* (trans. A. Blair), Cape, London (1952).

(4) D. Hume, *Dialogues concerning Natural Religion*, in *Philosophical Works of David Hume* (ed. T.H. Green and T.H. Grose), Longmans, London (1874) (first published 1775), vol. 2, p. 402. It is interesting to note that Cleanthes, into whose mouth the quoted sentence is put, is represented as locked in argument with an Epicurean who insists that the atomic theory makes any argument from design impossible.

Indeed, Darwin has been called by some the Newton of biology. Although in many ways the analogy is poor, in one sense it is certainly apt. For just as Newton may be credited with showing that the same laws applied to the supra and the sublunary spheres, a unification which would have astonished the ancients, so the Darwinian theory showed how the gulf between animate and inanimate creation was only apparent, not real. Perhaps, however, Darwin might be better compared with another of the great seventeenth-century physicists: Galileo Galilei. For just as Galileo sought to show that physical phenomena need not be accounted for in the occult terms of essence, final cause, and so on, so Darwin effected a similar revolution in biology. After 1859 it was no longer necessary to argue that the existence of a variegated living world demanded as a *sine qua non* an 'inner' vital, or psychic, force. It was no longer necessary to argue that the superbly designed bodies of living organisms implied *intentional* design, or that final causation, however much out of fashion in physics and chemistry, was indispensable in biology. Mystery was banished; the whole world was on one level; and as the ripples from the *Origin* spread through wider and wider tracts of thought, Keats' prophecy began to seem only too apt:

> Do not all charms fly
> At the mere touch of cold philosophy?
> There was an awful Rainbow once in heaven:
> We know her woof, her texture; she is given
> In the dull catalogue of common things.
> Philosophy will clip an angel's wings
> Conquer all mysteries by rule and line
> Empty the haunted air and gnomèd mine
> Unweave the rainbow.

Indeed Darwin himself might have agreed with this reaction. For it is well known how towards the end of his life he came to feel that all one side of his nature, the side responsive to poetry, painting and music, to 'the haunted air and gnomèd mine', had atrophied. 'My mind', he sorrowfully reports (5), 'seems to have become a kind of machine for grinding general laws out of large collections of facts...if I had to live my life again I would make it a rule to read some poetry and listen to some music at least once every week.' Looking back at the end of his life he is amazed at the industry he has displayed: the catalogues of books read, the indexes he has made, the facts he has tabulated. Perhaps it was only thus that the theory could be established. Perhaps it was essential that a single mind should distil the meaning from the multitudinous evidence, marshall the facts, conceive the theory. The theory's co-discoverer, Alfred Wallace, candidly confessed himself incapable of such labour: 'I have not the love of *work, experiment* and *detail* that was so preeminent in Darwin, and without which anything I could have written would never have convinced the world' (6).

(5) Darwin, *ibid.*, p. 54.
(6) A.R. Wallace, in a letter to Professor A. Newton in 1887; quoted in Darwin, *ibid.*, p. 200.

The principle of plenitude

It was at the marshalling of detailed observations, the construction of, precisely, 'catalogues of common things', at which Darwin was so preeminent. Although, as he says, he read widely as a young man, he was never enamoured of the abstract systems developed by metaphysicians: 'I read a good deal...on various subjects, including some metaphysical books, but I was not well fitted for such studies' (7). Perhaps this was just as well, for another of the consequences of the revolution which he initiated (in addition to emptying the poet's air and mineshaft) was the controversion and ultimate overthrow of a deeply held metaphysical position. This position dates from at least the time of Plato and has been said to follow from that philosopher's moral conviction that 'being good entails doing good' (8). The creator being, by definition, perfect had of necessity to *do* good, in other words to create, to create the universe. Another part of this metaphysical credo held that the cause was always superior to its effect. One can perhaps see the image of a hierarchical, static social system in this concept. Actions were initiated by the aristocrat and carried out by his underlings. This profoundly pervasive notion is succinctly expressed by Ralph Cudworth in *The True Intellectual System of the World* (1678): 'It is utterly impossible that Greater Perfection and Higher Degrees of Being should Rise and Ascend out of Lesser and Lower, so as that which is most Absolutely Imperfect of all things should be the First Fountain of All'. It is clear that the idea of evolution is diametrically opposed to this way of thought.

From the first axiom mentioned above another consequence seemed to follow. If God was good—and who would deny this?—and if, in consequence, creation was good, then it followed that the world was 'full', full not only of a number of things but of the greatest number of 'compossible' things. This consequence has been termed by Lovejoy (9) the principle of plenitude. Everything which could exist, whose existence did not contradict the existence of some other thing, did exist. Leibniz sets out this theory with geometrical rigour. He proposes that existents, in particular living organisms, are, in a mathematical sense, dense. That is, between any two existents there is always room for another, and that other will be found. This notion, as we shall see, was also held by Lamarck. For Leibniz it implied that there were no discontinuities in nature: from inanimate objects to man and beyond, the chain was unbroken. Leibniz's faith in this tenet was helped by the contemporary developments in microscopy. The merest drop of pond water was revealed as a miniature universe filled with a myriad living forms (10). From this axiom, too, Leibniz was convinced of the existence of

(7) Darwin, *ibid.*, p. 33.

(8) See, for instance, chapter 7, fifth section; for an account of the early development of this position, see J.M. Rist, *Plotinus*, Cambridge University Press (1967), p. 66 *et seq.*

(9) A.O. Lovejoy, *The Great Chain of Being*, Harper, New York (1960), chapter 2.

(10) G.W. von Leibniz, *Monadology and other philosophical writings* (trans. R. Latta), Oxford University Press, London (1898); *Monadology* (1714), paras. 65 - 70.

zoophytes, or plant-animals, many years before Abraham Trembley rediscovered them in the form of Hydra in 1739.

The concept of a rational and good creator also had another consequence. The creation was not only quantitatively optimal but also qualitatively optimal. Beings were created at all possible levels of perfection. Consequently it was possible for savants to arrange them in a scale of increasing excellence: the *scala naturae* of the Peripatetics. It will be recalled from earlier chapters how very influential had been the Aristotelian classification of living nature into three great divisions, depending upon their psychic powers, of plants, animals and men. In Aristotle's zoological compendium, *Historia Animalium*, we read of another, more detailed, scaling system. The animal kingdom is divided into eleven great groups with the zoophytes (so long to disappear from observational science) at the bottom and man at the top. In later centuries the scale was extended. It was suggested that there were as many orders of being above man as there were below him. Perhaps the best known expression of this vision of the universe is the often-quoted passage from *An Essay on Man* (1733):

> See, thro' this air, this ocean, and this earth,
> All matter quick, and bursting into birth.
> Above, how high progressive life may go!
> Around, how wide! how deep extend below!
> Vast chain of being, which from God began,
> Natures aethereal, human, angel, man,
> Beast, bird, fish, insect! what no eye can see,
> No glass can reach! from Infinite to thee,
> From thee to Nothing! – On superior pow'rs
> Were we to press, inferior might on ours:
> Or in the full creation leave a void . . . .

Bouleversement of the Aristocratic principle

The eighteenth century has many similar poetic expressions of this so-called 'principle of plenitude' (11). In Pope and in many of his contemporaries the orientation is still conservative: 'Whatever is, is best'. The poet is very far from being an advocate of social mobility, let alone of revolutionary change. Any displacement in the continuum, any attempt at betterment, would leave a 'void', and nature abhors a vacuum!

However, it is notorious that towards the end of the eighteenth century the old order began to disintegrate under the pressure of new forces. These pressures emanated from new technologies: technologies made possible by the investigations of the pneumatic chemists of the previous century. Concomitant with this loosening of the social order, mobility came increasingly to be seen in the *scala naturae*. For Emerson, in the nineteenth century, the worm, 'striving to be man', 'Mounts through all the spires of form'. And Charles Darwin's grandfather, Erasmus, had, as we shall see,

(11) See, for instance, B. Willey, *The Eighteenth Century Background,* Chatto and Windus, London (1965), and M.H. Carré, *Phases of Thought in England,* Clarendon Press, Oxford (1949).

similar expressions in the poetry which he wrote at the beginning of the same century.

The *scala naturae*, the great chain of being, had been imbued with a new dynamism. We are not yet fully in the universe of discourse which envisages a progressive transformation of species into fresh forms as time progresses (although this concept is certainly present in Erasmus Darwin and Lamarck), but far more frequently in the ambience of a more abstract, more 'poetic' notion, of a 'life force' which strives upward through the orders of animate creation. Robinet, for instance, in the mid-eighteenth century, sees the chain of being as the axis through which an active principle (as opposed to passive matter) works its way *up* towards man. But although we are not yet in a position to enter into the Darwinian objectivity, we are in process of leaving the static world order of the early eighteenth century behind.

We are also in process of manoeuvring the classic concept of a moral deity whose existence guarantees the rationality of the universe into an untenable position. The revolution in social organisation is mirrored by a revolution in metaphysics. The old idea that the cause must be better, more perfect, than the effect is dying. It is becoming replaced by what had formerly seemed blasphemous: by the notion that the more perfect beings are gradually developed from the less perfect, that the higher orders on the *scala naturae* arise from the lowlier.

Indeed one of the most vividly argued objections to evolutionary theory was, in fact, an aspect of this *bouleversement*. This was the notorious 'monkey theory' which caused such consternation at a famous British Association meeting. The descent (or, as some would have it, ascent) of man from an animal ancestry is a logical consequence of the Darwinian position. This displacement of man from his position as the divinely appointed master of all creation was likened by Haeckel to yet another Copernican revolution in human thought (12).

But the *bouleversement* penetrated more deeply still. With it we reach a climax to another of the great themes of our history. We have noted, intermittently, since the inception of biology among the Presocratics, the existence of a profound debate between the Democriteans and the Aristotelians. Does it make sense to apply the atomic theory of matter in the life sciences? For millennia Aristotle seemed to have the better of the argument: it seemed inconceivable that the superb fitness for purpose exhibited by living organisms could be accounted for by the fortuitous coming together of atomic particles. Here, it seemed, was the *pons asinorum* of the notion that higher orders of being could arise from lower. But now, in the nineteenth century, after the publication of *The Origin of Species*, even this 'mechanical hypothesis' became intellectually respectable. It became clear how organic forms could have arisen from inorganic matter 'without the

(12) E. Haeckel, *The History of Creation*, vol. 2 (trans. revised by E.R. Lankester), H. S. King, London (1876).

aid of the Gods'. And this, in turn, was to imply, when the thought had fully worked through, that the world was not necessarily rational, not necessarily predictable by the isolated human intellect. Chance, it became clear, has a great deal to do with natural selection. It accordingly became less and less fashionable for savants to follow Leibniz's example and attempt to legislate as to what might or might not be found in nature from *a priori* principles.

The days of creation

From yet another aspect the late eighteenth and early nineteenth centuries witnessed a profound alteration working through man's vision of himself and the universe. Although the change was, as we shall see, preceded by many hints and half-hints in previous centuries, it was not until the middle decades of the nineteenth century that the evidence became inescapable. This change was to do with his understanding of the duration of past time. One of the best-known and most influential expressions of the orthodox view before the development of evolution theory is that of John Milton. Indeed, T H Huxley remarked that evolutionists frequently found themselves more in conflict with Miltonic than with Mosaic cosmology. In *Paradise Lost* we read of how the Almighty

> . . . took the golden Compasses, prepar'd
> In God's eternal store, to circumscribe
> This Universe, and all created things:
> One foot he center'd, and the other turn'd
> Round through the vast profunditie obscure,
> And said, thus farr extend, thus farr thy bounds,
> This be thy just Circumference, O World,
> Thus God the Heav'n created, thus the Earth . . .

The poetry continues by describing the six days of creation, closely following the biblical account. The earth, the waters, the sky are formed. The plants, the fish, the water and land birds are created. Penultimately the terrestrial animals appear:

> . . . The Earth obey'd, and strait
> Op'ning her fertil womb teemed at a Birth
> Innumerous living Creatures, perfet formes,
> Limb'd and full grown out of the ground up rose . . .
> Among the Trees in Pairs they rose, they walk'd:
> The Cattel in the Fields and Meddowes green:
> Those rare and solitarie, these in flocks
> Pasturing at once, and in broad Herds upsprung
> The grassie Clods now Calv'd, now half appear'd
> The tawnie Lion, pawing to get free
> His hinder parts, then springs as Broke from Bonds,
> And Rampant shakes his Brinded main; . . .

and ultimately, of course, comes man, 'The image of God Express...'.

In this vision of the days of creation is an immediacy and concreteness whose grip must have been difficult to break. Given the prevailing set of religious beliefs we can still easily feel that, yes, this is how it must have been.

The imagery is on a human scale and we feel at home with it. But it was, of course, precisely this homely and meaningful picture of the world's beginnings which crumbled, and had to crumble, before the Darwinian moment could arrive. For Charles Darwin could not have begun to put together his theory without a far greater perspective in time than the Bible, taken literally, allowed. This he recognised quite explicitly: 'The belief that species were immutable productions was almost unavoidable as long as the history of the world was thought to be of short duration...' (13).

The breach in the walls of the temporally 'closed' universe was ultimately forced by the geologists; and forced only very shortly before Darwin set off on his momentous voyage in the Beagle. Among the books he took with him on his circumnavigation was a copy of Lyell's *Principles of Geology*. This volume, says Darwin, was studied attentively and, he writes, 'The very first place I examined, in the Cape Verde islands, showed me clearly the wonderful superiority of Lyell's manner of treating geology...' (14). For Darwin, Lyell's researches, and those of his fellow geologists, provided the essential dimension which his own theory required: that of time. Rather in the manner of Plato adjuring all those who would study his philosophy to first master geometry, Darwin writes: 'He who can read Sir Charles Lyell's grand work on the *Principles of Geology*, which the future historian will recognise as having produced a revolution in natural science, and yet does not admit how incomprehensibly vast have been the past periods of time, may at once close this volume' (15).

Catastrophism and uniformitarianism

Lyell's *Principles* summed up and completed the long process through which the veil that had formerly concealed man's enormous ancestry was withdrawn; for mid-nineteenth-century geology rested upon a tradition stretching back to the earliest times. Fossils had been discovered by several of the Presocratic philosophers, especially Xenophanes, and something of their significance understood. Later, during the mediaeval and Renaissance periods, this knowledge was very largely lost. The prevailing organismic paradigm caused men to see the earth's crust as if it were alive (chapter 12). It was believed that not only fossils but also stones 'grew' in the earth. Robert Hooke, for instance, describes a meeting of the Royal Society which he had attended in 1663:

> There happened an excellent good discourse about petrifaction; upon which several instances were given about the growing of stones: some that were included in glass vials; others, that laid upon the pasture ground; others that lay in gravel walks; which was known by putting a stone in at the mouth of a glass vial, through which, after a little time it would by no means pass. Next, the story of a field's being filled with stones every third year was confirmed by some instances. And that the stones in gravel

(13) C. Darwin, *The Origin of Species*, John Murray, London (first edition, 1860), p. 481.
(14) Darwin, *Autobiography and selected letters*, p. 29.
(15) Darwin, *Origin of Species*, p. 282.

walks grow greater, had often been proved by sifting those walks over again, which had formerly passed all through the sieve, and finding abundance of stones too big to pass a second time (16).

Hooke later changed his position on this mysterious power of petrifaction and moved to a more modern-sounding theory. The first, however, to decisively break with the notion that fossils were Nature's 'sports', ill-fated results of the earth's primaeval power to generate living forms (or even more interestingly, forms awaiting birth in the future), was Leonardo da Vinci. Uniting, as we have already seen, the keen observational powers of a painter with the intellectual preoccupations of an engineer, Leonardo was inevitably struck by the finding of fossil fish, oysters, corals and sea-shells well away from the shore and at a considerable altitude. He, like Xenophanes, proposed that the remains were those of marine creatures which had, over great periods of time, become trapped in mud (for Leonardo, alluvial mud). Although he was unable to devise a satisfactory theory to explain the elevation of these deposits, sometimes to the peaks of high mountains, Leonardo's methodology marks a real advance on that of the majority of his predecessors and contemporaries. Instead of appealing to mysterious formative powers or, alternatively, to extraordinary events in the past such as the Noachian flood, which many pious minds invoked as an explanation of the presence of marine fossils in high places, Leonardo attempted an explanation in terms of the forces which he could see around him still at work. And this, of course, is the essential principle of modern geology.

Leonardo, however, was in this, as in so much else, far ahead of his time. Before the uniformitarian movement could triumph, the biblical account had to be disposed of; and theologians fought back vigorously. In order to satisfy this opposition a compromise position, known as catastrophism, was first of all developed. This theory was put forward by Buffon in 1778. In an earlier work, *Theory of the Earth* (1749), Buffon's geologising had run foul of the Parisian Faculty of Theology, and he had been required to explain himself. Thus, when *The Epochs of Nature* was published in 1778, he had made a considerable effort to homologise geological and scriptural history. This he had done by dividing the history of the earth into seven great epochs. These epochs were the 'days of creation'. But instead of enduring a mere twenty-four hours, these Buffonian 'days' lasted for periods varying from 3000 years to 35 000 years. The earth had existed altogether some 75 000 years and, according to Buffon, there were another 93 000 still to come. *The Epochs of Nature* also suggested a reason for the gaps which were already becoming apparent in the fossil record. Between each 'epoch' Buffon suggested that a major geological upheaval had occurred. These cataclysms ensured that the old order changed and gave place to new.

Buffon's work, although at first unpopular in theological circles, was soon

(16) Quoted in *Forerunners of Darwin* (ed. B. Glass, O. Temkin and W.L. Strauss), Johns Hopkins Press, Baltimore (1959), p. 24.

seen to have the merit of providing the orthodox with an answer to questions about the age and prehistory of the earth. Perhaps it was no longer possible to hold that the earth and all its inhabitants were created in six solar days some 6000 years ago. But it was still possible to retain some of the grandeur of the Mosaic and Miltonic accounts by increasing the time scale. Whether the passage of twenty-four hours or of fifty thousand years occurred while '...the waters (brought) forth abundantly the moving creature that hath life, and the fowl that may fly above the earth in the open firmament of heaven' was, perhaps, not so very important. The biblical account could still be held to be essentially true.

This was also the view propounded by Georges Cuvier at the beginning of the nineteenth century. We noticed in chapter 18 the extremely influential position which Cuvier held in science and in French affairs. Cuvier, moreover, had the benefit of a further half-century's research over Buffon. His knowledge of palaeontology was therefore far greater. Some of this knowledge was the fruit of his own research. The Paris basin is particularly rich in fossil remains, and Cuvier was thus able to gain a very extensive first-hand acquaintance with the subject. Indeed, Nordenskjold regards the *Recherches sur les Ossemens Fossiles*, which he published in 1812, as having 'created the science of palaeontology in the modern sense' (17).

Cuvier's insight into the interdependence and correlation of the parts of an organism, which he derived from his deep knowledge of comparative anatomy (chapter 18), served him well in his study of fossils. From a very few fossil bones he believed himself able to reconstruct the animal, its habits and its habitat. But his great knowledge of the palaeontology of the Paris basin also included a rather striking fact. The fossiliferous beds showed very marked discontinuities. There was no evidence that he could find which suggested that one ancient organism ever gradually changed into another. Freshwater fossils alternated with marine fossils alternated with strata containing no fossils whatever. The conclusion seemed to Cuvier clear. There had been in the past a series of quiet, peaceful epochs during which life was fruitful and multiplied, separated by violent, catastrophic events. The uniformitarianism preached by Leonardo, Hutton and Playfair before him was insufficient explanation. These intermittent catastrophes were periods during which

> . . . the march of nature is changed, and none of the agents that she now employs were sufficient . . . whole continents were trapped in ice or inundated . . . corpses of vast quadrupeds [he refers to the recently discovered Siberian Mammoth] were seized by the ice . . . one and the same instant killed these animals and glaciated the countries they inhabited. This event was sudden, instantaneous, without any degree . . . (18).

(17) E. Nordenskjold, *The History of Biology* (trans. L.B. Eyre), Tudor, New York (1928), p. 337.

(18) G. Cuvier, *Discours sur les Révolutions de la surface du globe,* Paris (1825); facsimile reprint of third French edition published by Culture et Civilisation, Bruxelles (1969), p. 17; English translation, *Theory of the Earth* (trans. R. Jameson), Blackwood, Edinburgh (1827).

Alongside the catastrophists and, as we have already mentioned, antedating them to some extent were those who believed that the forces which shaped the planet's crust, and which were responsible for the fossiliferous deposits, were the same forces which we can still observe operating all around us today. Of these uniformitarians we have already mentioned Leonardo da Vinci. In the eighteenth century several investigators took up the same general position. Of these the best known are James Hutton and his friend and populariser John Playfair. Both assumed that, given sufficient *time*, there was no need to postulate extraordinary, unprecedented events. All the phenomena of the earth's crust could be accounted for by the operation, the almost infinitely long continued operation, of everyday unexceptional forces. Lyell quotes a passage from Playfair's *Illustrations to the Huttonian Theory* on the title page of his *Principles,* the book which Darwin carried around the world:

> Amid all the revolutions of the globe the economy of Nature has been uniform, and her laws are the only things which have resisted the general movement. The rivers, the rocks, the seas and the continents, have been changed in all their parts; but the laws which direct those changes, and the rules to which they are subject, have remained invariably the same.

With the gradual acceptance of the uniformitarian position during the nineteenth century, the age of the earth had perforce to be vastly increased, The Ussherian 4004 BC became absurd in the eighteenth century; in the nineteenth the Buffonian 75 000 years became similarly ridiculous. For Darwin the earth's age was to be measured in millions of years, for us today in thousands of millions of years. The alteration in our understanding of man's place in nature has been correspondingly gigantic. It can, as Lyell himself points out, only be compared to the alteration consequent upon the acceptance of the seventeenth century's 'New Philosophy':

> The senses had for ages declared the Earth to be at rest, until the astronomer taught that it was carried through space with infinite rapidity. In like manner the surface of this planet was regarded as having remained unaltered since its creation, until the geologist proved that it had been the theatre of reiterated change . . . . By the geometer were measured regions of space and the relative distances of the heavenly bodies; – by the geologist myriads of past ages were reckoned . . . (19).

If the Copernican, Keplerian, Galilean, Newtonian insights had burst the tight spatial bounds of the mediaeval universe, creating a cultural shock which reverberated throughout the ensuing century, so the geological revolution of the eighteenth and nineteenth centuries burst the comfortable time scale of Genesis. The cultural shock produced by this development was fully comparable to that produced by the earlier astronomical revolution. Received religion seemed forced into an untenable position, and men pondered whether

(19) C. Lyell, *Principles of Geology,* vol. 1, John Murray, London (tenth edition, 1866), p. 88.

God and Nature (are) then at strife
That Nature lends such evil dreams?
So careful of the type she seems
So careless of the single life;

That I, considering everywhere
Her several meaning in her deeds,
And finding that of fifty seeds
She often brings but one to bear,

I falter where I firmly trod . . . .

'So careful of the type?' but no.
From scarpèd cliff and quarried stone
She cries, 'A thousand types are gone:
I care for nothing, all shall go.

'Thou makest thine appeal to me:
I bring to life, I bring to death:
The spirit does but mean the breath:
I know no more.' (20)

Lamarck and Erasmus Darwin

An appreciation of this new dimension—time—which geological research had introduced into the scientific debate did not, however, force an immediate acceptance of the evolution theory. It was a necessary but not a sufficient condition. Lyell himself, in the first edition of his *Principles*, is quite clear that the case for the transmutation of species is not proven. Indeed, he seems to anticipate some of the acrimony of the latter part of the century when he writes that he who would accept the Lamarckian notion of the origin of species must face the consequence that '...in fine, he renounces his belief in the high genealogy of his species, and looks forward, as if in compensation, to the future perfectibility of man in his physical, intellectual and moral attributes' (21). Lyell shows in this passage how difficult it was for minds schooled in the old ways to accept the *bouleversement* discussed in the fifth section of this chapter. It required considerable mental flexibility to 'go the whole Orang' and recognise that man should no longer see himself as a fallen creature, excluded in the past from Paradise, a creature who must necessarily look back to happier times, a golden age, but instead perceive him as a creature 'faring forward' and for whom the best is, he hopes, 'yet to be'.

But Lyell's objection is based, of course, not on mere rationalisations but on what appeared to him to be the hard palaeontological facts. These were the facts which, as we have already seen, Cuvier believed he had disinterred from the fossil beds of the Paris basin. 'If', Cuvier writes (22), 'the species

(20) Tennyson was regarded by the Huxleys as 'the only poet who thoroughly understood the movement of modern science': *Life and Letters of T. H. Huxley* (ed. L. Huxley), vol. 1, Macmillan, London (1900), p. 315.
(21) Lyell, *ibid.*, (first edition, 1832), vol. 2, p. 386.
(22) Cuvier, *ibid.*, p. 103.

have changed we ought to find traces of these gradual modifications. Thus between the palaeotheria and our present species, we should be able to discover some intermediate forms; and yet no such discovery has ever been made.' Furthermore, Cuvier backs up his position by referring to the investigations made by Geoffroy de St Hilaire and himself on the mummified remains of animals and men brought back after the Napoleonic excursion into Egypt. These, he says, show that no anatomical change has occurred in five thousand years. Thus for Cuvier and his followers, and in this aspect of his work Lyell must be included among his followers, the transmutation of species had little positive evidence to support it and considerable negative evidence to disprove it. Lyell concludes: '...it appears that species have a real existence in nature; and that each was endowed, at the time of its creation, with the attributes and organisation by which it is now distinguished' (23).

Against this orthodox position, however, several voices were raised. The most famous in later years was that of Jean-Baptiste de Lamarck, who had long laboured as a colleague of Cuvier in the Musée d'Histoire Naturelle. Unlike Cuvier, however, who worked on the vertebrates and investigated dramatic forms like *Palaeotherium, Megatherium* and *Mastodon*, Lamarck's province in the museum included only the lower animals: the invertebrates. The results of his investigations were published in the seven volumes of his *Histoire Naturelle des Animaux sans Vertèbres* (1815—1822). His only other major work was the *Philosophie Zoologique* (1809), and it is in this volume that he develops his theory of evolution. It seems not unlikely that his ideas about the transmutation of species originated during the years spent classifying and anatomising the museum's vast collections of invertebrates. He writes in the *Philosophie Zoologique*, for example, that the idea that species are immutable 'is continually being discredited for those who have seen much, who have long watched Nature, and have consulted with profit the rich collections of our museums' (24). It becomes, he says, more and more difficult as more and more animals are discovered to draw sharp lines of demarcation between one species and another: 'everything is more or less merged into everything else'. But what, he asks rhetorically, is the cause of this continuity? He answers himself in the following terms:

> . . . in the course of time the continued change of habitat in the individuals . . . living and reproducing in . . . new conditions, induces alterations in them . . . thus, after a long succession of generations these individuals, originally belonging to one species become at length transformed into a new species distinct from the first (25).

A modern reader would find this unexceptional enough. Where he would differ from Lamarck, of course, is in the cause assigned to this

(23) Lyell, *ibid.* (1832), vol. 2, p. 442.
(24) J.B.P.A. de Lamarck, *Philosophie Zoologique,* tome 1, tome 2, Paris (1809); facsimile reprint of first edition by Culture et Civilisation, Bruxelles (1970); English translation, *Zoological Philosophy* (trans. H. Elliot), Macmillan, London (1914); Hafner Publishing, New York and London (1963), p. 36.
(25) *Ibid.* (1963), p. 39.

transformation. Notoriously, Lamarck believed that the transmutation of species was effected by forces internal to the organism. He insisted that the habitual effort made by an animal forces an anatomical change which is, in turn, transmitted to the offspring during reproduction. This notion is expressed in his second law:

> All acquisitions or losses wrought by nature on individuals, through the influence of the environment in which their race has long been placed, and hence through the predominant use or permanent disuse of any organ; all these are preserved by reproduction to the new individuals which arise, provided that the acquired modifications are common to both sexes, or at least to the individuals which produce the young (26).

Lamarck's most famous example is that of the giraffe. This animal, through the constant effort, extending over many generations, to reach the succulent leaves of trees, has lengthened both its forelegs and its neck so that it now stands some twenty feet in heighı (27). This constant, habitual, effort is, for Lamarck, an aspect of the animal's 'will'. Thus we see that, for Lamarck, Emerson's lines about the worm 'striving (upward) through all the spires of form' have a literal meaning. The evolutionary 'movement' is psychically motivated. The organism reacts 'creatively' to its environment.

Lamarck's *Philosophie Zoologique* does not restrict itself solely to zoological and anatomical matters. It also contains important sections on physiology and psychology. We are thus able to discover how, in Lamarck's view, the 'will' is able to influence the body's form. His theory makes full use of the contemporary neurophysiological picture of 'subtle fluids' coursing through hollow nerves (see chapter 17). He maintains that when an animal 'wills' a certain action, the nervous juice flows to the appropriate organ and brings about the movement or movements required. Lamarck then proposes that 'numerous repetitions of these organised activities strengthen, stretch, develop and even create the organs necessary to them' (28). And, as we have seen, these modifications are transmitted, it is not clear exactly how, to subsequent generations. It is not clear because Lamarck's view of biology, a word he helped to invent, is profoundly different to our own. Lamarck saw organisms in the light of all the ancient tradition: as *organisms*, escaping in some way the constraints of mere mechanism. Alteration of any part caused sympathetic changes in all other parts of the living unity. So it is not surprising that hypertrophy or atrophy of parts of the adult should affect the quality of the reproductive juices. This was enough to ensure that the offspring bore some trace of their parent's experience. Lamarck's mind was satisfied; process, Heracleitian flux was all, material properties blended, fluids intermingled.

A rather similar position was taken by Charles Darwin's grandfather, Erasmus. He, too, wished to set the sciences of life in the context of the

(26) *Ibid.* (1809), p. 113.
(27) *Ibid.*, p. 122.
(28) *Ibid.*, p. 124.

nascent chemistry of the time. He did not, however, like Lamarck venture far into the budding science of physiology: he did not attempt to account for the living process in terms of electric, magnetic, calorific or other 'subtle, invisible fluids'. His aim was to give an account of physical and chemical forces so that living organisms could be understood to arise 'without parent, by spontaneous birth' from an inorganic environment.

Erasmus Darwin's evolutionary theory is remarkably similar to that of Lamarck. Indeed according to Ernst Krause (29), his biographer, he was the true progenitor of Lamarckism. In the *Zoonomia* we find that Darwin derives his theory from some of Hartley's theological speculations. 'The ingenious Dr. Hartley, he writes (30),

> in his works on men, and some other philosophers, has been of the opinion that our immortal part acquires during this life certain habits of action, or of sentiment, which become forever indissoluble, continuing after death in a future existence; and adds, that if these habits are of the malevolent kind, they must render their possessor miserable even in heaven.

The world, for Hartley, is thus a moral training ground for eternity. In exactly the same way, thinks Darwin, the parental generation forms a training ground for the forthcoming filial generation. 'I would apply this ingenious idea', Darwin continues, 'to the generation, or production, of the embryon or new animal, which partakes so much of the form and propensities of the parent.'

Thus the habits, attitudes, abilities and disabilities of the parents are visited upon the next generation and the generations to come. Erasmus Darwin did not, however, consider evolution an entirely random process. This is a much more modern idea. For him, as for Lamarck and practically all his contemporaries, evolution had a privileged axis, it was orientated towards man. In appropriate environmental conditions:

> Where milder skies protect the nascent brood
> And Earth's warm bosom yields salubrious food
> Each new Descendant with superior powers
> Of sense and motion speeds the transient hours;
> Braves each season, tenants every clime,
> And Nature rises on the wings of Time (31).

The vital optimism of the early nineteenth century is very evident. The vision of nature is unflawed by the doubts which beset Tennyson at the end of the century. The progress is from good to better. The static *scala naturae* of the previous century has been imbued with a new dynamism. From small beginnings, in the fulness of time, came 'Imperious man, who rules the bestial crowd' (32).

(29) E. Krause, *The Life of Erasmus Darwin* (trans. W.S. Dallas), John Murray, London (1879).
(30) E. Darwin, *The Poetical Works,* 3 vols., Johnson, London (1806): *Zoonomia,* xxxix, 1.
(31) E. Darwin, *The Temple of Nature* (1806), II, 31.
(32) *Ibid.,* I, 309.

The origin of the Origin

It fell to Erasmus Darwin's grandson Charles to make the penultimate manoeuvre. We have seen in our account so far how the Mosaic and Miltonic concepts of creation crumbled before the accumulating evidence provided by geology; how the abyss of past time made it possible for the speculative mind to conceive of the progressive modification, indeed transmogrification, of species. Classical ideas about the evolution of life were resuscitated. With Bonnet and Robinet in the eighteenth century this progressive development of organic forms was conceived to be an inherent property of living matter: an indwelling principle which ensured that successive forms became more and more perfect. With Lamarck and Erasmus Darwin we find the firm recognition that the environment plays an important role. It forms one arm of a parallelogram of forces. The other arm is the 'will' or 'effort' of the organism itself. The resultant is the progressive fitting of the organism to its environment. The giraffe, through its centuries-long struggle to reach the succulent leaves growing high in the trees, becomes progressively taller. Charles Darwin, in *The Origin of Species* (1859), carried this movement of thought further. He emphasised the part played by the environment and de-emphasised the 'creative' activity of the organism. Indeed, in the *Origin* he has very little to say about how the variation, the raw material on which the selective forces act, arises. In chapter 1, which is devoted to this topic, we read much about the multifarious forms of variation which have been observed, but as to how it originates Darwin is uncertain:

> No one can say why a peculiarity in different individuals of the same species, or in individuals of different species, is sometimes inherited and sometimes not so; why the child often reverts in certain characters to its grandfather or grandmother or other more remote ancestor; why a peculiarity is often transmitted from one sex to both sexes, or to one sex alone . . . (33).

As Gillispie points out (34), there is an analogy here between Darwin and Isaac Newton. Like his great predecessor Darwin was not interested, in the *Origin* at any rate, in making hypotheses. Bodies gravitated, organisms varied: this was indisputable and all that was required for the theory.

The story of the origin of the *Origin* is well known. Darwin had returned from his circumnavigation in the Beagle and, after a few years in London, retired, because of continuing poor health, to Down in Sussex, and there he worked for two decades on biological subjects, including his notion of the way in which species originated. In 1858, the post brought shattering news: Alfred Russell Wallace, nearly twenty years his junior, had sent him an essay, on which he asked for comment, entitled 'On the tendency of varieties to depart indefinitely from the Original Type'. Darwin's life-work, the opus which he knew to be his major contribution to science, had been anticipated. 'All my originality, whatever it may amount to', he wrote despairingly to

(33) Darwin, *Origin of Species,* p.13.
(34) C.C. Gillispie, *The Edge of Objectivity,* Princeton University Press (1960), p. 318.

Lyell, 'will be smashed...' (35).

In the event, after much tortured heart-searching on Darwin's part ('...it would be dishonourable in me now to publish...I dare say it is all too late, I hardly care about it...It is miserable in me to care at all about priority'), his work, together with Wallace's, was read as a joint paper to the Linnean Society in 1858. Darwin then set about organising his material in a form suitable for publication, and the result of this labour—*The Origin of Species by means of natural selection or the preservation of favoured races in the struggle for life*—was published in November 1859.

The second part of the title tells all. Darwin had perceived that life was war. De Candolle in the eighteenth century had written about the warfare of plants: 'All the plants of a given country are at war with one another...' (36), and Darwin referred to this passage at the beginning of the abstract read to the Linnean Society in 1858. He also referred to Thomas Malthus' *Essay on the Principle of Population*, which contains the famous generalisation that whereas populations increase in geometrical ratio, their food supply increases only in arithmetic ratio, *ergo*, competition. These ideas had germinated in Darwin's mind as had, in fact, the latter in Wallace's. On these concepts, concepts which we may suspect had connections with the prevailing economic *laisser-faire* of the nineteenth century, Darwin rested his case (37). The origin of new species depends, as he says, on 'the struggle for life'. He goes on:

> Owing to this struggle for life, variation, however slight and from whatever cause proceeding, if it be in any degree profitable to an individual of any species, in its infinitely complex relations to other organic beings and to external nature, will tend to the preservation of that individual, and will generally be inherited by its offspring. The offspring, also, will thus have a better chance of surviving, for, of the many individuals of any species which are periodically born, but a small number can survive. I have called this principle, by which each slight variation, if useful, is preserved, by the term of Natural Selection, in order to mark its relation to man's power of selection (38).

In this passage Darwin encapsulates the essence of his theory: an essence, in hindsight, so obvious that T H Huxley mentally kicked himself for not having thought of it himself.

Strangely for us today, Darwin's work was construed by some of the most acute of his contemporaries as resuscitating the case for teleology. Asa Gray, for example, wrote in 1874: 'Let us recognise Darwin's great service to Natural Science in bringing back to it Teleology; so that instead of Morphology versus Teleology, we shall have Morphology wedded to

(35) C. Darwin, *Autobiography and Selected Letters*, p. 196.

(36) A. P. de Candolle, Géographie botanique, in *Dictionaire des Sciences Naturelles*, vol. 18, Strasbourg (1820), pp. 359 - 422.

(37) There has been much learned argument concerning the extent of Darwin's debt to Malthus' gloomy sociology. Sydney Smith (1960), in a close study of Darwin's notebooks, shows that the idea of natural selection was clear in Darwin's mind before he read Malthus' *Essay;* R. M. Young (1969), on the other hand, shows how thoroughly the general intellectual milieu of the 1840s (during which period the first drafts of the *Origin* were written) was permeated with Malthusian thought.

(38) Darwin, *Origin of Species*, p. 61.

Teleology' (39). T H Huxley had said much the same in 1869 (40).

The point seems to have been that the acceptance of evolution theory allowed morphologists to study their subject in a new way. Since the discrediting of *Naturphilosophie*, morphologists had been reduced to mere description; now they were able to perceive that each structure had its adaptive significance in the organism's struggle to survive. But teleology in a more pernicious sense still lurked within the Darwinian system. It lurked because Darwin possessed no valid idea of the cause of variation. In the *Origin*, as we have already noticed, variation was simply taken for granted. We drew a comparison between this phenomenalism and the Newtonian *hypotheses non fingo*. But whereas Newton remained agnostic, Darwin, like the eighteenth- and nineteenth-century physicists, eventually sought an explanation. The physicists, perturbed by the Newtonian implications that 'matter could act where it is not', invented subtle fluids, aethers of various sorts, and their efforts were eventually crowned by Einsteinian relativity theory. If Darwin is to be compared with Newton we can hardly expect him to have been his own Einstein. In the event he was not: this position, to continue the analogy, was filled by August Weismann and Gregor Mendel. Darwin's attempt to account for the inheritance of variation is notoriously Lamarckian. It is the theory of pangenesis.

(39) Darwin, *Autobiography and Selected Letters*, p. 308
(40) T. H. Huxley, The genealogy of animals (a review of Haeckel's Naturliche Schopfungs-Geschichte, *The Academy* (1869), in *Critiques and Addresses*, Macmillan, London (1873), p. 305: '. . . perhaps the most remarkable service to the philosophy of Biology rendered by Mr Darwin is the reconciliation of Teleology and Morphology, and the explanation of the facts of both which his views offer'.

20 THE MECHANISM OF HEREDITY

Pangenesis: a provisional hypothesis

The theory of pangenesis receives its most complete exposition in the book which Darwin wrote to provide the full evidence needed to sustain the argument of *The Origin of Species.* This book—*Animals and Plants under Domestication*—was first published in 1868, and a second edition appeared in 1888. In it we find the following passage:

> How again, can we explain the inherited effects of the use, or disuse of particular organs? The domesticated duck flies less and walks more than the wild duck, and its limb bones have become diminished and increased in a corresponding manner in comparison with those of the wild duck. A horse is trained to certain paces, and the colt inherits similar consensual movements. The domesticated rabbit becomes tame from too close confinement, the dog, intelligent from associating with man; the retriever is taught to fetch and carry; and these mental endowments and bodily powers are all inherited. Nothing in the whole circuit of physiology is more wonderful. How can the use or disuse of a particular limb or of the brain affect a small aggregate of reproductive cells, seated in a distant part of the body, in such a manner that the being developed from these cells inherits the characters of either one or both parents? Even an imperfect answer to this question would be satisfactory (1).

The 'imperfect answer' which Darwin proposes takes the form of a 'provisional hypothesis': pangenesis. He suggests that the cells of all parts of the body continually throw off minute particles or 'gemmules' which ultimately aggregate to form the reproductive cells. In this way the characteristics acquired during an organism's life are faithfully recorded in the germ cells and transmitted to the next generation. 'Hence', as he says, 'it is not the reproductive organs or buds which generate new organisms, but the units of which each individual is composed' (2). The new organism, as his grandfather would have said, is hardly new at all: it is merely a branch or extension of its progenitor.

Pangenesis did not appear to Darwin as a mere appendix to his work. On the contrary, he regarded it as one of his most important contributions. He wrote to Lyell in 1867 with characteristic understatement to say that he was 'inclined to think...it will be a somewhat important step in Biology' (3), and he was delighted when Wallace, who should have known better, echoed his own feelings: 'It is a positive comfort to me to have any feasible explanation

(1) C. Darwin, *Animals and Plants under Domestication,* vol. 2, John Murray, London (second edition, 1888), p. 367.
(2) *Ibid.,* p. 370.
(3) Darwin, *Autobiography and Selected Letters,* p. 281

of a difficulty which has always been haunting me and which I shall never be able to give up till a better one supplies its place, and that I think hardly possible' (4).

We thus see that what Darwin in fact came to believe is rather different from what he is commonly credited with saying. It would be an exaggeration to accept the epigram that, if Erasmus Darwin was the first Lamarckist, Charles Darwin was the last—but not a very great exaggeration; for Darwin's 1868 theory is in some ways Lamarckism shorn of its doubtful philosophising and primitive zoology and completed with the Manchester economics of the struggle for existence. He differs in not hinting what Lamarck hints: the eighteenth-century notion that there is an inbuilt complexifying force within organisms which ensures that generation by generation improvements are made, an inevitable progress towards 'perfection' accomplished. He is clear that the forces causing species to differentiate are selective forces. But he agrees in seeing the twin phenomena of individual response to environmental conditions and the inheritance of the characters consequently acquired as responsible for variation. And without variation, natural selection would be inoperative.

The theoretical significance of pangenesis

Why, then, did Darwin retreat from the clarity of his original position? The answer probably lies in the fact that, in the ten years following the first publication of the *Origin*, a fundamental objection had emerged. Darwin's extensive analyses of the records of stockbreeders, which had done so much to establish the possibility of evolution by natural selection in his mind, had also impressed him with the fact that hybrids tended to exhibit a mixture of the characteristics of both parental strains. In *Animals and Plants under Domestication* we read the following graphic example (5): 'When two commingled breeds exist at first in nearly equal numbers, the whole will sooner or later become intimately blended'. The example he gives is of interbreeding between the individuals of a colony consisting of 'an equal number of black and white men'. In three centuries, he calculates, 'not one hundredth part of the whites would exist'. The great majority would be mulattos blending the features of both races.

Clearly a genetic system with these properties would very rapidly ensure that all merely accidental variation between different individuals would very rapidly be extinguished. For Darwin's own theory to account for the origin of the variegated forms of animate creation, it was essential that some large-scale source of variation should be operating to make up for the continuous loss due to blending inheritance. It was this large-scale source which Darwin believed himself to have described in his 'provisional hypothesis'.

(4) *Ibid.*, p. 282.
(5) Darwin, *Animals and Plants under Domestication,* vol. 2, p. 398.

The germ plasm

It seems clear that if pangenesis, or something like it, is not at work, then Darwin's theory is untenable. This was explicitly recognised by Spencer in *The Principles of Biology* (1864): 'Close contemplation of the facts impresses me more strongly than ever with the two alternatives—either there has been inheritance of acquired characters, or there has been no evolution' (6). And Spencer was a convinced evolutionist.

Darwin too recognised that the nature of heredity was crucial to his theory. But the cases which he studied—the genealogies of sheep, pigs and cattle, of horses and dogs—tended to be too complicated for him to achieve a genuine understanding. Apart from his conviction that acquired characters were inherited, he also concluded that diet and environment were involved in the production of variation.

Blending inheritance, pangenesis, the environmental and nutritional production of variability: as August Weismann puts it at the beginning of his *Evolution Theory*, 'The *How*? of evolution is still doubtful, but not the *fact*.... The world of life, as we know it, has been evolved and did not originate all at once'. And in fact it was Weismann himself who did as much as anyone towards the end of the nineteenth century to establish the Darwinian theory on the firm foundations which we believe it to possess today. The essence of Weismann's contribution lay in the sharp distinction he drew between germ plasm and somatoplasm, between, to use a later terminology, genotype and phenotype.

Weismann's early researches on Daphnidean embryology had shown that one of the first events in the development of this organism was the separation of the germinal cells from the body, or somatic, cells. Weismann was inclined to sharpen this distinction and to apply it generally to all multicellular organisms. His distinction between reproductive and somatic cells prefigures the much later distinction between nucleic acid and protein.

In *The Evolution Theory*, published in 1902 but in fact based on lectures which he had delivered at Freiburg from 1880 onwards, he narrows down his definition of germ plasm from reproductive cells, to their nuclei, to nuclear chromatin, and finally to chromosomes: 'I have endeavoured to prove that the germ substance proper must be looked for in the chromatin of the nucleus of the germ cell, and more precisely in those ids or chromosomes which we conceive of as containing the primary constituents [*Anlagen*] of a complete organism' (7).

Weismann's insistence on a real material basis for the phenomena of heredity was salutary. Darwin, as we have seen, also proposed a material basis in the form of pangenes or gemmules. But when Galton showed by the simple procedure of interchanging the blood of differently coloured rabbits that the gemmules were undetectable, Darwin had not been particularly

(6) H. Spencer, *The Principles of Biology,* vol. 1, Williams and Norgate, London (1864), p. 621.

(7) Weismann, A. *The Evolution Theory,* trans. J.A. and M.A. Thomson, Edward Arnold Ltd., London, vol. 1, p. 345.

perturbed (8). Darwin's noncommittal reaction, however, suggested that his 'provisional hypothesis' was unfalsifiable and hence unscientific. Weismann, on the other hand, elaborated his 'material basis' in ways which were in some cases prophetic. The elementary units, or biophores, he conceived to be bundles of organic molecules. The biophores were assembled progressively into 'determinants' and the latter into 'ids'. The 'ids' were grouped linearly into 'idants', his term for the newly discovered chromosomes. We seem to be breathing the same atmosphere as that to which we have become accustomed since the Watson–Crick breakthrough of 1953. Weismann's model, moreover, allowed him to predict the necessary existence of a reduction division in the meioses leading to the production of germ cells.

But even more important than his insistence on a material substratum was his insistence that it was the germ cells alone, isolated from the rich metabolism of the soma, which were responsible for hereditary transmission. The poor eyesight which frustrated his experimental ambition forced a redirection of his energies into theoretical polemic. He, more than any other worker in the later nineteenth century, maintained the original purity of the Darwinian insight (9). He was convinced that the vicissitudes of the soma in no way affected the germ plasm. He was convinced that there was no evidence that acquired characters were ever inherited.

Yet, although Weismann had focused attention on the germ plasm isolated, like the Stoic's soul, from the buffetings of the environment, he was still obsessed with the classical examples of evolutionary change: the gradual disappearance of disused organs, the gradual enhancement of useful characters. Indeed he conceived that the variation in the germ plasm, upon which evolutionary change depended, was brought about by minute fluctuations in the quality and quantity of the 'food stream' reaching the reproductive cells. Hence the structures determined by the 'determinants' might wax or wane, hypertrophy or atrophy, as the generations passed. In the event it required a radically different orientation to break the hereditary cipher. Instead of studying variation *per se*, instead of pondering the progressive change of an animal or plant species in geological time, it was necessary to concentrate on stability amid the variability, on the continuing reappearance of an unchanging feature. And this, of course, is precisely what occupied Gregor Mendel during his eight years of experimental breeding.

(8) Dawin, *Animals and Plants under Domestication*, vol. 2, p. 398.
(9) See, for example, H. Spencer, *ibid.*, (second edition, 1898), vol. 1, p. 559: '. . . concerning the fact of evolution there is agreement, concerning its causes there is disagreement. The cause which Mr Darwin first made conspicuous has come to be regarded by some as the sole cause; while, on the part of others there has been a growing recognition of the cause which he at first disregarded but afterwards admitted. Professor Weismann and his supporters contend that natural selection suffices to explain everything. Contrariwise, among the many who recognise the inheritance of functionally-produced changes, there are a few, like the Rev. Prof. Henslow, who regard it as the sole factor'.

Mendel's analysis

Mendel published his paper 'Versuche ueber Pflanzenhybriden' (Experiments in plant hybridisation) in 1866, seven years after *The Origin of Species* and two years before *Animals and Plants under Domestication*. Yet for over a third of a century its significance was not understood. Moreover this does not seem to be, as is often suggested, because it was published in a particularly obscure journal. The *Proceedings of the Brunn (Brno) Natural History Society* were sent to over 120 libraries, including those of the Linnean and Royal Societies in London. Mendel received forty reprints of his paper and sent them to several prominent biologists. The learned world seems, as suggested in the previous section, to have been unready to accept the Mendelian analysis; it was so orientated that the importance of Mendel's discoveries was not understood.

However, by the turn of the century things had changed. Several investigators were unwittingly working along lines similar to those pursued by the now dead Mendel. In 1900 three scientists, Correns, de Vries and Tschermak, arrived at results analogous to those which, they found to their surprise and perhaps disappointment, had been obtained by Mendel in the 1860s. As Correns writes:

> I thought I had discovered *something new*. But then I convinced myself that the Abbot GREGOR MENDEL in Brunn, had, during the sixties, not only obtained the same result through extensive experiments with peas, which had lasted for many years, as did DE VRIES and I, but had also given exactly the same explanation, as far as that was possible in 1866 (10).

Mendel's results are nowadays well known to all biologists. They do indeed show the extraordinary constancy of inherited characteristics. Mendel, indeed, emphasises that 'transitional forms were not observed in any experiment' (11). His pea plants were *either* tall *or* dwarf, the pods *either* spherical *or* wrinkled, the seed coats *either* white *or* greyish-brown, and so on. These features appeared unchanged generation after generation, and, moreover, the different characters were inherited quite independently of each other. In place of blending, modification, fusion, we have a far more atomic conception. In place of progressive amplification or diminution of a characteristic we have constancy. The continuum is broken; the organism is seen to consist of a mosaic of independently governed features which may be independently selected for or against. Mendel's factors, discovered by statistical procedures, naturally lent themselves to mathematical analysis.

Nevertheless, Mendelism remained unappreciated until 1900; and by the turn of the century cytologists had progressed far in the analysis of the nuclear events attending cell division. The chromosomes had been described

(10) C. Correns, G. Mendels Regel über das Verhalten der Nachkommenschaft der Rassenbastarde, *Berichte der Deutsche Botanische Gellschaft,* **18** (1900), 158–68; trans. trans. L. K. Pitternick in *Genetics,* **35** (1950), suppl. 33–41.

(11) G. Mendel, Versuche ueber Pflanzenhybriden, *Verh. naturforsch. Verein Brunn,* **4** (1866), 3–47; trans. Royal Horticultural Society of London (1901) and reprinted in *Experiments in Plant Hybridisation,* Oliver and Boyd, London (1965).

by Strasburger in 1880 and subsequently named as such by Waldeyer. Their movements during cell division, both mitotic and meiotic, had been described. It remained only for Bateson and then, more definitely, for Sutton (1903) to recognise the analogy between these movements and the movements of Mendel's hypothetical factors for the science of cytogenetics to be born (12).

The development of cytogenetics has since been rapid and continuous. In 1910 T H Morgan and his collaborators at Columbia University began to publish a series of papers on the genetics of the fruit fly *Drosophila melanogaster*. It soon became clear that the Mendelian factors, renamed genes by Johannsen in 1909, were indeed carried on the chromosomes. The point was finally established by the investigation of sex-linked mutants by Bridges in 1916 (13). Subsequent work in cytogenetics has largely consisted in an ever more precise localisation and characterisation of the gene. In 1953 Watson and Crick were able to describe its physicochemical constitution, and in 1969 a group of genes was isolated. After 1953, as Watson points out (14), 'the gene was no longer a mysterious entity whose behaviour could be investigated only by breeding experiments. Instead it quickly became a real molecular object...'. Thus, in the second half of the twentieth century, we may return affirmative answers to both of Weismann's questions: not only the fact of evolution, but also the 'How?' of evolution has been established.

Darwinian objectivity

In the wide perspective of this book the triumph of Mendelian genetics, not much earlier than 1920, marks a very decisive moment. Another of the great arguments which had enlivened the biological world seemed to have come to an end. To Isaac Newton's query, 'How came the bodies of animals to be contrived with so much art...?' there was now an answer: an answer which involved no mysterious hidden force wrapped within an organism's body, or produced by divine fiat, but only the commonplace forces with which we are from day to day familiar. Erasmus Darwin and Lamarck had argued for the doctrine of descent, and Charles Darwin had not only amassed the evidence which convinced the world that descent had indeed occurred but had also suggested a plausible mechanism. August Weismann had vigorously argued the case against the inheritance of acquired characters, and Mendel had shown how characteristics were in fact inherited. Morgan and his colleagues had shown how Mendelism perfectly accounted for the complicated genetics of the fruit fly. Subsequent geneticists had extended the analysis to other organisms, including man. The answer to Newton's question was plain. The Mendelian factors, the genes, occasionally mutated. The discovery of the

(12) W. S. Sutton, The chromosomes in heredity, *Biol. Bull.*, 4 (1903), 231 - 51.
(13) C.B. Bridges, Non-disjunction as proof of the chromosome theory of heredity, *Genetics,* 1 (1916), 1 - 52, 107 - 63.
(14) J.D. Watson, *Molecular Biology of the Gene,* Benjamin, New York (1965), p. 66.

molecular structure of DNA allows us to understand the causes and nature of this mutation. Weismann's theoretical speculations do not seem to be too far out. The gene *is* a complicated molecule; the nucleotide bases of which it is built *are* liable to alter in response to external influences—radiation or chemical. These mutations have, however, no necessary connection with the organism's situation in life. The phenotypic changes which result are, so far as is known, entirely random. On this random variability in a population, natural selection acts. The 'fittest' tend to survive (it is a tautology) and consequently perpetuate their characteristics in succeeding generations. The 'unfit' are eliminated. In this way environmental forces ensure that organisms are closely adapted to the habitats in which they customarily find themselves and for the types of life they habitually live. The evolutionary process can be seen to be as 'natural' as any other physical process. No mysterious inbuilt complexifying force is involved. no *élan vital*, no privileged axis directed towards a *point omègu*. It is every bit as natural as, for example, the second law of thermodynamics which, it will be remembered, says that natural processes tend to a maximum 'mixed-upness'.

Once the evolutionary doctrine is accepted it follows, as Darwin and his contemporaries very clearly saw, that man can no longer be regarded as standing over against nature, as theologians had customarily taught (see p.130), but must be seen as intimately bound up with nature, as, indeed, part of nature. Thus the evolutionary theory of the mid-twentieth century also answers another of the philosophical questions which had troubled earlier scientists. In the sixteenth century Simplicius teased Galileo with the following conundrum: 'You account the understanding the chief distinction of man, who is made by Nature...therefore you must say that neither did Nature understand how to make an understanding that understands' (15). But so long as 'understanding' is defined in behaviouristic terms, the evolutionary theory can do just this. It can, moreover, plausibly explain the origin of the Kantian categories, the filters through which, according to his philosophy, we must needs 'see' the world. For it would be surprising if, during the thousand million years of animal evolution, the sensory apparatus had not become adapted to filter out all but the relevant aspects of the impinging environmental energy.

But the Darwinian viewpoint can do more even than this. As the theory has grown more subtle, more sophisticated, it has become clear that the Malthusian analysis is not always correct. Not always does a population expand to the uttermost limits of its food supply, but instead the selective process acts on the breeding system, or even on the group, to maintain optimum conditions. This, Darlington and others suggest, accounts for several features of human societies and social psychology. It explains, for example, the careful mechanisms, the totems and taboos, which ensure outbreeding in early human societies. Many primitives have very complicated

(15) Galileo, *Great World Systems* (trans. T. Salusbury) (ed. G. de Santillana), p. 114.

mating dos and don'ts, and spend much time and effort in determining family connections and interrelationships. The incest taboo still remains deeply rooted in the human psyche. Selective forces can also plausibly account for the prevalence of 'initiation ceremonies' and other sexually-orientated rituals (16). Thus twentieth-century biology allows us to 'see' the activities of primitives from an entirely different stance to that adopted in the early chapters of this book. Instead of sympathetically appreciating their experience from, as it were, the 'inside', we are now observing their behaviour 'objectively', from the 'outside'. Those societies which have developed adequate outbreeding systems survive in the struggle for existence, while those less well adapted become extinct. A similar 'objective' analysis may, of course, be extended to account for certain features of present-day so-called civilised societies. Ritual and taboo are, after all, not the sole prerogative of the primitive. The Darwinian analysis may thus be applied to account for the origin and survival of ethical systems and aesthetic experience.

And here, perhaps, we can see as clearly as anywhere the 'deep structure' of our history. In chapter 7 we noticed how the Socratic endeavour, which shifted philosophical interest from the macrocosm to the microcosm, was an endeavour to establish immutable 'forms' of the good, the just, of wisdom and virtue, analogous to the eternal forms of mathematics. The social significance of this endeavour, we suggested, lay in an attempt to counter the ethical relativities generated by the civil strife of the time. Plato was able to establish on the moral certitude thus obtained the structure of his ideal, unified, state. We saw, also, how the concomitant teleological metaphysics helped Aristotle buttress his 'biological' world view. But here, with the triumph of Darwinism, this world view is finally demolished. With the triumph of Darwinism the essence of Thrasymachus' argument, so summarily dispatched by Socrates in the *Republic*, is resurrected. The exponents of Darwinism would maintain that ethical systems, like all other biological phenomena, have selective value. The value in this case is social cohesion. Thrasymachus argued, as we saw in chapter 7, that 'justice is nothing more than that which is advantageous to the stronger'. How very similar is the implication of modern Darwinism. There is no absolute moral code. Morality is valuable only in so far as it contributes to the survival of the group. How very similar, too, is the social background which has allowed this modern version of Thrasymachus' argument to gain the ascendancy. Karl Marx referred to Malthus as 'the mountebank-parson': the reason for his animus is not difficult to understand.

(16) C.D. Darlington, Genetics and society, *Past and Present,* **43** (1969), 3–37: 'the rules for breeding . . . give a coherence to all primitive communities. They also give them an interest in kinship and descent which extends irrationally to the most advanced peoples today where it still supports the institutions of aristocracy and monarchy.' See also Darlington's more extensive survey, *The Evolution of Man and Society,* Allen and Unwin, London (1969).

21 A CLOCKWORK EMBRYO?

The absurdity of generatio aequivoca

In chapter 18 we noticed that Immanuel Kant had stumbled across an idea which belonged by rights to the nineteenth century. He believed that by a 'daring venture of reason' he could, perhaps, detect a mechanism—evolution, no less—through which the manifold forms of life originated. In the same chapter of the *Critique of Judgment* he insists, however, that such a hypothesis could never account for th development of an embryo. The 'generatio aequivoca', he wrote 'by which is understood the production of an organised being through the mechanics of crude unorganised matter', is an impossibility.

Kant was not alone in this opinion. In spite of the early success of the scientific-technological revolution and the consequent spread of the mechanical hypothesis throughout science, the development of the embryo still seemed something quite different. Mechanistic accounts of morphogenesis seemed, especially to biologists, next to meaningless. This repugnance is well expressed by de Maupertuis in the following passage from Système de la Nature (1756):

> An attraction uniform and blind, diffused in all matter, could not serve to explain how those particles became arranged to form even the simplest body. If all have the same tendency, the same force to unite with one another, why are these going to form an eye and these an ear? Why this marvellous action? And why don't they unite pell-mell? One will never explain the formation of any organised body solely through the physical properties of matter: and from the time of Epicurus to that of Descartes one has only to read the writings of all the philosophers who have attempted it to be persuaded of the fact (1).

The principle which de Maupertuis expresses in the passage above seemed, until very recently, applicable to all physicalist 'explanations' of embryonic development. The profound absurdity of a 'clockwork' embryo formed the core, for example, of Driesch's argument in his famous Gifford Lectures: ***The Science and Philosophy of Organism*** (1907, 1908). We shall return to Driesch's argument and the solution he proposed in a later section of this chapter.

(1) P.L.M. de Maupertuis, *Système de la Nature, Oeuvres,* vol. 14, Lyon (1756); quoted by B. Glass, *Forerunners of Darwin,* The Johns Hopkins Press, Baltimore (1959), p. 44.

William Harvey's theory of epigenesis

Before discussing twentieth-century developments, however, let us look briefly at the types of explanations advanced in the more distant past. We have already noticed that while Aristotelianism prevailed, embryology posed no sharp question. As we have seen, the Stagirite was inclined to regard the whole world as a giant embryo; the explanations which he proposed for the macrocosm, ludicrous as they appear to present-day and indeed to Renaissance physicists, accounted well enough for the development of the microcosm. Once the paradigmal revolution was accomplished, however, embryological phenomena passed from being the standard by which all was to be explained to being the most powerful exception to prove (in the sense of test) the new paradigmal rules. The reader will recall from chapter 14 that William Harvey, the perhaps unwitting instigator of a mechanistic physiology, remained profoundly Aristotelian in the field of embryology:

> Nor are they lesse deceived who make all things out of Atomes, as Democritus, or out of elements, as Empedocles. As if (forsooth) Generation were nothing in the world, but a mere separation, or Collection, or Order of things. I do not indeed deny that to the Production of one thing out of another, these forementioned things are requisite, but Generation her self is a thing quite distinct from them all. (I finde Aristotle in this opinion) and I myself intend to clear it anon, that out of the same White of Egge (which all men confesse to be a similar body, and without diversity of parts) all and every of the part of the chicken whether they be the Bones, Clawes, Feather, Flesh, or what ever else, are procreated and fed. Besides, they that argue thus assigning only a material cause, deducing the causes of Natural things from an involuntary or causal concurrence of the Elements, or from the several dispositions or contriving of Atomes; they do not reach that which is chiefly concerned in the operations of Nature, and in the Generation and Nutrition of animals, namely the Divine Agent, and God of Nature, whose operations are guided with the highest Artifice, Providence and Wisdom, and doe all tend to some certaine end, and are all produced for some certaine good (2).

In spite of this conviction that teleological reasoning is fundamental to an understanding of embryology, Harvey is usually held to have initiated the modern phase in the study of development. ***Exercitationes de Generatione Animalium*** contains a mass of detailed observations on the development of a wide variety of animals; not only mammals, not only vertebrates, but the whole animal kingdom provided organisms for the Harveyan analysis. His exact account of the development of the chick embryo remains, according to Needham (3), unrivalled to the present day. His perception that all living forms derived originally from eggs, ***omne vivum ex ovo***, although of necessity theoretical in his time (for want of the compound miscroscope) is a seventeenth-century anticipation of Virchow's equally famous nineteenth-century declaration: ***omnis cellula e cellula***. With the exception of a difficult period at the end of the seventeenth and the beginning of the eighteenth

(2) W. Harvey, *On the generation of Animals* (1653), p. 51.
(3) J. Needham and A. Hughes, *A History of Embryology*, Cambridge University Press (second edition, 1959), p. 136.

centuries, the progress of embryological research initiated by the publication of *De Generatione Animalium* in 1651 has been continuous.

The preformation theory

William Harvey's approach to the phenomena of embryology was that of an epigeneticist. The organs of the adult cock or hen are progressively formed from, or emerge from, an originally undifferentiated, homogeneous 'White of Egge'. But this was not the opinion of many of his immediate successors. Harvey, as we have already mentioned, carried out his researches without the aid of the compound microscope. Early examples of such instruments were, however, beginning to become available in the second half of the seventeenth century, and embryologists were quick to take advantage of them. Their observations started a controversy which was to last nearly a century.

The quest for an understanding of how organisms reproduce themselves is, of course, of great antiquity. Exactly how does a new individual arise from the union of his parents? Is it the case, as Aristotle and his scholastic successors maintained, that the male semen induced form on the undifferentiated matter of the female? Or did the mother, as Hippocrates, Empedocles and Democritus maintained, contribute equally with the father in the formation of their offspring?

Throughout the revival of anatomy in the fifteenth and sixteenth centuries, the reproductive system had been investigated with acute interest. Leonardo's researches (figure 14.3) are nowadays well known; Fallopius not only gave his own name for the Fallopian tubes but also introduced the terms vagina and placenta (*Anatomicae Observationis*, 1561); Fabricius, Harvey's teacher, wrote on the foetus and also on the development of the eggs of birds. Indeed, Harvey is believed to have derived some of the impetus for his own investigations from these works. It is perhaps significant that the illustration which forms the frontispiece of *De Generatione Animalium* bears the legend *omnia ex ovo*.

This motto in fact shows considerable courage. For in 1651 the human egg had not been observed. The ovarian follicles (long mistaken for ova) were first reported by Steno and de Graaf in 1672. Four years later, in 1676, Leeuwenhoek reported his observations of 'vast numbers of living animalculi' in human seminal fluid to the Royal Society. The name 'spermatozoa' was only coined years later, in 1826, by von Baer.

These achievements of the new science of microscopy raised profound and difficult questions. Although the actual fertilisation of an ovum by a spermatozoan was not observed until 1875 (Hertwig), it was quickly realised after the communications of de Graaf and Leeuwenhoek that such an event must lie at the root of sexual reproduction. But what then became of the Aristotelian and Scholastic notions of conception?

When and how did the soul enter to shape the developing embryo? If the semen consisted of teeming myriads of animalculi, did each and every one carry a soul? And if so, whence did they obtain so precious a cargo? Were they perhaps ensouled in the original act of creation? And had they remained awaiting their chance of life in the macrocosm ever since? Or perhaps it was the ovum, after all, which contained the vital principle. But if so, the same perplexities still obtained.

These conundrums were sharpened by microscopists interested in entomology. Swammerdam in the *Biblia Naturae*, completed in 1669 although not published until 1737, convinced himself that the perfect insect lay fully formed within the skin of the nymph and chrysalis. 'In reality', he writes (4), 'the caterpillar, or worm, is not changed into a nymph or chrysalis; nor, to go a step further, the nymph or chrysalis into a winged animal...', but the same animal, casting its skin, becomes first nymph or chrysalis and then imago. And, he goes on, it is the same with chicks and tadpoles: the adult is formed by the 'expansion of the parts already formed'.

Thus we see emerging from this strange amalgam of theology, entomology and early microscopy the very influential group of ideas summed up in the terms preformation, preexistence, *emboîtement*. So strong a hold did these ideas achieve that more than one microscopist believed himself able to see the unborn man, complete but miniaturised within the spermatozoon (figure 21.1). Others (ovulists rather than animalculists) maintained that miniscule people were to be observed within the follicles. How useful an example this is for those who suggest that we are inclined to force our presuppositions and prejudices upon the facts! Indeed Bonnet, one of the strongest adherents of the theory, openly regarded it as 'une des plus belles victoires que l'entendement pur ait remporté sur les sens': one of the greatest victories which the understanding has achieved over the senses (5). In this we can perhaps hear an echo of Galileo's praise of the heliocentric astronomers Aristarchus and Copernicus (6).

It is clear that the concept of preformation is closely related to that of *emboîtement*. We have already seen how Swammerdam's researches into insect nymphs, caterpillars and chrysales seemed to support the latter concept. Towards the end of the seventeenth century Malebranche, in the second chapter of *La Recherche de la Vérité* (1672), gives an excellent account of both preformation and *emboîtement*: 'a single apple pip', he writes 'contains all the apples and all the apple trees of an infinity or near infinity of centuries' (7). Leibniz, at the beginning of the next century, published much the same idea (8). For some, the first female of each species

(4) Quoted in T.S. Hall (ed.), *A Source Book in Animal Biology*, McGraw-Hill, New York (1951).
(5) C. Bonnet, *La Palingénésie Philosophique*, Geneva (1769).
(6) Galileo, *Dialogue on the Great World Systems:* 'I cannot find any bounds for my admiration, how that reason was able in Aristarchus and Copernicus, to commit such a rape upon the senses, as despite thereof to make herself mistress of their credulity.'
(7) Le P.N. Malebranche, *Recherche de la Vérité*, Paris (1672), p. 74.
(8) G.W. von Leibniz, *Monadology*, Leipzig (1714), paras. 74–6.

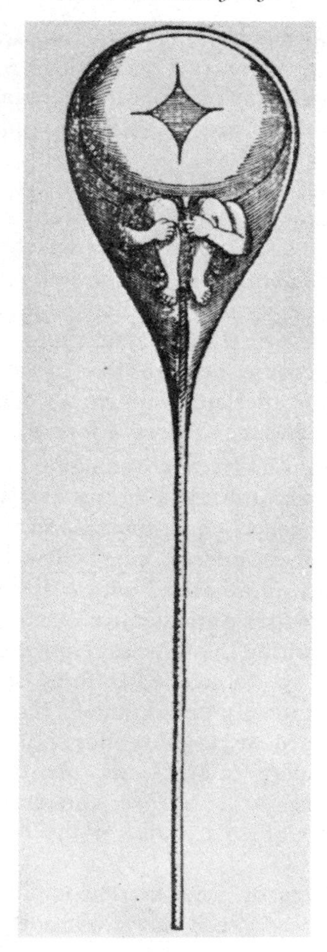

Figure 21.1. Drawing of a spermatic homunculus published by Hartsoeker in 1694.

contained within herself, Chinese-box fashion, all the subsequent individuals of that species. 'I would say', writes Malebranche 'that the females of the first animals were, perhaps, created with all those of the same species to which they have given birth and to which they will give birth until the end of time' (9). In this way, for the theologically inclined, was the doctrine of original sin justified (10).

The influence of the great names of Malebranche and Leibniz may have been responsible for the wide currency which *emboîtement* and preformation achieved. For, on reflection, the concepts contained rather serious internal

(9) Malebranche, *ibid.*, p.5.

(10) For instance, Swammerdam in *Miraculum Naturae* (1672): '. . . there is no place in

difficulties. At the level of theology, where preformation had seemed so valuable in saving the creative initiative of the deity, a considerable problem appeared. If each animalculus, each ovum, which the microscopist observed contained in fact a perfect preformed man or woman merely awaiting engrossment in order to take part in the affairs of the great world, what became of the countless animalculi and ova which failed to find each other? Were the tiny men and women doomed to die without ever being born? This seemed to some in the optimistic times at the beginning of the eighteenth century altogether too cruel a fate. Fantastic theories circulated which envisaged supernumerary ova and spermatozoa escaping into the atmosphere, there to hibernate until called forth once more to take their chance in the competition for a place on earth (11).

Other difficulties, of a more mundane variety, also became apparent. How, with the theory of preformation and *emboîtement*, could one account for hybrids? How, indeed, could one even account for the obvious fact that offspring resemble both parents? Perrault invoked the maternal imagination, suggesting that this could in some way affect the flow of nourishment to the enlarging homunculus (12). This explanation gained wide acceptance at the end of the seventeenth century.

Another obstacle to the smooth progress of the preformation theory was provided by the facts of regeneration. If a lizard's tail is amputated, it grows again, if a crab loses its leg, it is regenerated; if the zoophytes which Trembley had rediscovered in 1740 were transected, two new polyps developed. How could the theory account for these observations? Were there perhaps preformed germs so small as to be invisible distributed all over an animal's body: in the lizard's tail, the crab's legs, the hydroid's stem? Or was there perhaps a divisible soul paralleling the divisible body? Réamur (1741) toyed with the idea but in the end concluded that the problem was insoluble (13).

A third difficulty was presented by the birth of monsters. We noticed in chapter 8 that Aristotle was concerned with just this problem as long ago as the fourth century B.C. His solution, in essence, was to say that the hand of the potter sometimes slips. But how can the preformationist account for

the nature of things for the generation of parts but only for their propagation or growth . . . . If this be true, and of this I have no doubt, it is very easy to explain . . . in what way Levi is said to have paid his tithes long before he was born . . . and . . . in the judgment of a very learned man, to whom our efforts and experiments are immediately accesible, even the foundation of original sin itself would have been discovered, since the whole of mankind would have been concealed in the loins of Adam and Eve, and to this could be added as a necessary consequence that when these eggs have been exhausted the end of mankind will be at hand.' Quoted in H.B. Adelmann, *Marcello Malpighi and the Evolution of Embryology*, Cornell University Press, Ithaca, N.Y. (1966), vol. 2, p. 908.

(11) See, for example, J. Cooke, *The New Theory of Generation*, London (1762).

(12) C. Perrault, *De la Méchanique des Animaux*, Paris (1680).

(13) See J. Roger, *Les Sciences de la Vie dans la Pensée Française du XVIIIe Siècle*, Colin, Paris (1963), p. 395ff.

monstrous births? Is it, perhaps, an accident of the enlargement process or, perhaps, the consequence of disease? But this affronts belief in the deity's omnipotence, a belief which provided much of the psychological force of the preformation theory itself.

The theory nevertheless held sway for nearly a century. From the discovery of spermatozoa in 1672 until the sixties of the next century it dominated the minds of embryologists. Its heuristic merits were small. Viewing development as a mere engrossment of what was already there, biologists tended to look elsewhere for their research problems. In consequence rather little advance was made in embryology during this period. Biologists were inclined, as Needham points out (14), to assume that the structure and organisation of the adult were already present in the embryo although contemporary microscopes might not be able to detect it. It is not difficult to see, however, that the theory chimed well not only with predestinarian schools of theology but also with the prevailing mechanising tendencies of those years. A clockwork embryo may appear absurd to us today, as it did to Harvey and the Aristotelians. But if embryology merely consists in the scaling-up of an originally miniaturised machine, then the absurdity is somewhat less obvious. There would, in fact, to use the Kantian terminology, be no *generatio aequivoca* because the embryo is just as highly organised as the adult from the very beginning. The design and creation of the clockwork could still be allowed the creator; the working of the machinery during historic time could still be allowed the mechanical philosophers.

Unfortunately for this neat scheme the facts, as we have seen in other instances, proved recalcitrant. They just could not be forced into the theory's framework. In chapter 16, where we looked at the initial post-Cartesian development of physiology, we saw that towards the middle of the eighteenth century the mechanical philosophy began to decline. This withdrawal of the mechanising tide seems to have occurred throughout the realm of the sciences. In its place developed the Romantic philosophies of nature which were outlined in chapter 18. Nowhere was the ebbing of the tide more apparent than in the world of embryology.In place of the concept of inert matter blindly obeying the clearly understood laws of mechanics, there developed once again the idea that matter contained a plastic, vital, even divine, principle continuously at work. This reaction in turn eventually overreached itself in the speculation of the *Naturphilosophie* movement of the early nineteenth century (chapter 18). But, as Roger puts it (15), the image of life provided by the mechanical philosophy of the eighteenth century was deeply unbiological; the clarity which it was intended to introduce proved illusory; the biologist was turned away from the study

(14) J. Needham and A. Hughes, *ibid.*, p. 238.
(15) Roger, *ibid.*, p. 452.

proper to his subject; it was necessary for science to learn once more the virtues of doubt, humility and patient observation.

Nineteenth-century controversies

The preformation theory found one final champion before it began to decline. Albrecht von Haller was the most prominent physiologist of his generation. His *Elementa Physiologiae Corporis Humani* was, as mentioned in chapter 17, the first modern textbook on the subject. His influence was thus profound. Yet in the last volume of this work, published in 1766, we read, 'Epigenesis omnino impossibilis est': epigenesis is absolutely impossible.

It is arguable that von Haller believed in preformation with especial tenacity because he was a convert from the other camp. In section XIII of his important work ***Sur la Formation du Coeur dans le Poulet*** (1758), he writes:

> I have in my works shown well enough that I leaned towards epigenesis and regarded it as the point of view most fully in accord with experience, but these matters are so difficult and my experiments on the egg have been so numerous that it is with less repugnance that I propound the opposite opinion, which is beginning to appear to me the more probable. The chick has furnished me with arguments in support of unfolding which I believe I ought to submit to the judgment of the reader (16).

The only reason, he goes on, why the parts of the adult cannot be seen in the embryo is that they are too fluid and transparent. But they are certainly there. 'One does not see the wind'.

In 1759, only one year after Haller had brought out his work on the development of the chick's heart, Caspar Freidrich Wolff published his *Theoria Generationis.* Although Haller gave Wolff's work a generous review, it is nowadays seen as the publication which more than any other destroyed the preformationist position. Wolff's work exhibits a curious mixture of tendencies: couched in a metaphysical form was an essay which looked back to Aristotle and Harvey and away from the mechanical philosophy of the early eighteenth century. Yet Wolff sought to establish these more ancient ideas by valid scientific procedures, most importantly by microscopic examination of the chick embryo. In contrast his preformationist opponents, notably Albrecht von Haller, adopted the almost scholastic position of insisting that if Wolff could not see the miniaturised chicken in the egg this was due to nothing more than the low resolving power of contemporary microscopes. And because Haller's authority was far greater than Wolff's, the preformationist view for a time prevailed. Thus the full recognition of the value of Wolff's work, especially that of his later treatise *De Formatione Intestorum* (1768), did not occur until after his death in 1794.

The kernel of Wolff's argument against the preformationists may be summarised in the following way. If, he said, the embryo is indeed preformed

(16) Quoted in Adelmann, *ibid.*, p. 878.

within the egg but invisible because too small then, when it does first become visible, it should appear already fully formed, perfect. This was evidently open to experimental investigation. Wolff accordingly looked at the development of the blood vessels in the chick embryo. He found that they developed by the coalescence of a number of primordially independent 'blood islands'. Even more convincingly he showed in *De Formatione Intestorum* that the chick's intestine, far from first appearing fully formed, developed by the rolling up of a sheet of tissue on the ventral surface of the embryo. Wolff generalised these and other observations and argued that, far from being preformed, all the organs of the adult body developed by a similar epigenesis.

At this distance in time it can be seen that the theory which Haller derided was, far from being backward looking, destined to form the basis of the modern subject. Furthermore, and somewhat paradoxically, it was also to form a significant source for the Romantic biology of the nature-philosophers in the early nineteenth century (17). Its philosophical orientation contradicted the view held by many preformationists that the body was some kind of machine wound up at creation and left to tick away through the generations until Armageddon. The machine concept in the eighteenth century was far narrower, far less rich, than the concept which has emerged from twentieth-century technology. In the eighteenth century it was inadequate, as we have several times remarked, to serve as an explanatory paradigm for the living process.

As embryology entered the nineteenth century it finally became inescapable that, whatever notions philosophers might like to hold, the facts of the matter were that structure just simply did originate in apparently structureless material. In the middle of the eighteenth century Haller observing a two-day-old chick embryo, and seeing an already pulsating heart may have felt himself justified in championing preformation. For, as we have seen throughout our history, one of the most consistently held general ideas has always been that 'nothing can come of nothing', 'ex nihilo nihil fit'. But a century of observation, a century of steady improvement in microscopical technique, forced embryologists to abandon this seemingly self-evident position. The derided Aristotelian ideas of the pre-microscopical Harvey who, with the naked eye, or perhaps with a simple hand lens, had observed the chick's anatomy to materialise from the formless matter of the egg, seemed, after all, unavoidable. Preformation, the *emboîtement* principle, received a final *coup de grâce* at the hands of Karl Ernst von Baer in the early nineteenth century. The first volume of his major work, *Ueber Entwicklungsgeschichte der Thiere*, was published in 1828 and the second volume in 1837.

(17) Lorenz Oken, for example, in the *Lehrbuch der Naturphilosophie* (preface), mentions Wolff's work on the intestines as having anticipated some of his own theory.

Von Baer's friend and colleague, H C Pander, had published in 1817 a monograph on the development of the chick in which he had shown that one of the earliest events was the differentiation of the blastoderm into three layers: nowadays called endoderm, maesoderm and ectoderm (18). Von Baer generalised his friend's germ layer theory so that it illuminated the development of the vertebrates at large. So comprehensive was von Baer's treatise that it is usually regarded as forming the foundation of classical embryology. The germ layer theory has formed the basis of comparative embryology from that day to this.

But von Baer's analysis went further. He emphasised that development always proceeds from the general to the particular. The chick begins simply as a vertebrate, then becomes an air-breathing vertebrate, then a bird, then a terrestrial bird, then a gallinaceous bird and finally a domestic chicken (19). Clearly we are in the near vicinity of the group of ideas which Haeckel later so forcibly expressed in the form of his recapitulation theory. Von Baer, however, was not a Haeckelian, for he never accepted the evolution theory. From the point of view of this chapter, however, von Baer's 'biogenetic law', whereby the embryo 'differentiates' from a generalised beginning, is of considerable significance; for it clearly contravenes any belief in preformation, any belief that a perfectly differentiated homunculus lies latent in the egg (or sperm) only awaiting the appropriate stimulus for enlargement.

But, as is so often the case in the history of science, the victory for the epigeneticists' cause was not final. In an earlier chapter we noticed how the Aristotelian technique for solving a problem seems to have been by way of a number of ranging shots, some passing over the target, some falling short. Gradually the ranging improves, the shots falling ever closer on each side of the target until hits begin to be scored. Something similar may be said about the approach to a solution of a major scientific issue. And nowhere is this better exemplified than in the case we are considering: the analysis of development. For in 1888 Wilhelm Roux performed an experiment which seemed to prove the preformationists had been right after all; seemed to show that in the argument between Haller and Wolff the former had after all been justified in his scepticism.

Wilhelm Roux is chiefly remembered today as the founder of analytical embryology or, to use his own term, *entwicklungsmechanik*. He was dissatisfied with what seemed to him the merely descriptive embryology of his time, with the tracing out of phylogenies and the pursuit of the germ layers throughout the whole extent of the animal kingdom. He believed, with Aristotle, that knowledge was knowledge of causes. But, *contra* Aristotle, Roux had no faith in final causes, in teleology; he was interested only in efficient and material causes. Hence the name which he gave to the subject

(18) These terms were in fact introduced by Remak in 1845.
(19) Von Baer, *ibid.*, vol. 1. p. 140. This finding had wide implications, as we noticed in chapter 18, section 5.

which he conceived himself to have founded: developmental mechanics.

But let us return to Roux's 1888 experiment. It was simple. He destroyed one cell of a two-celled frog embryo. He found that the remaining cell developed into a half-embryo (20). This Roux interpreted to mean that below the reach of microscopical investigation the anatomy of the future organism was already mapped in the substance of the zygote and perhaps, indeed, in the substance of the egg. It required only triggering to unfold. Division of the zygote divided the latent homunculus.

The Drieschean entelechia

However recrudescence of the preformationist theory in this form proved short-lived. It was soon shown that if the injured blastomere was removed from contact with its neighbour the latter would develop into a complete embryo, albeit only half the size of the normal. Some time before Roux's error was discovered, moreover, Driesch had performed experiments whose results pointed to quite a different interpretation (21). He had been able to obtain quite normal embryos from single blastomeres taken from the 2, 4, 8, 16 and even 32-cell stages of sea-urchin blastulae. Thus the victory seemed to swing back once again into the epigeneticists' camp. As Driesch points out in his 1892 paper, Roux's theory of 'organ-forming germ areas (must)...be discarded, at least in its general form'.

In the years immediately following 1891, Driesch performed many other experiments on sea-urchin embryos in an attempt to elucidate the mechanisms at work in development. He altered the relative positions of the blastomeres, showing *inter alia* that prospective ectoderm and endoderm cells were interchangeable; he changed the directions along which the cleavage furrows developed and again demonstrated that normal embryos resulted. At first he struggled hard to provide a mechanistic explanation for his observations. In 1891 he published a small book entitled *Mathematico-mechanistic investigations of the problem of morphology in Biology*. But continued contemplation of the results of his experiments caused him to have second thoughts. It began to seem to him inconceivable that any mechanism could possibly account for the results emerging from his experimental work. Driesch, like so many nineteenth-century biologists, had a profound interest in philosophy. His perplexity when faced with the sea-urchin embryo made him eager to find some general system of ideas to replace the mechanistic

(20) W. Roux, Contributions to the developmental mechanics of the embryo. On the artificial production of half-embryos by the destruction of one of the first two blastomeres and the later development (postgeneration) of the missing half of the body, *Virchows Archiv path. Anat. u. Physiol. u. Kl. Med.*, **114** (1888), 113 - 53, in *Foundations of Experimental Embryology* (ed. B.H. Willier and J.M. Oppenheimer), Prentice-Hall, New Jersey (1964), pp. 2–37.

(21) The experiments were performed in March and April 1891. They were reported in 1892: *Zeitschrift für wissenschaftliche Zoologie*, **53**, 160–78, 183–4.

philosophy which he believed himself to have shown to be so woefully inadequate. Driesch eventually left biology altogether to profess philosophy at Leipzig.

The results of Driesch's philosophical endeavours received their best-known statement in the 1907—1908 Gifford Lectures: *The Science and Philosophy of Organism*. In these lectures he proved to his own satisfaction that no conceivable mechanism could account for the phenomena of morphogenesis. A machine, for Driesch, was 'a typical configuration of physical and chemical constituents by the acting of which a typical effect is attained' (22). Such configuration, Driesch insisted, would be destroyed by his experiments on the *Echinus* embryo and could hardly be at work in the regeneration of limbs, and so on. Life for Driesch was not, could not be, 'a specialised arrangement of inorganic events', and biology was therefore not 'applied physics and chemistry', it was something apart, an 'independent science'. And the factor with which living organisms were endowed and which no inanimate object possessed was termed by Driesch, notoriously, 'entelechy'.

The psychological paradigm and its challengers

Hans Driesch's account of entelechy is extraordinarily obscure. Like the concept of mind, to be discussed in the next chapter, it was an immaterial something which somehow acted upon material bodies. Driesch's attempts to say what it was came down mostly to saying what it was not. Entelechy, we read (23), 'may set free into actuality what it has *itself* prevented from actuality, what it has suspended hitherto'. And, in a letter to H S Jennings, he writes:

> Two systems absolutely identical in every physico-mathematical respect may behave differently under absolutely identical conditions if the systems are living systems. For the specificity of a certain entelechy is among the complete characteristics of a living organism and about this entelechy knowledge of physico-chemical things and relations teaches absolutely nothing.

In the 699 pages of Driesch's *Science and Philosophy of Organism* we see one of the last major flare-ups of vitalism before the concept faded into the light of common day. For although Driesch hoped that his concept of entelechy would become meaningful through its use, as the axiomatic concepts of physicists—mass, energy, force, and so on—have become meaningful, it was not to be. Scientific biology remained true to its Cartesian impulse: the way forward in morphopoiesis, as in all other aspects, lay in a progressive reduction, a progressive reinterpretation, into the languages of physics and chemistry.

(22) H. Driesch, *The Science and Philosophy of Organism,* Adam and Charles Black, London (1908), vol. 1, p. 139.
(23) *Ibid.,* vol. 2, p. 180.

For many years, however, a direct and unequivocal connection between embryology and the exact sciences remained elusive. Embryologists, while in general spurning Driesch's concept, found themselves unable to put any other of similar generality in its place. They were forced to continue using terms which had merely descriptive connotations. We read of 'potency' and 'determination', of 'induction' and 'competence', of primary, secondary and tertiary 'organisers', of embryonic 'fields'. How these terms were to be reinterpreted into the languages of physics and chemistry is nowhere made clear. Indeed, the distance separating embryology and chemistry is made strikingly evident in the concluding remarks of a volume of Silliman lectures written by one of the most eminent of twentieth-century embryologists, Hans Spemann:

> Again and again terms have been used which point not to physical but to psychical analogies. This was meant to be more than a poetical metaphor. It was meant to express my conviction that the suitable reaction of a germ fragment, endowed with the most diverse potencies, in an embryonic 'field', its behaviour in a definite 'situation', is not a common chemical reaction, but that these processes of development, like all vital processes, are comparable, in the way they are connected, to nothing we know in such a degree as to those vital processes of which we have the most intimate knowledge, viz, the psychical ones (24).

Nevertheless, in spite of the conceptual difficulties, experimental and analytical embryology continued its advance. Long before Spemann wrote his Silliman lectures an equally great authority, Theodor Boveri, had pointed the way to a mechanistic theory of epigenesis. He had recognised the fundamental importance of the chromosomes in development, and in his 1902 paper 'On multipolar mitosis as a means of analysis of the cell nucleus' (25) foreshadowed, in a qualitative way, the concepts of nucleo-cytoplasmic feedback and control which some sixty years later became familiar in the theories of Jacob and Monod. Furthermore, he points out that his own experiments 'teach us that the nucleus whose structure may have any degree of complexity, behaves just as Driesch demands from a machine' (26). Driesch, he observes, never divided the nucleus, and if he had he would have found his entelechy divided also. Finally Boveri, as a student of morphology, is fully prepared to see his subject taken over by the chemists. Indeed he looks forward eagerly to the day when 'morphological analysis (is) carried forward to the point where (its) ultimate elements would be specific chemical entities' (27). This day, as is well known, arrived a full half-century later when Watson and Crick published their solution to the structure of DNA,

(24) H. Spemann, *Embryonic Development and Induction,* Yale University Press, New Haven (1938), pp. 371–2.
(25) In *Foundations of Experimental Embryology* (ed. B.H. Willier and J.M. Oppenheimer), Prentice-Hall, New Jersey (1964).
(26) *Ibid.*
(27) T. Boveri, *Ergebnisse über die Konstitution der chromatischen Substanz des Zellkerns,* Gustav Fischer, Jena (1904); quoted in J.M. Oppenheimer, *Essays in the History of Embryology and Biology,* MIT Press, Massachusetts and London (1967), p. 81.

and Kendrew and Perutz published structures for the first globular proteins: myoglobin and haemoglobin.

The evolution of matter-theory

The coming together of morphopoiesis and chemistry which is imminent in the last third of the twentieth century owes as much to the development of chemistry as it does to that of biology. In particular it has depended upon the development of an understanding of how atoms join together to form molecules—the most elementary case of morphopoiesis. In the twentieth century a deep understanding of the nature of chemical bonds was achieved. The achievement of this insight went far towards destroying the ancient antagonism between atomic physics and biology. It did so in rather an interesting way; for it resolved at the same time the age-old paradox of the existence of a void.

It would be inappropriate to follow out the development of this understanding in these pages. Suffice it to say that electrons were discovered by J J Thomson in 1897 and that by 1911 Rutherford was able to show that they orbited a central, positively charged, atomic nucleus. In 1913 Bohr published his views on the electronic structure of atoms which included the notion of quantised orbits. In a qualitative sort of way it was possible to account for many of the combinatory proclivities of atoms in terms of Bohr's model.

Bohr's atom, however, remained something of an enigma until the 1920s. For it seemed quite arbitrary to restrict electrons to certain orbits and to no others. A genuine understanding of the substructure and hence the chemical activity of atoms awaited revolutionary developments in fundamental physics. In 1923 de Broglie made the bold suggestion that, since light had for some time been known to possess both undulatory *and* particle properties, then, for the sake of symmetry, particles of matter should also exhibit wavelike properties. If this were the case, if electrons could also be treated as waves, then might it not be that Bohr's quantised orbits were in fact positions where stationary waves could occur? This would account very nicely for the restricted number of orbits permitted. In 1926 Schroedinger was able to develop de Broglie's idea into a precise mathematical theory: wave mechanics.

Wave mechanics, as the name indicates, marks a very radical change in the physicist's account of the world. It unifies two great and formerly disparate fields of physics: particle physics and wave physics. It rests on the inescapable finding that all matter shows the properties of both waves and particles. Only, however, at the subatomic level does the duality become obvious. At this level phenomena must be interpreted in some cases as due to underlying particles, in other cases as due to underlying waves. This perception clearly sends tremors throughout the structure of physics and back through the history of physics as far as Parmenides. For what becomes of the ancient dichotomy between atoms and the void, between, indeed, the one and the many, if the basic stuff of which the world is composed is for some purposes best treated as a continuum and for other purposes as a multitude of particles? What becomes of the Abderan void, introduced to save the phenomena, if a particle can, for many purposes, be considered to extend in space like a wave?

This transformation of the bases of physical thought has vastly important implications for biology; for it is at the level of wave mechanics that the bonding between atoms can be understood. In the twentieth century it is no longer necessary to propose crude models of atoms resembling miniature billiard balls. After Schroedinger, the classical Newtonian picture of 'solid, massy, hard, impenetrable moveable Particles' could no longer be taken seriously. Instead the crude picture in the mind's eye is far more that of a drop of liquid. When atoms join, the electron waves associated with the valency electrons fuse together to enfold the nuclei of both the atoms concerned. This notion chimes in well with the morphologist's understanding of organic form. The molecules of life can be visualised not as a lattice of discrete impenetrable particles but as complex three-dimensional forms varying in density from part to part. A model of such a molecule is shown in figure 21.2. The molecule shown is that of myoglobin, and it is the first globular protein to have had its three-dimensional structure solved. Perutz, who shortly afterwards succeeded in determining the structure of the considerably more complex protein, haemoglobin, writes of protein molecules in general that they 'may be compared to an animal in having a three-dimensional anatomy laid out to a definite plan, rigid in some parts and flexible in others, with perhaps some minor variation in different individuals of the same species' (28).

The morphologist and the anatomist can thus rest content that the Democritean picture of the world has undergone so radical a transformation that a science of biophysics is no longer a fatuous dream. But what of the embryologist? Does the early twentieth-century revolution in fundamental physics also resolve his equally long-standing and even more profound antagonism to atomism? Does it answer the persistent question of Aristotle, of Galen, of Harvey, of de Maupertuis?

Self-assembly

The answer appears once again to be in the affirmative—with the qualification that the additional mechanism of Darwinian evolution is required.

It can, for instance, be shown that globular proteins automatically coil

(28) M.F. Perutz, *Proteins and Nucleic Acids,* Elsevier, Amsterdam, London and New York (1962), p. 15.

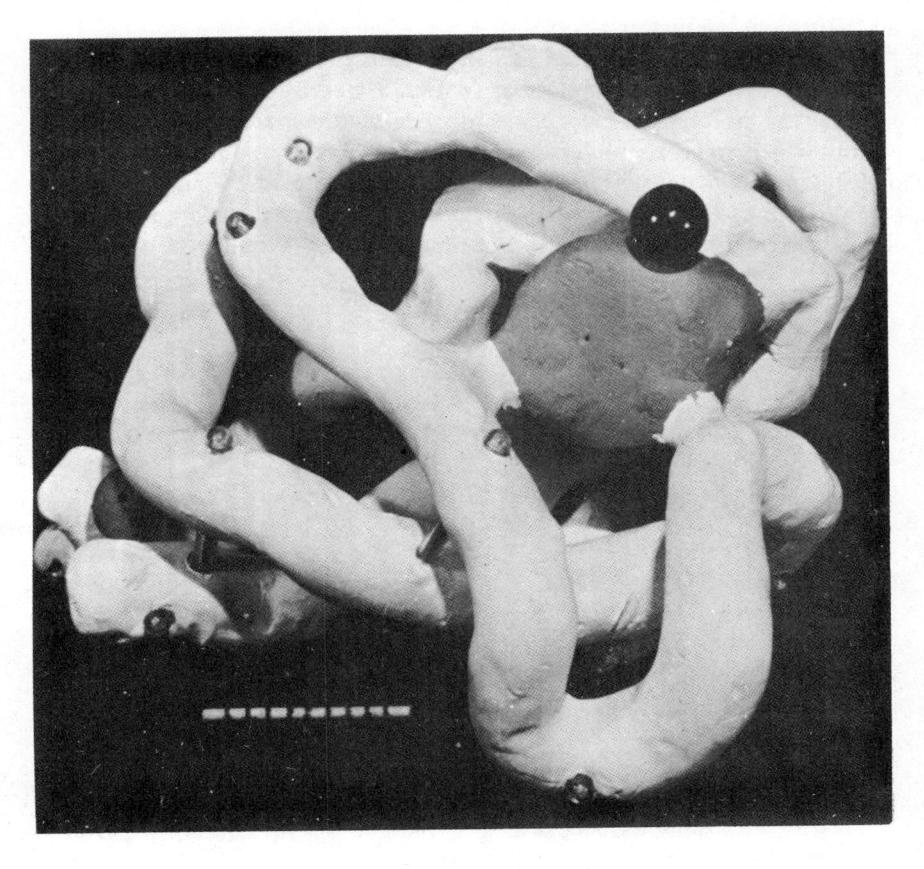

Figure 21.2. Plaster of Paris model of the myoglobin molecule constructed by J C Kendrew (1958). The scale is marked in Angstrom units (1 Å = 10^{-8} cm).

themselves into their complex three-dimensional configurations when placed in the correct physicochemical environment. Morphopoiesis, in this case, is the outcome of the interaction of multitudinous short-range coulombic forces.

Next it can be shown that many important biological molecules such as, for example, haemoglobin consist of a number of subunits. Haemoglobin consists of four subunits: two identical α-chains and two identical β-chains. Now it is found that these subunits are synthesised independently of each other in the maturing blood cell and come together, once again automatically, to form the complete haemoglobin molecule. The picture is of large numbers of subunits jostling together until complementary surfaces 'stick' to each other. The stickiness is, of course, due once again to short-range physical forces. And, once again, morphopoiesis emerges from solely physical interactions.

The next step on this modern *scala naturae* is the formation of structures consisting of two different molecular species. We arrive at this step when we consider the morphogenesis of cell components such as ribosomes and those entities which stand at the threshold of life—the viruses. In both cases the molecular species involved are proteins and nucleic acids. In both cases spontaneous assembly can be demonstrated if the constituents are suspended in the correct physicochemical environment. Perhaps the most revealing

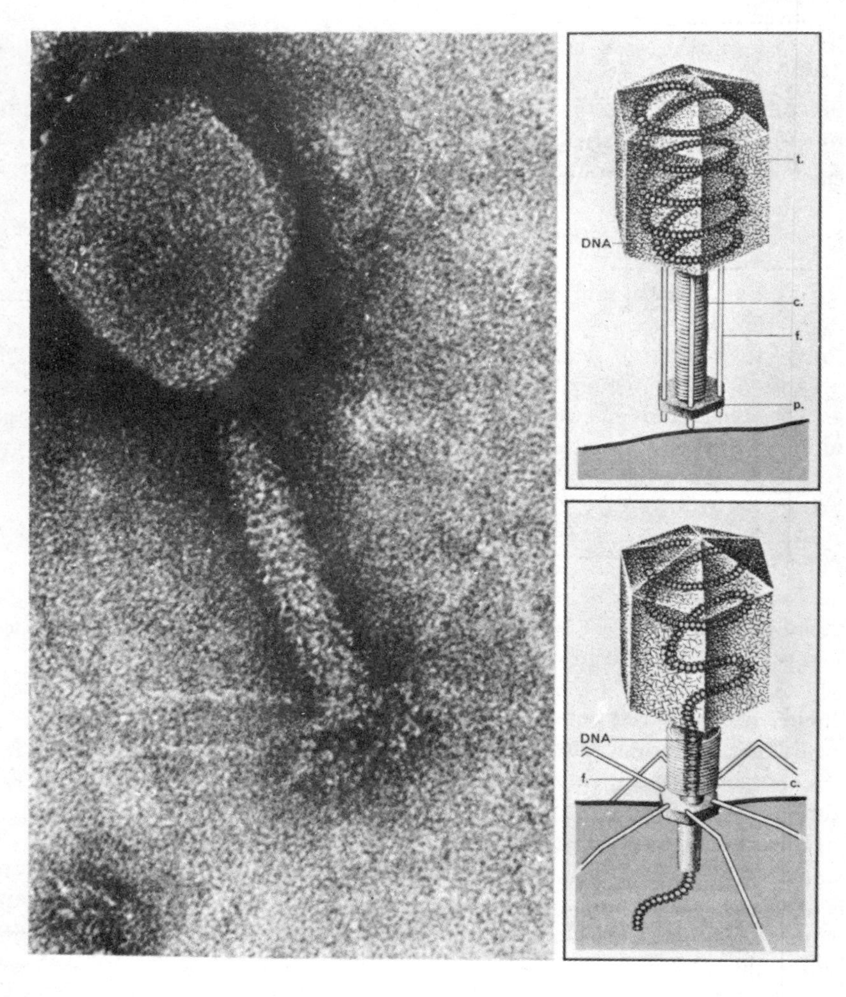

Figure 21.3. On the left is an electronmicrograph of a T4 bacteriophage. On the right are two diagrams showing the way in which it enters a bacterium whose boundary membrane is represented by a dark line at the bottom of both diagrams. The structural complexity of this type of bacterial virus is plain.

investigation of this power of spontaneous morphopoiesis is that carried out by Wood and Edgar on the T4 bacteriophage (29). Figure 21.3 shows that this bacterial virus has quite a complex morphology. It consists of a head, tail, tail-plate and tail fibres. The head consists of a spiral of DNA contained within a capsule of globular proteins. When a bacteriophage parasitises a bacterium its DNA is injected into the bacterial cell. Once within the cell the phage DNA is able to programme the bacterial protein synthesising machinery so that fresh bacteriophage components are synthesised. Of the hundred or so genes believed to be present on the phage DNA, at least forty have been shown to be concerned with morphopoiesis. The majority of these genes programme the synthesis of bacteriophage components. It can be demonstrated that many (although not all) of these components assemble together spontaneously. Once again, and at an even more complex level, we seem to have an instance of morphopoiesis due to mere physicochemical forces. We at last have a riposte to Galen's second-century scorn for those who considered that organic form could arise from the coming together of 'primary corpuscles'.

But Galen might still not be entirely satisfied. Aristotle might still insist upon the presence of a 'final cause'. For is it not still the case that the components of a virus are exquisitely designed to fit together to form the organism? They are hardly the 'primary corpuscles' of the atomists. And is it not the case that they are produced sequentially at precisely the right moment in the morphopoietic process? And, finally, is there not a great gulf fixed between even the most complex of viruses and the simplest genuinely living organism?

Without pretending that we are as yet anywhere near fully understanding the process of morphopoiesis, it can nevertheless be seen that these are defensive questions, reminiscent of the 'God of the Gaps' employed by theological apologists. The breach in the Aristotelian system has been made, and the molecular mechanists are busy exploiting their new-found land.

The response of the modern biologist to questions about the exquisite design of morphopoietic components is to investigate the mechanisms of molecular evolution. The perfectly tailored components of bacteriophages are, after all, no more mind-baffling than, for example, the superbly designed photoreceptors of mammals. The answers emerge from fairly straightforward extensions of the Darwinian principle into the molecular world and involve no 'new principle'.

The delicate timing of the morphopoietic process begins to find an explanation in terms of the complex network of molecular control circuits which has been evolved within the living cell. These circuits operate at several different levels within the cell. The feedback mechanism associated with the

(29) W.B. Wood and R.S. Edgar, Building a bacterial virus, *Sci. Amer.*, **217** (1967), 1, 60–75.

names of Jacob and Monod controls the expression of genetic information inscribed upon the cell's DNA. Other controls operate at the level where this information is 'translated' into the structure of protein molecules and yet other controls govern the quantity of metabolites synthesised. This intricate system ensures not only that morphopoiesis is a well-organised process but also that the mature cell operates as a closely integrated system. It ensures not only that the comparatively simple morphopoiesis of the bacteriophage is smoothly accomplished but also that the development of the cell itself is similarly well organised. A host of cellular control factors are at present being discovered in the molecular biology laboratories of the world. In consequence the teleological character of the living process which so impressed generations of biologists is now beginning to find an explanation in terms of cybernetics and control theory. This approach has, of course, received a considerable impetus from the technological advances of our times. It seems likely that, once again, the necessities of earning our living will throw up a symbolism which we can use in our attempt to understand the nature of the living process.

Still outstanding is the question which points to the gulf between virus particles and living organisms. All viruses are obligate parasites of living cells. In the absence of the latter they are unable to reproduce themselves. The mechanist's account, however, does not depend upon solving the relationship between viruses and cells. The morphopoiesis of virus particles has been studied merely because they are extremely convenient systems to investigate. The same principles of automatic self-assembly can be shown to obtain in the morphopoiesis of such universal cell organelles as ribosomes. The previous paragraph, also, has indicated the type of mechanistic cell biology now becoming available.

Finally what has become of the great controversy between the preformationists and the epigeneticists? Nowadays we can see that both sides had some of the truth. The germ cells do not, of course, contain a fully formed homunculus. They do, however, contain, inscribed upon their DNA, a representation of the ground plan of this homunculus. The embryo does not emerge from a homogeneous and structureless protoplasm. It does, however, differentiate from indifferent material under the control of instructions built into the DNA molecules. What the eighteenth and nineteenth centuries lacked was an appreciation of computer and information science. Saturated as we are today in these sciences and their concomitant jargon, we need no longer regard the processes of embryology as a miracle which escapes the framework of physical science.

22 DEACTIVATION OF THE MIND

A possible symmetry in the history of ideas?

In chapter 21 we noticed that the great twentieth-century embryologist, Hans Spemann, believed that the phenomena which most closely resembled the phenomena of morphopoiesis were 'those vital processes of which we have the most intimate knowledge, viz. the psychical ones'. That this is no new idea the reader of this book will realise. He will recall that we have many times pointed out that this was the great analogy upon which Aristotelian science was founded. In subsequent chapters we have seen how the sixteenth- and seventeenth-century scientific revolution ridiculed this analogy in physics and astronomy. We have also noticed how the Aristotelian paradigm was slowly driven from the realm of biological science, even from the realm of embryology, its original power base. Now in the last chapter of this book we come finally to the concept of mind. Can the Galilean paradigm, the correlation of 'accidents', replace the Aristotelian in its own country? Or is there a symmetry in the history of ideas, and will it be the case that just as Aristotelianism was broken by the existence of projectiles, so Galilean science will be broken by the existence of consciousness? Part of the motive for writing this book has lain in the hope of gaining sufficient perspective, of gaining perhaps an Archimedean fulcrum, to suggest some answers to these questions. First, however, let us briefly review and develop some of the material covered in earlier chapters.

Ventricular psychology

It will be recalled that we started our story on the shores of Asia Minor. Socrates' hopes, we may remember, had been raised by Anaxagoras' doctrine that the mind impelled and governed all: '...all things that were to be, all things that were but are not now, all things that are now or shall be, Mind arranged them all, including this rotation in which are now rotating the stars, the sun and the moon...' (1). And further back still, at the beginning of the tradition, Thales had insisted that the initiator of movement in the world was a self-moving mover — the soul.

Socrates had been disappointed to find that Anaxagoras conceived 'mind'

(1) Quoted in G. S. Kirk and J. E. Raven, *The Presocratic Philosophers,* Cambridge University Press (1971), p. 373.

to be a very subtle sort of matter and his own theory, as developed by Plato, was more penetrating. He remained convinced, however, that the mind was posterior to the physical universe and responsible for its behaviour. This profound dualism remained in Western thought for millennia: remained, however, in a somewhat debased form, for spiritual substance was quite customarily regarded, as we have seen Anaxagoras regarded it, as merely a very 'subtle' and 'active' form of matter. The main structure of the classical philosophy was, however, retained: the mind, or soul, whatever its existential status, was the body's active principle and was responsible for its behaviour.

The notion that the soul, like Gaul, was divided into three parts we first met in the *Timaeus*. The idea seems to have been accepted by many ancient thinkers. That the rational soul was itself tripartite seems to have been a later accretion. The further notion that the three parts of the rational soul were to be located in the three cerebral ventricles seems to have originated with Nemesius of Emesa (*fl. c.* AD 390). He was of the opinion that the three important faculties—imagination, intellect and memory—were the functions of psychic spirit contained in the lateral, third and fourth ventricles respectively. An elaboration of this psychic schematum is found in the very influential writings of Avicenna (AD 980 – 1037), and as the mediaeval period drew to a close we find precisely the same idea in another widely read work:

(a)

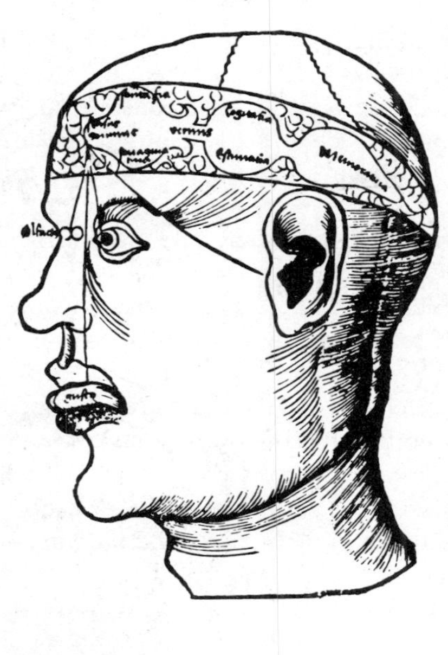

(b)

Figure 22.1. Ventricular psychology. (a) This illustration from Albertus Magnus' *Philosophia Naturalis* (originating about 1250) shows the traditional psychophysiology. (b) This figure from Reisch's *Pearl of Wisdom* (1503) shows the traditional ventricular theory in more detail. The lines drawn from the eye, nose, tongue and ear show that information from these organs is delivered to the front of the first ventricle. This ventricle is consequently devoted to association or 'sensus communis' and more posteriorly to 'fantasia' and 'imagination'. The choroid plexus (vermis) connects the first ventricle to the second which is devoted to cogitation and judgment. The third ventricle is concerned with memory.

Gregor Reisch's *The Pearl of Wisdom*. The ventricular theory lent itself to diagrammatic representation, and figures similar to figure 22.1(b), reproduced from Reisch's book, are not uncommon.

But, as the fifteenth century gave way to the sixteenth, students were not quite so ready to accept without question the teachings of tradition. At about the turn of the century Leonardo da Vinci developed a technique for investigating the cerebral ventricles which consisted in filling them with molten wax and then, after allowing time for solidification, digesting away the surrounding brain. In this way he was able to obtain a reasonably accurate representation of these putative seats of the soul. The contrast between Leonardo's results and the traditional picture is shown in figure 22.2

Leonardo's work was not, however, known until the nineteenth century (chapter 14), and Vesalius during his student days was exposed to the same catalogue of false ideas and theoretical dissections which generations of mediaeval physicians had suffered before him. In *De Humanis Corporis Fabrica* (1543) we may read his ironic comments:

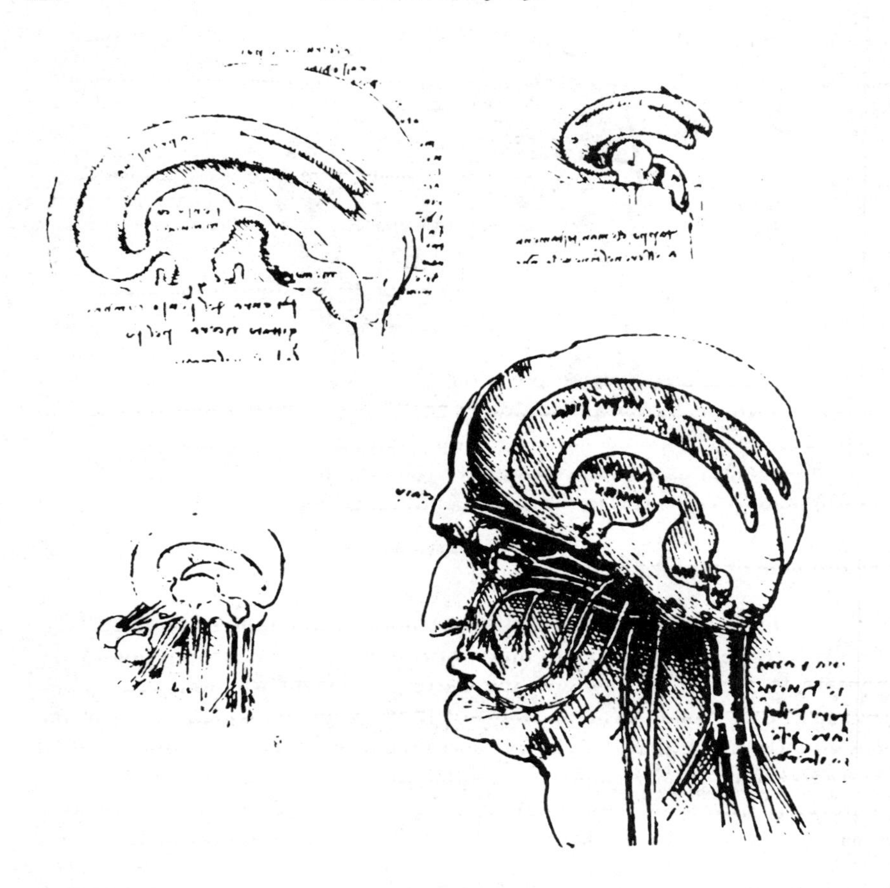

Figure 22.2. These drawings from Leonardo's notebooks show casts of the cerebral ventricles, The anterior chambers (lateral ventricles) are labelled 'imprensiva' (perception), the middle chamber (third ventricle) is labelled 'sensus communis', and the posterior chamber (fourth ventricle) is labelled 'memoria'.

I well remember how when I was following the philosophical course at the Castle School, easily the leading and most distinguished school of the University of Louvain, in such commentaries on Aristotle's treatise, *On the Soul,* as were read to us by our teacher, a theologian by profession and therefore, like the other instructors at that school, ready to introduce his own pious views into those of the philosophers, the brain was said to have three ventricles (2). The first of these was anterior, the second middle and the third posterior thus taking their names from their sites; they also had names according to function. Indeed those men believed that the first or anterior, which was said to look outward towards the forehead, was called the ventricle of the sensus communis because the nerves of the five senses are carried to it from their instruments, and odours, colours, tastes, sounds and tactile qualities are brought into this ventricle by the aid of those nerves. Therefore, the chief use of this ventricle was considered to be that of receiving the objects of the five senses, which we usually

(2) This theory is not, in fact, to be found in the Aristotelian treatise.

> call the common senses, and transmitting them to the second ventricle, joined by a passage to the first so that the second might be able to reason and cogitate about those objects; hence cogitation or reasoning was assigned to the latter ventricle. The third ventricle (our fourth) was consecrated to memory, into which the second desired that all things sufficiently reasoned about those objects should be sent to be suitably deposited. The third ventricle, as it were moist or dry, either more swiftly or more slowly engraved them as into wax or a harder stone.... Furthermore, that we might more aptly consider each thing that was thus taught, an illustration was shown us taken from some Philosophic Pearl [figure 22.1 (b)] presenting to our eyes the aforesaid ventricles which each of us studied very carefully as an exercise and added a drawing of it to our notes (3).

Vesalius' irony was well merited. In chapter 14 we quoted his account of anatomical teaching as he found it when a student. The lecturer, notoriously a theologian, read appropriate passages from a classical text while an unlettered menial struggled to point out the structures described. Vesalius changed all this. Nevertheless, although healthily sceptical about ventricular psychology, he did not dispute that the ventricles contained a psychic pneuma (4). And, as we saw in chapter 15, a variant of this ancient theory was still held by Descartes in the seventeenth century.

The problem of perception

The Cartesian neurophysiology was, as we noticed in chapter 15, extremely ingenious. It pointed the way to much of contemporary neurophysiology. In chapters 16 and 17 we saw how the Cartesian insight was followed by the eventual development of a mechanistic physiology in which the psychic pneumata were transformed first into nerve juices, then into animal electricity and finally into the potential changes across lipoprotein membranes with which we are familiar today. These developments were largely the work of students of the peripheral nervous system. But Descartes also had his successors among investigators of higher nervous functions.

One of the central problems for scientists interested in the function of the brain was the problem of perception. It will be recalled from chapter 15 that, for Descartes, objects in the outside world were focused on to the pineal gland within the brain. This was, of course, no revolutionary idea. We have seen in the Vesalius quotation above how the traditional physiology regarded the anterior cerebral ventricle as the 'viewing room' to which objects detected by the five senses were brought; and in previous parts of this book we have noticed that the classical theory envisaged the continuous emanation of tenuous 'effluvia' from objects (much as scented objects emanate scents and noisy objects emanate noise), and that these effluvia were focused by the eye into the hollow of the optic nerve. The optic nerve delivered these thin, subtle 'ghosts' to the brain where they were recognised (like recognises like) by the

(3) Trans. E. Clarke and C.D. O'Malley in *The Human Brain and Spinal Cord,* University of California Press (1968), p. 468.
(4) See C. Singer, *Vesalius on the Human Brain,* Oxford University Press, London (1952).

similarly thin and subtle matter of the soul.

Although its origins are largely forgotten, this para-optical theory has remained to puzzle the philosophically naive, among whom must be numbered many scientists, for centuries. Post-Cartesian philosophers, from Leibniz through Kant to Sartre, have, however, been quick to recognise the confusion. In the *Monadology*, for instance, Leibniz provides a famous rebuttal:

> ...and supposing there were a machine, so constructed as to think, feel and have perceptions, it might be conceived as increased in size, while keeping the same proportions, so that one might go into it as into a mill. That being so, we should on examining its interior, find only parts which work one upon another, and never anything by which to explain a perception (5).

And, a little further on, he concludes that 'perception and that which depends upon it are *inexplicable on mechanical grounds*'.

This cannot be gainsaid so long as we are considering our own experience of perceptions and sensations. But when we set out to describe the phenomenon of perception in others, the case is different. It is self-contradictory to suppose that we can experience another's aching tooth or perception of a sunset. On the other hand, we may very well hope to find in them something analogous to Leibniz's mill which is responsible for the behaviour, verbal or otherwise, by which we know that a perception has occurred (6). And, unlike the majority of their continental colleagues, the British empiricist philosophers attempted to do just this. It is to this tradition, the true heir of scholastic psychophysiology, that we must therefore turn next.

Towards an experimental physics of the mind

Thomas Hobbes (1588 – 1679) was a contemporary of Descartes and was indeed acquainted with the French philosopher. Like the succeeding British empiricists, he was concerned to get rid of the doctrine that the mind was born primed with certain inbuilt ideas. Against Descartes he wished to argue that the mind came innocent into the world, a *tabula rasa*. All that was later to be found had entered through the gates of the senses, *nihil est in intellectu quod non prius fuerit in sensu*, 'there is no conception in a man's mind which hath not at first, totally, or by parts, been begotten upon the organs of sense. The rest are derived from that original' (7). This was, of course, an assumption on Hobbes's part. In the seventeenth century it would have been difficult to provide a rigorous proof. It was an assumption, however, which harmonised well with his general philosophical position. Hannah Arendt

(5) G.W. Leibniz, *Monadology,* 17, 1714; trans. R. Latta, Oxford University Press, London (1898).
(6) See, for example, H. Feigl, *The 'Mental' and the 'Physical',* University of Minnesota Press, Minneapolis (1967).
(7) T. Hobbes, *Leviathan,* part 1, chapter 1, London (1651).

refers to Hobbes as 'the only great philosopher to whom the bourgeoisie can rightly and exclusively lay claim' (8), and Macpherson claims that the citizens of Hobbes's social philosophy inhabited 'a possessive market economy' (9). Society consisted of a congeries of individualists, each hoping to better his own selfish lot. The aristocratic principle of hereditary rank and inborn knowledge was anathema to the working of such a system. The intellectual connections of Hobbes's sociology and the prevalent corpuscular hypothesis of the seventeenth century have also often been remarked. Indeed, many passages in his works make the connection quite explicit. For example, he bases his philosophy on an axiom which would not have been strange to Newton: 'That when a thing lies still, unless somewhat else stir it, it will lie still for ever...when a thing is in motion, it will be eternally in motion, unless somewhat else stay it...' (10).

When he turns to psychology, he remains firmly within the same physicalist tradition: 'All the qualities called *sensible* are in the object that causes them but so many several motions of the matter by which it presseth our organs diversely. Neither in us that are pressed are they anything else but divers motions; for motion produces nothing but motion' (10a). We are aware not only of Newton but also of Galileo in the background. The analysis of sensory qualities into aspects of motion was, as we saw in chapter 14, characteristic of the Italian. Hobbes carries this analysis a step further when he explains exactly how it is that we become aware of colours, sounds, scents, and so on: 'The said image or colour is but an apparition unto us of the motion agitation, or alteration which the *object* worketh in the *brain*, or spirits, or some internal substance in the head' (11). 'An apparition unto us', proceeding from some alteration in the substance of the brain: we can hardly do better than that today!

From sensations, those motions within the skull, Hobbes passes to imagination and memory. For Hobbes these are very much the same thing. Both are traces left behind by sensations, or, to use his materialistic terminology, the continuance of movements in the brain after the stimulating object has been withdrawn. For, says Hobbes,

> We see in water though the wind cease, the waves give not over rolling for a long time after: so also it happeneth in that motion which is made in the internal parts of man....For after the object is removed, or the eye shut, we still retain an image of the thing seen, though more obscure than when we see it....Imagination therefore is nothing but decaying sense (12).

Finally, and importantly, he expounds his conception of 'mental

(8) Quoted in *Hobbes Studies* (ed. K.C. Brown), Blackwell, Oxford (1965), p. 185.
(9) *Ibid.*, p. 187.
(10) Hobbes. *ibid.*, part 1. chapter 2.
(10a) T. Hobbes, *Leviathan*, London (1651), part 1, chapter 2.
(11) T. Hobbes, *Human Nature* (1650) chapter 2, 4; ed. W. Molesworth, John Bohn, London (1840).
(12) Hobbes, *Leviathan, ibid.*, chapter 2.

discourse', or thinking. Here we meet one of the pioneering accounts of the association of ideas. Again it is best to allow the philosopher to speak for himself:

> When a man thinketh on anything whatsoever his next thought after is not altogether so casual as it seems to be. Not every thought to every thought succeeds indifferently....The reason whereof is this: all fancies [i.e. images] are motions within us, relics of those made in sense; and those motions that immediately succeed one another in the sense continue also after the sense; insomuch as the former coming again to take place and be predominant, the latter followeth by coherence of the matter moved, in such manner as water upon a plain table is drawn which way any part of it is guided by the finger (13).

Hobbes presents a clear and trenchantly argued dual-aspect theory of the mind. The corpuscularian happenings within the head *appear* to the individual whose head it is as sensations and perceptions, imaginings and memories. In Hobbe's theory we can see the germ not only of the associationist school of psychology but also of the Pavlovian theory of conditioned reflexes. We can also perhaps see at work the influence of his social atomism. There is nothing in the psychology which Hobbes outlines to play the part of an integrating soul. Disparate sensations collide with one another, decay or reenergise each other: the mind, like the chemist's gas, is treated under the sign of the corpuscular hypothesis.

Hobbes's successor Locke has the greater philosophical reputation. This is strange, for his work owes much to Hobbes and his exposition is far more laboured. Nevertheless, according to d'Alembert, 'Locke created the science of metaphysics in somewhat the same way as Newton created physics....In one word, he reduced Metaphysics to that which it ought to be, viz., the experimental physics of the mind' (14).

Locke's method, as d'Alembert implies, was thoroughly empirical. He was not interested to study the books of previous metaphysicians. Indeed, he was contemptuous of the scholasticism still flourishing at Oxford, his own university. Instead he determined to enquire by introspection 'into the original of those ideas, notions or whatever else you may please to call them which a man observes and is conscious to himself he has in mind' (15). This is admirable. But, as d'Alembert hints, Locke had an ulterior purpose. He wished to imitate Newton. He wished to treat mental phenomena in the same sort of way that the physicists had so successfully treated physical objects. The analogy, however, is dangerous. Our feelings, hopes, fears, aspirations, and so on, are, as Descartes recognised, quite different things from the atoms and aethers with which the physical scientists were concerned. Although

(13) *Ibid.*, chapter 3.
(14) J.L. d'Alembert, *Preliminary Discourse to the Encyclopedia of Diderot,* Paris (1751); trans. R.N. Schwab, Bobbs-Merrill, Indianapolis, New York (1963), pp. 83–4.
(15) J. Locke, *Essay Concerning Human Understanding,* London (1690), introduction.

Locke at the outset of his work explicitly denies any interest in the physical correlatives of mind—'these are speculations which however curious and entertaining, I shall decline'. He systematically uses metaphors derived from the sciences of ***res extensa***. Indeed, the central concept of his philosophy is expressed in terms which imply this false analogy:

> Let us then suppose the mind to be, as we say, white paper, void of all characters, without any ideas; how comes it to be furnished, whence comes it by that vast store which the busy and boundless fancy of man has painted on it with an almost endless variety? (16)

We read elsewhere in the *Essay* of characters being 'stamped on the mind', of ideas being 'in men's minds', of 'the senses convey(ing) into the mind', and so on. And in book 2, chapter 33, where we meet once again an exposition of the doctrine of the association of ideas, we can read an explicit fragment of psychophysiology:

> Custom settles habits of thinking in the understanding, as well as determining in the will, and of the motions of the body; all of which seem to be but trains of motion in animal spirits, which, once set a-going, continue in the same steps they have been used to, which by often treading, are worn into a smooth path and the motion in it becomes easy, and as it were natural.

It is interesting to notice how strong a hold the theory of animal spirits still possessed at the end of the eighteenth century. In general Locke's concept of how the mind 'works', of how it comes to be furnished with ideas, seems to be but a shadow-play of the mechanisms imputed to the physical world by contemporary scientists. Simple ideas, for example, are combined within the mind into several sorts of compound or complex ideas very much as contemporary chemists saw elements joining together to form 'mixts'. This combination, however, still seemed to Locke to be an *activity* of the mind. His philosophical successor David Hume, however, carried the argument to a logical conclusion:

> For my part, when I enter most intimately into what I call myself, I always stumble on some particular perception or other....I never catch myself at any time without a perception and can never observe anything but the perception... [our minds] are nothing but a bundle or collection of different perceptions, which succeed each other with an inconceivable rapidity, and are in a perpetual flux and movement (17).

We have already noticed the importance of Hume's work in awakening Kant from his 'dogmatic slumbers'. However, perhaps more influential in the development of brain science were the roughly contemporaneous publications of David Hartley. Indeed, R M Young maintains that Hartley may justly be regarded as 'the founder of the physiological psychology of the higher nervous functions' (18). Hartley's major work, *Observations on Man*,

(16) *Ibid.*, book 2, chapter 1.
(17) D. Hume, *Treatise of Human Nature,* London (1739), book 1, part 4, section 6.
(18) R.M. Young, *Mind, Brain and Adaptation in the Nineteenth Century,* Cambridge University Press (1970), p. 97.

was first published in 1749. In it he attempts to combine some of Sir Isaac Newton's ideas with some of those of John Locke. In particular he wishes to develop a hint which Sir Isaac published in the 'General Scholium' to the second edition of the *Principia* (1713). Newton writes as follows:

> And now we might add something concerning a most subtle spirit which pervades and lies hid in all gross bodies; by the force and action of which spirit the particles of bodies attract one another at near distances and cohere, if contiguous; and electric bodies operate to greater distances, as well repelling as attracting the neighbouring corpuscles; and light is emitted, reflected, refracted, inflected and heats bodies; and all sensation is excited, and the members of animal bodies move at the command of the will, namely, by the vibrations of this spirit, mutually propagated along the solid filaments of the nerves, from the outward organs of sense to the brain, and from the brain to the muscles.

Hartley takes up this hint (although he admits that he does not grasp exactly what it is that Newton has in mind) and suggests that when external objects impinge on the sensory nerves vibrations are set up in the aether present within the pores of the neuronal substance. These vibrations are transmitted along the nerves to the brain. Once within the brain, says Hartley, the vibrations may persist for a little way through its substance keeping their original directions. Now when such vibrations are very often repeated they may beget miniature vibrations (vibratiuncles, according to Hartley) having the same direction and location. These vibratiuncles will also arise when the substance of the brain is heated or when neighbouring arteries pulsate. In this way Hartley believes himself to have accounted for the apparently spontaneous origin of ideas. Next he goes on to show that his aetherial vibratiuncles can easily provide a physical basis for Locke's association theory. The constant association of sensations A, B and C result in the constant association of vibratiuncles a, b and c and hence, after a time, whenever sensation A occurs (vibration A), b and c are elicited. The form of his argument is clear. Indeed, to moderns it is strongly reminiscent of Pavlovian conditioned reflex theory. Hartley, however, goes further than this. He makes use of his theory to account for the origin of compound ideas. For Hartley, as for Locke, general ideas are merely compounds formed by the aggregation of numerous simple ideas—in physical terms, vibratiuncles. The ancient notion of the mind as an active unifying and coordinating power has in this theory quite disappeared. If it can still be said to exist at all, it forms, like its physical substratum the brain, only an arena, a receptacle, in which the coming together and falling apart of the simple ideas (vibratiuncles) may occur.

Hartley's detailed physiology need not detain us. It was not taken seriously even by his contemporaries. His general theory was, however, very influential. *Observations on Man* was republished in 1775 and 1790. There is no need to emphasise once again how similar the Hartleyan psychology is to the prevalent corpuscular hypothesis. Indeed, as if to drive the point home, none other than Joseph Priestley made himself responsible for republishing the *Observations*, the author having died in 1757. It is interesting, too, to notice that clear lines of influence run not only back to the bourgeois social philosophy of Thomas Hobbes but also forward, as we saw in chapter 19, to the evolutionary theory of Erasmus Darwin and ultimately to Charles Darwin. At the end of chapter 20 we noticed how radically different were the Aristotelian and Darwinian visions. Here we see the emergence of a view of the mind as a mechanism which, once again, the Stagirite would have found unintelligible: a view of the mind, moreover, which appears to share the same sociological roots as the Darwinian theory. Can this be merely coincidence? It is a view of the mind, however, which lends itself to experimental investigation. Is it, for example, possible to localise Hartley's vibrations and vibratiuncles within the brain? The philosophers could thus pass to the physiologists the task of determining how far an 'experimental physics of the mind' was possible.

Phrenology

The Hartleyan association theory has the advantage of eliminating the necessity of positing an active yet immaterial soul. In effect, Hartley unmasked the true import of Descartes' neurophysiology. The residual dilemma left by Descartes (perhaps due to his temporising) was how mind could act on matter and *vice versa*. The British empricists solved this dilemma at a stroke by rejecting one side of the dichotomy. Sensations, and the like, were in some way aspects, concomitants, of vibrations and vibratiuncles within the substance of the brain.

Within this philosophical framework, scientists were able to advance their understanding of the brain. In chapter 15 we saw that Descartes had been content to subscribe to the ancient notion that the cerebral ventricles and their contents played a predominant role in cerebral physiology. Towards the end of the seventeenth century, Thomas Willis directed attention away from the ventricles to the substance of the brain. The details of his system need not detain us but it is interesting to notice that he, like Descartes before him, was strongly influenced by the current interest in optics. Galileo's telescope and Newton's prism seem so to have impressed the intellectual world that Willis, who was not a deeply original physiologist in his own right (19), propounds the following physiological basis of perception: 'images or representations of all sensible things, sent in through the passages of the nerves, like tubes or narrow openings, first pass through the *corpora striata* as through a lens; then they are revealed upon the *corpus callosum* as if on a white wall, and so

(19) See Sir M. Foster, *History of Physiology during the Sixteenth, Seventeenth and Eighteenth Centuries,* Dover, New York (1901), p.277.

induce perception...' (20). This concept of a 'medullary screen' upon which images of the outside world are displayed is not uncommon among physiologists in the late seventeenth and early eighteenth centuries.

Attempts to understand the physiology of the brain made little further progress until the beginning of the nineteenth century. At this time, Gall and Spurzheim introduced their well-known doctrine of phrenology (21). Gall believed there to be some 26 or 27 distinguishable behavioural propensities and that each of these was dependent upon a particular region of the brain. Hypertrophy of these cerebral areas resulted in both a prominent exhibition of these propensities in an individual's behaviour and also in an expansion of the cranium immediately above. In consequence, the phrenologists believed it possible to read a person's character from the detailed shape of his skull. Figure 22.3 shows a phrenological topology.

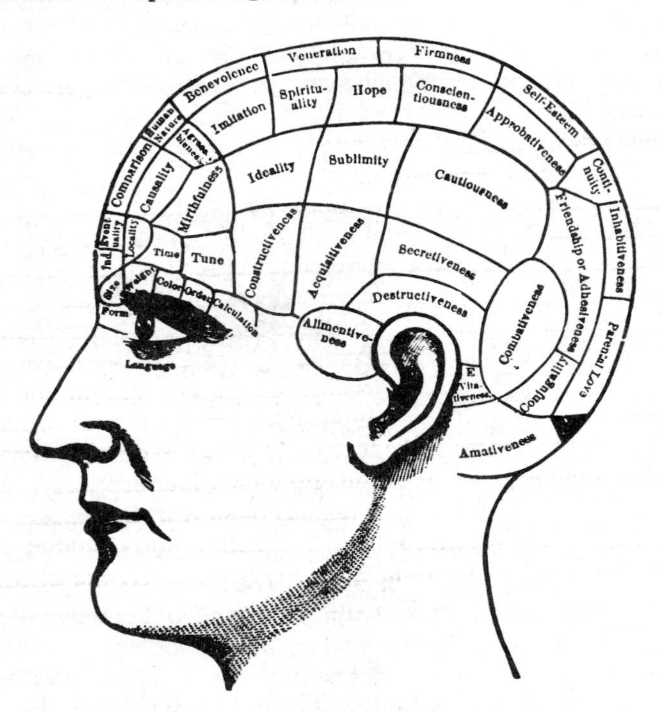

Figure 22.3. Map of the phrenological organs of the brain.

(20) T. Willis, *De Anima Brutorum,* London (1672); trans. E. Clarke, and C.D. O'Malley, in *The Human Brain and Spinal Cord,* University of California Press (1968), pp 43–4. The term 'corpus callosum' was used by Willis to denote all the white matter of the brain: not just the tract of fibres designated by the name today.

(21) Gall became interested in the localisation of faculties and feelings in the brain in the late 1790s. His major work, *Anatomie et physiologie du système nerveux en générale, et du cerveau en particulier,* 4 vols, was published between 1810 and 1819.

Although phrenology did not stand up to detailed experimental investigation, it did have the virtue of concentrating scientific attention once more on to the localisation problem; of emphasising that distinctions in human personality must, if the materialist theory is valid, have a basis in cerebral structure. Phrenology at the beginning of the nineteenth century, like the Cartesian beast – machine at the beginning of the seventeenth, was, however, vastly premature; for at the beginning of the nineteenth century rather little was known about the structure of the brain. It is clear that any attempt to localise function in the brain's anatomy was futile before the latter had been thoroughly elucidated. One of the consequences of Gall's phrenology was, in fact, to provide a strong motive for this elucidation.

Subsequent investigation of the brain's structure in the nineteenth century proceeded at two rather different levels. Anatomists concerned themselves with tracing out the ontogeny and phylogeny of the convolutions of the cerebral cortex which, before the work of the phrenologists, had been regarded as a mere random chaos. Outstanding names in this work are those of Richard Owen, François Leuret and, more recently, Ariëns Kappers, who with Huber and Crosbie has gathered much of the material into two volumes entitled *The comparative anatomy of the nervous system of vertebrates including man* (1936). Secondly, microscopists investigated the arrangements of the brain's cells and the pathways of its fibres. This work had, perforce, to await the advent of the achromatic microscope (see p. 211). Before the 1830s the fine structure of the cerebrum was virtually unknown. Since the 1830s, however, a continuous accession of knowledge has occurred culminating, at the turn of the century, in the work of Betz, Golgi and Cajal (see p. 296).

The three streams of research—localisation of function in the brain, morphology and pattern of the convolutions, and microscopical structure—came together at the beginning of the twentieth century to form the discipline of cell and fibre architectonics. It is this discipline, founded largely by the work of Karl and Cécile Vogt and their famous pupil Karl Brodmann, which provides the anatomical substratum for twentieth-century studies of the localisation of function in the brain. Even today, however, the attempt to identify cortical areas responsible for behavioural function remains confused and controversial.

In the early years of the last century the prospects for establishing a science of character on a neuroanatomical foundation consequently seemed quite forlorn. Accordingly it lost the respect of mainstream physiologists. An idea ahead of its time, as we have many times remarked, is just as invalid as an idea behind its time. Nevertheless, as R M Young shows (22), phrenology remained a strong background influence, and through Herbert Spencer affected the work of Hughlings-Jackson and other prominent neurologists at the end of the century.

(22) Young, *ibid.*

The concept of a reflex 'arc'

Meanwhile more orthodox physiologists were beginning to piece together the fragments of a physiology of the central nervous system. In 1811 Charles Bell published a short book entitled *Idea of a New Anatomy of the Brain, submitted for the observations of his friends.* He, too, was interested in establishing the functions of the different parts of the brain. This concern to assign functions to different cerebral regions is strongly reminiscent of Bichat's contemporary endeavour to localise the functions of the body in different tissues (see chapter 17). So far as the brain is concerned, Bell saw the cerebrum as the centre of sensation and motion, while he assigned to the cerebellum control over involuntary activities. In particular he believed that he would be able to establish the functional anatomy of the brain by experimenting on the spinal nerves. For he still retained, as was common at the beginning of the nineteenth century, the classical idea that the spinal cord was a rope of nerve fibres originating in the brain. He believed, therefore, that by stimulating these nerves and observing the behavioural effect, he could gain an insight into the functions of the brain. And it was in the course of this mistaken investigation that he established the distinction between the anterior and posterior roots of the spinal nerves, showing that only the former contained motor fibres. Bell's functional distinction between the anterior and posterior roots did not, however, create much immediate stir in the physiological world. He was loth to follow up his work for, as he wrote in 1822, 'I cannot convince myself that I am authorised in nature or in religion to do these cruelties' (23). He searched for other preparations and finally decided to investigate the nerves of the face. Here he was able in the 1820s to confirm his earlier observations. Some nerves he found to be sensory, others motor.

The clear distinction between the functions of the anterior and the posterior roots of the spinal nerves, called by Longet (1842) 'the most beautiful discovery of modern times', provided a strong impetus for the further investigation of reflex behaviour. We saw in chapter 15 that René Descartes had developed a clear though entirely theoretical account of reflex activity in the early seventeenth century. Indeed, at the beginning of the previous century, Leonardo da Vinci had demonstrated that a decapitated frog retained many behavioural automatisms so long as the spinal cord

(23) François Magendie did not share Bell's scruples. In 1822 he described how he had succeeded in sectioning both anterior and posterior roots and had demonstrated that whereas the anterior roots had a motor function, as Bell had shown before him, the posterior roots were sensory. 'At first', he writes, 'I thought the member corresponding to the cut nerves was entirely paralysed; it was insensible to the strongest prickings and pressures, it seemed to me also incapable of moving; but soon, to my great surprise, I saw it move in a manner very apparent, although sensibility was entirely extinct' (*J.Physiol. exp. Path.*, 2 (1822), 276–9). The rule that the posterior roots are sensory and the anterior motor is consequently termed the Bell – Magendie law.

remained intact. But in spite of a continuing interest (24), little exact analysis was achieved before Bell's breakthrough. After Bell it was probably Marshall Hall's work which did most to establish the modern concept.

Before Hall's work was published, and indeed for some time after as well, the movements to be observed in a 'spinal animal' were supposed to be due to a spinal 'soul'. It is interesting to note that as late as the mid-nineteenth century AD the classical notion of 'souls' as self-moving movers still lay embedded in physiological thought. Edouard Pflüger, in particular, insisted that a spinal frog because it is able to behave in so complex and integrated a manner must still possess a soul and this, of necessity, had to be located in the spinal cord. Pflüger developed his theory with relentless logic. Divide the cord into two parts, he says, and the soul is divided into two. These and similar ideas attracted the attention of T H Huxley in his celebrated essay 'On Animal Automatism' (1874). Huxley, however, drew the opposite conclusion. A spinal frog must be an automaton; hence why should we not treat an intact frog as a mechanism too?

Returning, however, to the 1830s and to Marshall Hall's work on spinal reflexes, we find that his paper in the *Philosophical Transactions of the Royal Society* (1833) establishes the modern concept of the reflex arc. The central concept is expressed in the following passage where, speaking of muscular actions mediated through the spinal cord, he writes:

> In this kind of muscular motion, the motive influence does not originate in any central part of the nervous system, but at a distance from the centre: it is neither spontaneous in its action, nor direct in its course; it is, on the contrary, excited by the application of appropriate stimuli, which are not, however, applied immediately to the muscular or neuro-muscular fibre, but to certain membranous parts, whence the impression is carried to the (spinal) medulla, reflected, and reconducted to the part impressed, or conducted to a part remote from it, in which muscular contraction is effected (25).

Hall carried out innumerable experiments on animals as various as turtles, frogs and snakes. He believed that his analysis of reflex activity, to which he claimed to have devoted some 25 000 hours of his life, was comparable in importance to William Harvey's discovery of the circulation of the blood. Indeed, there are certainly some similarities. The nerve 'impulse' does 'circle' from receptor through the central nervous system to the effector. Hall introduced the term reflex *'arc'* to bring out this analogy. The connection between sensory input and motor output was made within the central nervous system, still largely *terra incognita* when Marshall Hall was writing. Harvey, too, we may remember, did not know exactly how blood from the arterial system passed across into the venous system. The capillaries were discovered

(24) Perhaps the best-known eighteenth century exponent of reflex action was Robert Whytt. His *Essay on the vital and other involuntary motions of animals* (1751) did much to stimulate research into the physiology of the central nervous system.

(25) M. Hall, On the reflex function of the medulla oblongata and medulla spinalis, *Phil. Trans. Roy. Soc.*, **123** (1833), 635–65.

by Malpighi in 1661, four years after Harvey's death. Finally the reflex concept ousted all notion of spinal 'soul' or 'vital force', just as Harvey's theory eliminated the Galenical notion that the left ventricle was the seat of a 'dark' living fire, the source of the body's heat. The reflex concept, by the end of the nineteenth century, had come to form a basis for the understanding of the physiology of the central nervous system in somewhat the same way as Harvey's great discovery had come to form the basis for the whole of mammalian physiology.

The establishment of the neuron doctrine

The man who played Malpighi to Marshall Hall's Harvey was Ramón y Cajal. His preeminence is not, however, as clear cut as Malpighi's. At least two other prominent workers must share his position. Both Wilhelm His and August-Henri Forel published the results of histological work in 1887 which showed that the central nervous system consisted of multitudinous *separate* units, the nerve cells or neurons. Cajal, however, working independently of these two investigators, provided very extensive and exact evidence for the independence of neurons and neuroglia within the central nervous system. The publication of his results in 1888 marked the beginning of the end for the alternative and very influential doctrine propounded by Golgi and others that the brain consisted of a great reticulum of branching and anastomosing fibres: it marked the end of the beginning of the neuron doctrine.

The transmission of impulses through the spinal cord from sensory to motor nerve could thus not be continuous, as was the flow of blood through the capillaries from the arterial to the venous side of the circulation. If Cajal was right and the central nervous system consisted of a multitude of discrete functional units, the neurons, then impulses would have to 'jump' from one neuron to the next as they circulated through a reflex arc. If this were indeed the case, then these transmission zones between neurons might be of considerable functional significance.

This consideration did not escape C S Sherrington who, in 1906, wrote:

> If there exist any surface of separation at the nexus between neurone and neurone, much of what is characteristic of the conduction exhibited by the reflex arc might be more easily explicable....Such a surface might restrain diffusion, bank up osmotic pressure, restrict the movement of ions, accumulate electric charges, support a double electric layer, alter in shape and surface tension with changes in difference of potential, alter in difference of potential with changes in surface tension or in shape, or intervene as a membrane between dilute solutions of electrolytes of different concentration or colloidal suspensions with different sign of charge. It would be a mechanism where nervous conduction, especially if predominantly physical in nature, might have grafted upon it characters just such as those differentiating reflex-arc conduction from nerve trunk conduction... (26).

(26) C.S.Sherrington, *The Integrative Activity of the Nervous System*, Cambridge University Press (1906), p.17.

Sherrington had himself introduced the term 'synapse' to describe this all-important junction in 1897.

C S Sherrington is, of course, of far greater importance in the history of neurophysiology than merely as the coiner of the term 'synapse'. He it was who demonstrated the full subtlety of spinal reflexes. He emphasised that the single isolated reflex was to some extent an abstraction. Like the neuron, the unit of neural anatomy, reflexes were never in ordinary life discrete and unitary. Always they occurred in permutations and combinations whose overall effect was to adapt the organism to the situation in which it found itself. The isolated reflex, the pathologist's 'knee-jerk' reflex, or the Babinski reflex, was artificial: in the normally behaving animal, single reflexes were merely the functional units from which the total behaviour was built. It was the peculiar function of the spinal cord to integrate, as the title of Sherrington's famous Silliman lectures (27) indicates, this multichannel outflow. Sherrington was, moreover, quick to admit that the reflex concept, although fully adequate for the spinal cord, might well prove inappropriate when the physiology of the brain, especially the physiology of the cerebrum, came to be analysed.

Sherrington's analysis of the physiology of the spinal cord has been largely responsible for the orientation of twentieth-century neurophysiology. His approach was always couched in terms of the neuron theory and he came to place great emphasis, as the quotation above shows, on synaptic transmission. This approach has energised much twentieth-century work on the physiology of the central nervous system. Highly sophisticated techniques for stimulating and recording from single neurons in the central nervous system have gone far towards verifying the Sherringtonian vision. Extensive investigation of synapses and synaptic transmission has indeed shown that these minute but multitudinous junctions are crucial elements in the physiology of the brain.

Physiology of the brain

Physiological analysis of the spinal cord was not, however, the only line of advance in the latter part of the nineteenth century. Another line of investigation also takes its origin from the reflexologists of the mid-nineteenth century. Thomas Laycock (1812–1876), a Scottish physician, and I M Sechenov (1829–1905), in Russia, both asked the next question: if the reflex concept is fruitful in accounting for the behaviour of the spinal cord, might it not also prove fruitful in accounting for the physiology of the brain (28)? Might there not be cerebral reflexes analogous to the spinal

(27) *Ibid.*

(28) T. Laycock, On the reflex functions of the brain, *Brit. and Foreign Medical Rev.*, 19 (1845), 298–311. I.M. Sechenov, *Physiogische Studien über die Hemmungsmechanismen für die Reflexthatigkeit im Gehirne des Frosches,* Berlin (1863); trans. S. Belsky, in *Selected Physiological and Psychological Works* (ed. G.Gibbons), Foreign Languages Publishing House, Moscow (1962).

reflexes of the cord? As in Hartley's theory, emotions and thoughts might perhaps accompany these higher reflexes. We have already noticed that T H Huxley's famous essay 'On Animal Automatism', originating as an address to the 1874 meeting of the British Association, persuasively argued a similar point.

Laycock's ideas awaited the arrival of John Hughlings-Jackson, whom we shall discuss in the next section of this chapter, and Sechenov's theory awaited the emergence of another Russian physiologist, I P Pavlov (29). Pavlov (1849 – 1936) was a slightly older contemporary of Sherrington, and at the turn of the century these two scientists provided the growth points of central nervous system physiology. Yet, strangely, there was little, and there has since been little, cross-fertilisation. Pavlov, apparently, found Sherrington's approach to the brain incomprehensible. He was unable to understand how a neurophysiologist could, at the end of some sixty years of intensive study, still agree with du Bois Reymond that the answer to all questions about the relationship of mind to matter must remain 'ignorabimus': it can and will never be known. He suspected Sherrington of being an idealist, a term of abuse in the Russia of 1934, and even of being senile (30).

Vice versa, Western physiology has never been able to fully incorporate the Pavlovian tradition; and, apparently, for a very paradoxical reason—it is too idealistic! Western physiologists find it very difficult to understand how the Pavlovian interpretation of higher nervous activity can be grounded in the known physical structure of the brain. Indeed, as a reviewer of Pavlov's work in *Nature* remarked, a psychological interpretation in terms of 'Association, distraction, interest, consciousness, attention, memory etc.' seems much more appropriate (31).

Pavlov came to his researches in cerebral physiology by way of investigations of the digestive tract. Indeed, it was for the latter work that he was awarded the 1904 Nobel prize. His original interest in digestive physiology is reflected in the reflex he chose to devote the latter part of his life to: the secretion of saliva. This reflex has the great advantage that it is easily quantifiable and that it can be coupled to many previously neutral stimuli. Thus salivation could be induced not only by the famous bell but also by tactile and visual stimuli. Pavlov believed that by his widely ramifying investigations of this reflex he was examining not simply reflex salivation but

(29) Pavlov makes his position quite explicit: 'Our starting point has been Descartes' idea of the nervous reflex...the whole activity of the animal should conform to definite laws... machine-like...reflexes are like the driving belts of machines of human design': *Conditioned Reflexes* (trans. and ed. G. V. Andrep), Oxford University Press, London (1927), p.7.

(30) I.P. Pavlov, *Selected Works* (trans. S. Belsky, ed. J. Gibbons), Foreign Languages Publishing House, Moscow (1955), pp. 563–9.

(31) D.Denny-Brown. Physiology of the higher functions of the brain. *Nature,* **121** (1928), 662–4

the physiology of the cerebrum itself. He conceived himself to be using the conditioned reflex as a lens to examine the working of the brain, much as contemporary geneticists were using ***Drosophila*** to investigate the nature of heredity. The concept was clearly brilliant, but the analysis which made no attempt to relate the findings to underlying nervous structure and function, has not so far proved fruitful. In spite of attempts, like those of Konorski (32), to ground the Pavlovian theory on Sherringtonian neurophysiology, the Russian school has remained somewhat apart from the main line of advance.

This main line, as has already been indicated, lies in the analysis of brain function in terms of its microscopic units. It attempts to derive the observed behaviour from the known neuroanatomical and neurophysiological substratum. Like the Pavlovian theory, it nevertheless remains firmly within the corpuscular/associationist paradigm which we reviewed in the third section above.

In addition to the Sherringtonian tradition originating in the researches of Bell, Magendie and Marshall Hall, it is, however, possible to trace a parallel tradition. This tradition stems from the phrenological speculation of Gall and Spurzheim at the beginning of the nineteenth century. These two traditions merge, as we shall see, in contemporary brain science.

The tradition stemming from the work of Gall and Spurzheim consists in a gradual elucidation of a relation between the brain's anatomy and its function. In recent years it has been possible to show that certain areas, often quite small areas, control certain bodily movements or functions, or are excited by the stimulation of defined parts of the body. This approach promises to provide the firm anatomical substratum which the Pavlovian theory lacks.

We noted in the fifth section above that the phrenological theories of the beginning of the nineteenth century did not survive experimental investigation. They were based entirely on descriptive evidence. This evidence was in some cases very widely based. Gall, for example, made a famous collection of skulls. But it *was* only descriptive, and in some cases only anecdotal. The theory was, however, symptomatic, as indicated in the fourth section, of a widespread belief that the psychological faculties were seated in distinct and separable localities of the brain (figure 22.3).

Probably the most widely accepted arguments against phrenology in the early part of the nineteenth century were those provided by Pierre Flourens. Flourens carried out a great number of experiments on the cerebral hemispheres of mammals and birds. His procedures included the surgical removal of portions of the cerebral hemispheres and also the pricking of

(32) J.Konorski, *Conditioned Reflexes and Neuron Organisation* (trans. S. Garry), Cambridge University Press (1948), and *The Integrative Activity of the Brain,* Chicago University Press (1967).

various regions of the brain: in both cases the behavioural effects of these operations were carefully observed. He concluded that

> the cerebral lobes are the exclusive sites of sensation, perceptions and volitions. All these sensations, perceptions and volitions concurrently occupy the same area in these organs. Therefore the ability to feel, to perceive, and to desire constitute only one essentially single faculty (33).

Clearly this conclusion is in direct conflict with the mosaic theory of the phrenologists. It was, however, very convenient for Flourens' own metaphysical position; for Flourens was a convinced follower of Descartes. He believed, like the philosopher, that introspection reveals an inward unity; and if the mind was a unity, was it not likely, *contra* the phrenologists, that its material substratum, the brain, was so also? 'Unity', he writes, 'is the outstanding property which rules. It is everywhere, it dominates everything. The nervous system therefore forms but a single system' (34).

Although Flourens' work went far to discredit the ideas of the phrenologists among the scientific community, Gall was not himself unable to reply. He pointed out that Flourens' operative procedures were contrary to the organisation of the cerebrum. Flourens was accustomed to slice through the cerebrum progressively and horizontally. Gall writes that 'he mutilates all the organs at once, weakens them all, extirpates them all at the same time' (35). In fact this criticism was not effective against all of Flourens' experiments, but for some it was certainly valid.

In the first half of the nineteenth century, however, the opprobrium in which phrenology was held turned interest away from attempts at cerebral localisation. Nonetheless several physicians, especially in France, persisted. Both Bouillaurd and Auburtin in the mid-nineteenth century attempted, for example, to locate the centre or centres concerned with speech. Bouillaurd, in 1825, pointed out that post-mortems frequently showed that loss of speech was associated with damage to the anterior lobes of the brain. Later, Bouillaurd's son-in-law, Auburtin, also argued for the same theory. However, the clinching evidence was provided by a third French physician, Paul Broca. In the early 1860s, Broca produced evidence which strongly suggested that the 'seat of articulate language' was almost certainly to be found 'in the posterior third of the third frontal convolution' (36).

Broca had achieved his results by a combination of observing pathological behaviour (an aphasic patient) and anatomical investigation (of the deceased patient's brain). This clinical approach was shared by John Hughlings-Jackson; and it is to Hughlings-Jackson and his junior contemporary David

(33) P. Flourens, *Recherches expérimentales sur les propriétés et les fonctions du système nerveux dans les animaux vertébrés*, Paris (1824); Clarke and O'Malley, *ibid.*, p. 487.
(34) *Ibid.*, p. 488.
(35) F.J. Gall (1835), quoted in Young, *ibid.*, p. 61
(36) P.P. Broca, Siège du langage articulé, *Bull. Soc. Anthrop.*, Paris, 4 (1863), 200–2; quoted in Clarke and O'Malley, *ibid.*, p. 497.

Ferrier that we owe our present concepts of cerebral localisation. Hughlings-Jackson explicitly regarded the disease process as an experiment made upon the brain: '...I have considered convulsion as *a symptom of disease of the brain;* and also for the purposes of anatomy and physiology, *as an experiment made upon the brain by disease*' (37).

Hughlings-Jackson, David Ferrier and cerebral localisation

John Hughlings-Jackson had studied medicine at York, where he was associated with Thomas Laycock. Laycock was later to become Professor of Medicine at Edinburgh, and we have already noticed the significance of his ideas on cerebral reflexes. We shall soon see that these ideas exerted a considerable influence on the young Hughlings-Jackson.

In 1859, at the age of 24, Jackson arrived in London from York, determined to give up medicine in favour of philosophy. Fortunately for neurology, he was argued out of this intention. However, he always retained, as will be apparent from the following account, an interest in philosophy. This interest was nourished by his close friendship with Herbert Spencer. He pays tribute to his friend's philosophy in many places in his published writings. He also pays tribute to Thomas Laycock. In the 1875 paper already mentioned he commences with a quotation from his old colleague:

> Four years have elapsed since I published my opinion, supported by such facts as I could then state, that the brain, although the organ of consciousness, *is* subject to the laws of reflex action; and that in this respect it does not differ from the other ganglia of the nervous system . . . (38).

Jackson accepts this position. He regards the physiology of the cerebrum as merely a more complex case of the physiology of the cord. In the cord the Bell–Magendie law is plain for all to see. Environmental energies impinging on the organism lead through the inbuilt structure of the reflex arc to a stereotyped reaction. In the cerebrum it is only the complexity which obscures the same phenomenon: a complexity which Jackson saw, under Spencer's influence, as having evolved from the simpler primordia of the cord. In the cerebrum, however, the response is of the entire organism, not simply of a small group of muscles as is sometimes the case with spinal reflexes; and the response, moreover, is often made to a very precise stimulus, not to the rather unspecific stimuli to which the cord often responds. Jackson writes as follows:

> In all centres the connection of sensation and motion is, I think, necessary.
>
> In the lower centres there is direct adjustment of few and simple movements to few and simple peripheral impressions. In the very highest centres also there is a similar adjustment, but then it is of exceedingly special movements (representing movements of the whole organism) to the most special impressions from the environment. So far for resemblances; there is reflex action in each case. Now for certain differences. The difference is not that

(37) J.Hughlings-Jackson, On the anatomical and physiological localisation of movements in the brain (1875), in *Selected Writings*, vol. 1, p.37; italics in the original.

(38) T. Laycock, *British and Foreign Medical Review,* 19 (1845), 298.

> reflex action is the characteristic of the lower centres, but that exact and perfect reflex is characteristic of them. In the higher centres as Laycock long ago insisted, reflex action occurs, but it is imperfect. In the simple reflex actions of lower centres the movements follow the afferent excitations with little or no delay. In the highest centres, if my speculations as to motor and sensory regions be correct, the movements will not be immediate. For the speculation is, that there is in the cerebral hemispheres wide geographical separation, and thus probably some delay in action (39).

The Bell—Magendie law, Marshall Hall's concept of the 'reflex arc', gave an anatomical substratum to the psychological laws of the mind. Jackson puts it this way: 'sensori-motor processes are the physical side of, or, as I prefer to say, form the anatomical substrata of, mental states' (40). Thus it made sense once more to search for character in the architecture of the brain. In the light of these ideas Hughlings-Jackson made a fresh approach to the study of epilepsy. He no longer concerned himself with cataloguing whether or not there was loss of consciousness, an onset of tongue-biting, or relaxation of the sphincters: he was far more concerned to establish the nature of the spasm's onset, its progress through the body's musculature and the condition, paralysed or otherwise, in which the patient was left. Like Broca before him, he was intent upon connecting the behavioural symptoms to the pathology of certain specific regions of the brain. When, in 1873, David Ferrier completed the first of his pioneering brain stimulation experiments, Jackson was delighted to find his theories confirmed.

David Ferrier had indeed undertaken the seminal experiments at the West Riding Lunatic Asylum with the direct intention of subjecting Hughlings-Jackson's concepts to experimental test.

> The objects I had in view in undertaking the present research were twofold: first to put to experimental proof the views entertained by Dr. Hughlings-Jackson on the pathology of epilepsy, chorea and hemiplegia by imitating artificially the destroying and discharging lesions of the disease which his writings have defined and differentiated—and secondly to follow up the path which the researches of Frisch and Hitzig (who have shown the brain to be susceptible to galvanic stimulation) indicated to me as likely to lead to results of great value in the elucidation of the functions of the cerebral hemisphere, and in the more exact localisation and diagnosis of cerebral disease (41).

Ferrier later collected the results of his researches into his book, *The Functions of the Brain* (1876). His technique, as indicated above, was to subject the brains of experimental animals (monkeys) to electrical stimulation and to observe their subsequent behaviour. He quickly found that stimulation directly in front of the fissure of Rolando resulted in 'certain definite and constant movements of the hands, feet, arms, facial muscles, mouth and tongue etc.'. He had discovered what we now call the motor cortex: as Sir Charles Sherrington wrote in his obituary (1928), 'He

(39) Hughlings-Jackson, *ibid*, (1875), pp. 60–1.
(40) *Ibid.*, p.49.
(41) *West Riding Asylum Reports*, 3 (1873), 30.

established the localisation of the motor cortex very much as we know it today. He showed that its focal movements were obtainable with such definition and precision that "the experimenter can predict with certainty the result of stimulation of a given region" (42). Ferrier's stimulation experiments investigated not only the motor cortex but the whole cerebrum. He attempted to establish centres for hearing, vision, olfactory and tactile sensations, and so on.

In addition to being an experimental physiologist, Ferrier was also a clinician. In consequence the human significance of the results he was obtaining from monkeys was of great importance to him. He went so far as to transfer the map of cerebral localisations which he had drawn for the monkey unaltered to the human brain. The result (figure 22.4), which is depicted in

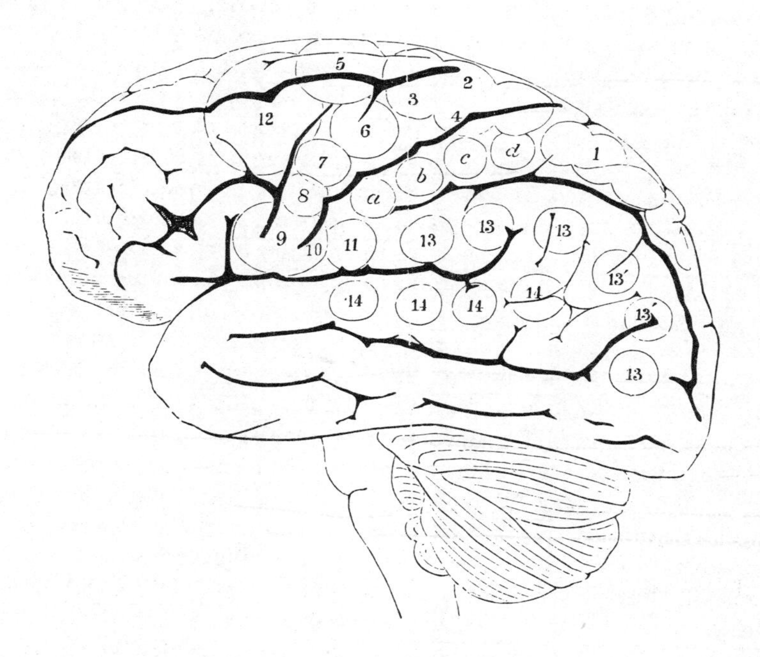

Figure 22.4. Lateral view of the human brain, from David Ferrier, *The Functions of the Brain*, (second edition, 1886), figure 130. The numbers signify functions, such as: 1, position of centres for movements of opposite leg and foot . . . ; 2, 3, 4, centres for complex movements of arms and legs . . . ; 5, centre for extension forward of the arm and hand . . .

(42) C.S.Sherrington, Obituary notice of Sir David Ferrier, *Proc. Roy. Soc. B*, **103** (1928), viii–xvi.

his book, *The Functions of the Brain*, is strikingly reminiscent of the phrenological diagrams constructed by Gall and Spurzheim at the beginning of the century!—but with this all-important difference: instead of mapping psychological proclivities such as benevolence, conscientiousness, mathematical capacity, and so on, Ferrier mapped physiological characteristics such as movements of the arms and legs, centres for vision and taste, and so on. There is no suggestion that these regions vary in size from person to person, or that their nature can be ascertained by palpation through the skull.

The comparison of brains to the artefacts of contemporary technology

With the work of Ferrier, neurology enters its modern phase. Once again the tradition was funnelled through Sherrington, whose book *The Integrative Activity of the Nervous System* is dedicated to Ferrier. The combination of cerebral localisation and the Sherringtonian emphasis on analysis in terms of unit neurons has subsequently proved very fruitful. Much interesting electrophysiological work has examined the response of single cerebral neurons to patterned stimuli presented to the organs of special sense. *Vice versa*, the single unit recording technique in the hands of Sherrington's pupil, J C Eccles, has proved its worth in the analysis of motor output. It has become possible to develop plausible schemata to account for the organisational functions of the cerebellum; to show how, working as a species of 'neuronal computer', it manages the orchestration of innumerable skeletal muscle actions to produce smooth, aesthetically satisfying, behaviour.

In the 1970s, very much still remains to be learnt about the structure and functioning of the brain. However, there seems every reason to believe that the Cartesian approach, the treatment of the brain as an extraordinarily complicated machine, will in the end prove successful. Descartes, it will be recalled, compared the brain to the mechanical and hydrodynamic machinery of his own day. The analogy was very strained. In the last third of the twentieth century our technology has provided the computer for our comparisons. Modern students of artificial intelligence would scoff at Vaucanson's flute player. Compared with the robots which NASA plans to send to the surface of Mars, Vaucanson's automaton is as a spinal frog to the intact animal. The situation in neurobiology thus resembles the situation in morphology and morphopoiesis outlined in the last chapter. Biological analysis and technological advance approach each other rather like teams of excavators building a tunnel beneath the Alps. In the seventeenth century the analogy between brains and computers was so far-fetched as to be almost absurd. Little was known about neurophysiology and automation had hardly begun to develop. Three centuries later physiologists know far more about how the brain works and technologists are capable of building far more lifelike artefacts. What in 1640 seemed laughable must today be taken seriously.

The 'mental' and the 'physical'

This brings us very nearly to the end of our long argument. The circle is nearly closed. In the opening chapters of this book we saw the world through the eyes of the early philosopher – scientists as a living organism; now we see organisms, including our own bodies and brains, through the eyes of contemporary scientist – philosophers as machines. We have seen how this perceptual alteration has developed in parallel with the development of our technology. It seems that in order to feel ourselves thoroughly to have understood anything we have first to have built it ourselves (43). The profound understanding which follows on this activity can then provide models by means of which the phenomena of the natural world can be organised.

A special aspect of this intellectual movement from the organismic to the mechanistic world-picture has been the decline and fall of the concept of the soul. This source of activity, of creativity, perhaps derived from the muscle-powered technologies of antiquity, disappeared first from the astronomical world, then from the world of chemistry, then from the biological world and lastly, as we have just seen, from the psychological world. The historian of psychology, E G Boring, sums this progression up in a pithy phrase: 'Psychology first lost its soul, then its mind, and then it lost its consciousness, but it still has behaviour of a kind...' (44).

This historical movement is mirrored by the structure of this book. In chapter 1 we discussed the Coleridgean view of perception and imagination. We saw how Coleridge, following Kant, sought to treat them both as genuinely creative processes. In this way the Romantics hoped to escape the deadening influence of the associationist philosophy in which the mind was conceived to consist of but a 'bundle' of discrete, externally generated, perceptions (45). But in the present chapter we see that the mechanistic view has finally prevailed. The brain has come to be regarded as a machine, albeit enormously complicated, but nevertheless a machine—an object. It is, according to the tradition of Descartes, of Hobbes, Locke, Hume and Hartley, of Laycock and Sechenov, a reflex mechanism responding in a highly complex manner, perhaps, to the environmental energies which impinge upon it. It is, in the last analysis, no more capable of independent, 'creative' activity than is the leg when the patellar ligament is tapped. The mind, according to the tradition a mere epiphenomenon of the brain, must

(43) Strauss (1965) argues that this provides the reason for the high epistemological status of mathematics: for here we have even constructed the axioms ourselves.
(44) Quoted in E.R.Taylor, A new view of the brain, *Encounter,* **36** (1971), 35.
(45) The deadening influence of this philosophy seems to have played a large part in the nervous breakdown suffered by J.S.Mill, perhaps the foremost exponent of the 'technologico Benthamite' cause. It is interesting to note that Mill consoled himself with the Romantic poetry of Wordsworth and Coleridge and the equally Romantic music of Weber.

similarly be regarded as nothing more than the shadow of a reflex mechanism. We started out with an 'interior' account of our experience, in particular of our experience of doing science. We saw how the apparently creative activity of the mind allows us to order our experience in different ways, according, perhaps, to different paradigms. We have ended by eliminating the subjective aspect altogether. We seem to have eliminated the possibility of knowledge itself.

But have we? Does contemporary brain science really have this implication? The answer to this question turns, of course, on the relationship of our conscious experience to the tangle of neurons within the skull. To pick up the beginnings of this answer let us return to John Hughlings-Jackson. It will be recalled that in 1859 when he first arrived in London he had to be persuaded that neurology was a better prospect than philosophy. It is thus not surprising to find that the man who is usually regarded as the founder of modern neurology has also been a seminal influence on the development of ideas relating mind and brain. Hughlings-Jackson's position is made quite clear in the paper from which we have already quoted. He insists, in 1875, on a species of the mind – brain identity theory which has, a hundred years later, again become popular among metaphysicians (46). His position is well epitomised in the passage which we quoted in the seventh section of this chapter: 'Sensori-motor processes are the physical side, or as I prefer to say, form the anatomical substrata of, mental states' (47).

This position is not to be confused with the psychoneural parallelism mentioned above. Hughlings-Jackson does not suppose, as Sherrington seems to have supposed (48), that there were two things, two substances, body and mind, and that in some mysterious way activity in one is always accompanied by concomitant activity in the other. Jackson is quite explicit. He contrasts the parallelism of Tyndall with the identity theory of Lewes:

> Tyndall writes that the passage from the physics of the brain to the corresponding facts of consciousness is unthinkable. Granted that a definite thought and a definite molecular motion occur simultaneously we do not possess the intellectual organ, nor apparently any rudiment of an organ, which would enable us to pass by a process of reasoning from one phenomenon to the other. They appear together we know not why (49).

Many other authorities thought similarly. Lewes, however, as Hughlings-Jackson points out, disagreed:

(46) See, for example, *The Mind–Brain Identity Theory* (ed. C.V.Borst), Macmillan, London (1970).
(47) Hughlings-Jackson, *ibid.*, p. 49.
(48) Sherrington writes as follows: 'That our being should consist of two fundamental elements offers, I suppose, no greater inherent improbability than that it should rest on one only.' Introduction to second (1947) edition of *The Integrative Activity of the Nervous System.*
(49) Hughlings-Jackson, *ibid.*, p. 41.

> That the passage of motion into a sensation is unthinkable, and that by no intelligible process can we follow the transformation I admit; but I do not admit that there is any such transformation....I ask, who knows this? On what evidence is the fact asserted? On examination it will appear that there is no evidence at all for such a transformation; all the evidence points to the very different fact that the neural process and the feeling are one and the same process viewed under different aspects. Viewed from the physical or objective side, it is a neural process; viewed from the psychological or subjective side, it is a sentient process (50).

Jackson is fully in agreement with this analysis. He is not interested to pursue the philosophical question further. 'I do not trouble myself', he writes, 'about the mode of connection between mind and matter...indeed so far as clinical medicine is concerned, I do not care' (51). He is, however, concerned to keep the two aspects of his underlying unity separate. As he says in another paper, some people, conflating the two linguistic systems, 'speak of a brain as if it were a "solid mind"' (52). And, indeed, he does not regard himself to be always quite blameless in this respect. For does not the term 'sensori-motor process' which, as we have seen, he takes to be the 'anatomical substratum' of a mental process, itself appear ambiguous? The term 'sensation' is often, as he points out, applied both to the physical stimulus and to the resulting mental state. This need not, however, be so. It is quite possible to define sensory, or afferent, nerves and centres without reference to the psyche. Furthermore, it is no more and no less mysterious that activity in these centres has a psychological aspect than that activity in the motor centres should also have a psychological accompaniment; for it is Hughlings-Jackson's contention that it is activity in these efferent centres which is the physical aspect of language and thought.

Jackson's account of the mind – body problem derives, as he many times acknowledges, from the philosophical work of Bain, Lewes and Spencer. In Jackson's case, however, the philosophy is grounded in a profound study of neuropathology. For Jackson, as we have already noticed, every disease of the brain was an experiment. Every epilepsy helped him to further his analysis of the brain's physiology, helped him to confirm the theoretical framework he had borrowed from the philosophers. And the psychoneural identity theory, or, as some authorities phrase it, the 'dual aspect theory', remains today the working hypothesis of most neurobiologists. For the scientist the philosophical puzzle is fruitless. This is just the way things are. If pressed he might recall Galileo's apophthegm (or Bacon's variant): 'I have always accounted extraordinarily foolish those who would make human comprehension the measure of Nature' (53).

(50) G.H. Lewes, *Problems of Life and Mind,* Trubner, London (1879), p.459.
(51) Hughlings-Jackson, *ibid.,* p. 52.
(52) J. Hughlings-Jackson, On the scientific and empirical investigation of epilepsies (1874), *Selected Writings,* vol. 1, p. 170.
(53) G.Galileo, *Dialogues on the Two Great Systems of the World,* Day 1.

He would also point out that the philosophical misgivings of Coleridge and others are nowadays obsolete. For these misgivings, it will be remembered, were occasioned by the empiricist theory of mind. A 'bundle of sensations', a concatenation of impressions, 'entirely loose and separate', could not, it was felt, describe the mind (54). The reader may, however, perceive an analogy here. For did not the Aristotelians deride the Democritean biologists for a very similar reason? How could a mere congeries of atoms be conceived to constitute an organism? The development of the Darwinian theory and the twentieth-century advances in chemistry, biochemistry and molecular biology have very clearly shown how this *is* possible.

Similarly with the 'bundle of sensations': the development of neurophysiology has shown how the information which reaches the brain is selected for its relevance to the individual's situation. The work of Hubel and Wiesel and others has, in other words, begun to show us how we can discern a physiological counterpart to Coleridge's 'primary imagination' (chapter 1). The brain is analogous to the mind in being perpetually active, organising and reorganising the patterns of incoming impulses: it, too, is far from being a passive *tabula rasa*. Furthermore, there is no good reason to suppose that our introspected mental 'coherence' with which the Romantics made so much play does not have an analogous physical aspect. Although, as we have seen, many facets of behaviour have been shown to have localised 'control centres' in the brain, the total activity may well be highly integrated—as is, for example, that of the multicellular body. Mental states may well correspond to complex 'global' brain states. We should not allow ourselves to be led astray by the residue left behind by the associationist psychology: the notion of ideas as discrete, colliding, attractive corpuscles; of 'thoughts' waiting in an antechamber to be summoned separately to the bar of consciousness (55). Instead we should envisage the brain as a complex three-dimensional system whose state varies in innumerable ways from moment to moment. The brain state corresponding to a 'thought', an 'idea', a 'feeling' or a 'memory' may thus arise from the cooperative interaction of innumerable neurons spread throughout large regions of the brain (56). The neurophysiological aspect of the 'association of ideas', of 'recognition', of 'searching for the right word' is thus very different from that imagined by David Hartley and resembles much more closely our subjective experience of these phenomena.

The notion, in other words, that one inspects one's thoughts, one's

(54) See, for example, T.H.Green's classical statement of this view in the introduction to his edition of Hume's *Treatise of Human Nature.*

(55) See, for instance, the account given by Poincaré in Mathematical Creation, *The Creative Process* (ed B.Ghiselin), University of California Press (1955).

(56) See, for example, *Science*, 177 (1972), 850–64, where E.R.John writes that his findings suggest that 'coherent temporal patterns in the average activity of anatomically extensive neural assemblies may constitute the neurophysiological basis of subjective experience'.

memories, one's after-images seems to be merely a hangover from scholasticism and from the para-optical theories of the seventeenth century: from Descartes' pineal images and from Willis' medullary screens, and so on. It is another instance of the 'linguistic drag' which we examined in chapter 3. The true situation is much more interesting. Jean-Paul Sartre has pointed out that the consciousness we reflect upon is an object like other objects in the world. The consciousness which does the reflecting can only be described metaphorically. Sartre writes that it is like a 'rent in the tissue of being' (57). As we can see it is the opposite pole of the 'global brain state': it is the opposite aspect of the dual aspect theory of the mental and the physical.

Thus, to sum up, there seems to be no justification for criticising the dual aspect theory because of faults in an out-of-date empiricist doctrine of the mind. There is no reason to suppose that our introspected mental unity, our powers of 'co-adunation', do not possess a perfectly adequate physical aspect in the mechanisms discovered, and being discovered, by contemporary neurophysiologists.

An ineradicable duality

We may thus conclude our journey through the historical and philosophical hinterlands of modern biology in search of an answer to Shelley's question by emphasising the ineradicable duality of our present outlook. At the beginning of this chapter we asked whether there might not be a 'symmetry' in the history of ideas; whether it is as inappropriate to treat mentality in terms of post-Galilean science as it is to treat the projectile in terms of Aristotelian science. We can now see that this is a false trail. The historical process is not best figured as a pendulum swinging from one invalid extreme to another but, as we stressed in some of the earlier chapters of this book, as a progressive separation and demarcation of an objective from a subjective realm. This modern dichotomy was not clear to the ancients, and it is not clear in Aristotle's system. It has only become inescapable in very recent times.

Our contemporary position emerges from an analysis stretching over two thousand years and more of a once far more holistic and unparticularised experience. The analysis, as we have seen, has not been continuous. In some periods it has hardly progressed at all; in other periods it developed more rapidly than the known facts warranted. The rate of advance, as we also noticed in earlier chapters of this book, seems to be correlated in a complex way with the social milieu. In modern times the manipulative *zeitgeist* of Western economies has allowed the 'objective' treatment of organisms and

(57) J-P. Sartre, *Being and Nothingness: an essay in phenomenological ontology* (trans.H.E. Barnes), Methuen, London (1957), p. 617. Wittgenstein *(Tractatus Logico-Philosophicus,* 5, 632) makes much the same point: 'The subject does not belong to the world: rather it is the limit of the world'.

brains to gain a very considerable momentum.

This operational or nominalist outlook makes it possible for us to use the analogies derived from our technology to describe the living world in terms of inanimate objects responding to efficient causation. On the other hand, however, we remain inescapably aware of our own selves as psychological entities struggling to achieve objectives. Thus whereas we may, on the one hand, treat organisms as machines, doctor them with chemicals or tinker with their genes, we may also, on the other hand, 'see' them as analogous to ourselves: sentient beings.

Indeed, one might perceive here an analogy to the complementarity principle so celebrated among quantum physicists. An entity, it will be recalled from chapter 21, may for some purposes be treated as a particle, for others as a wave. Similarly with living creatures. As the *scala naturae* is ascended, from microorganism to man, the frame of reference changes progressively from mechanistic to mentalistic. Whereas one may treat *E. coli* as a chemical machine and manipulate it accordingly, this would often be counterproductive when dealing with the behaviour of *H. sapiens*. This biological 'complementarity principle' applies, moreover, not only to the *scala naturae* but also to the different levels of organisation within the 'higher' animal. It would clearly be appropriate to treat the tissues of the body mechanistically, whereas such treatment might well be inappropriate for cases of behavioural derangement (58).

Finally this profound duality in our view of the world does not cut us off from our most precious possession: the perception that we are goal-seeking, creative persons. Although such characteristics have their objective correlative in the cybernetic concepts of Claude Bernard and his successors; and although the origin of a cybernetic mechanism, such as a human being, has been accounted for by Charles Darwin and *his* successors as yet one more event in crustal geochemistry, no more, no less, surprising than any other—yet, nevertheless, our 'internal' perception of these processes remains equally valid. 'The teleology we are *aware* of in the mind', writes C S Sherrington, 'is no greater, no less, than that which we *observe* in the organism's body...our psychical lives are tissues of purposes' (59). We are free (60), creative, beings, and with Andrew Marvell can assert that

> The mind creates exceeding these
> Far other worlds and other seas.

(58) See, for example, issues of *J. Psychosomatic Res.*, especially **15** (1971), no. 4, Psychosomatic disorders of movement.

(59) C.S. Sherrington, *The Integrative Activity of the Nervous System*, Cambridge University Press (1947).

(60) That the dual aspect theory does not eliminate human freedom and hence responsible action is well shown by D. M. MacKay in *Freedom of Action in a Mechanistic Universe*, Cambridge University Press (1967), and elsewhere.

APPENDIX 1

Significant dates 585 B.C. – A.D. 1953. Low magnification survey [scale 100y ≈ 8mm].

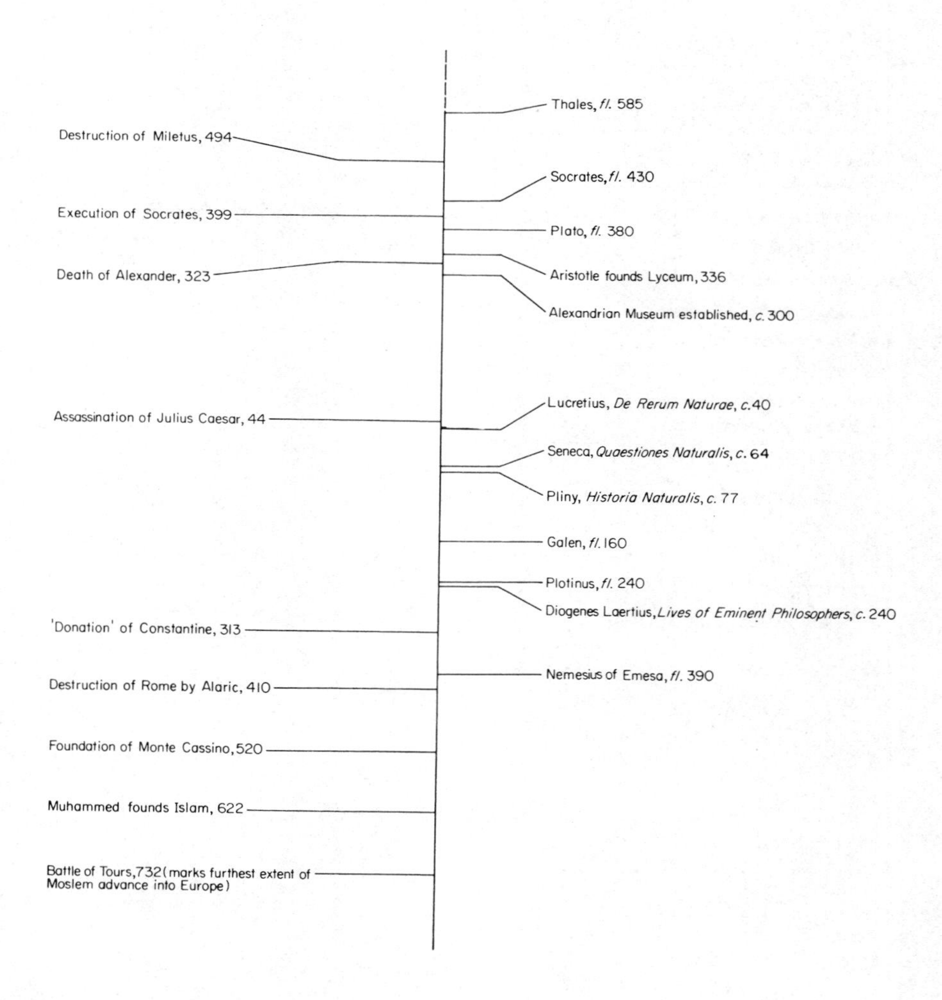

Appendix 1 (continued)

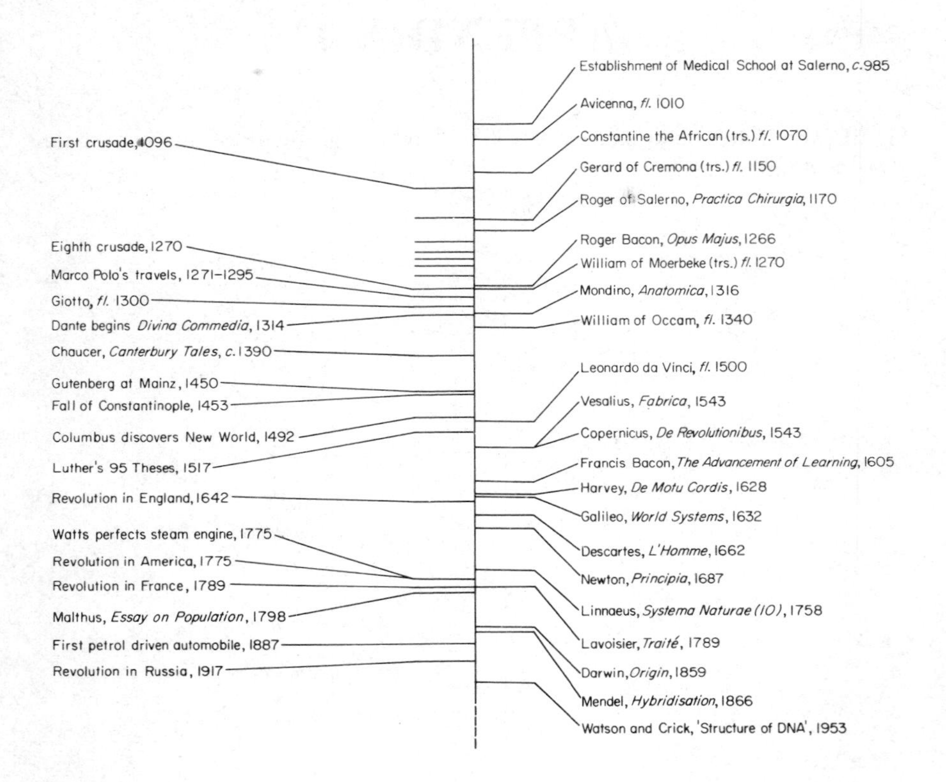

APPENDIX 2

Significant dates A.D. 1450 — A.D. 1957. Higher magnification [scale 100y ≈ 37 mm]

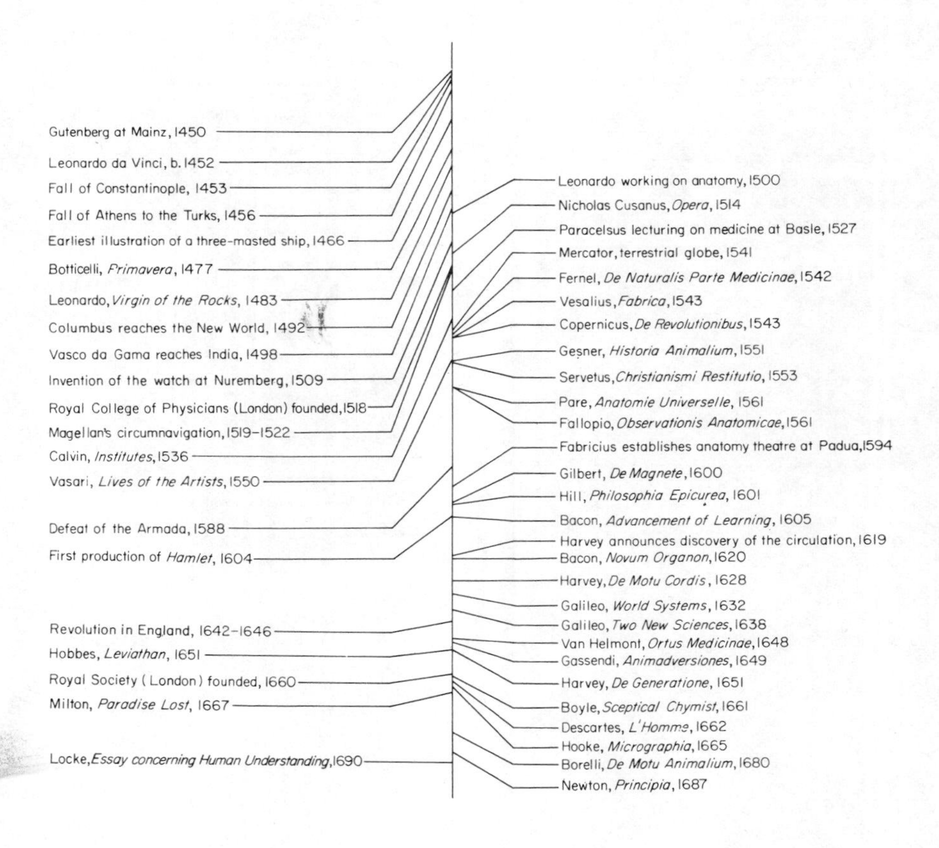

Appendix 2 (continued)

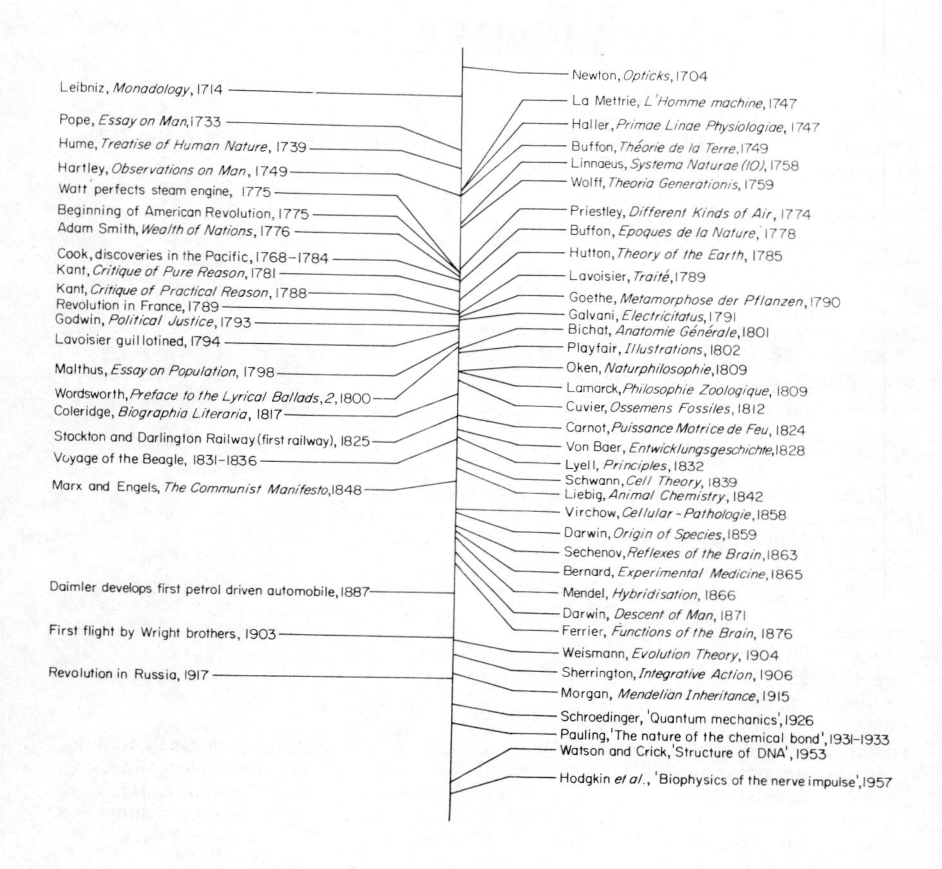

BIBLIOGRAPHY

Although it has been argued that all that is living in the work of our predecessors has been incorporated into the structure of contemporary science, it seems to me that there is no substitute for reading the original work (or its facsimile) if one wishes to grasp the full depth and force of the originating mind. I hope that if this book has succeeded in nothing else it will have succeeded in raising questions which will send the reader back to examine the work of our intellectual ancestors. To facilitate this, the bibliography collects together most of the books and papers on which the foregoing pages have been based. I have endeavoured, wherever possible, to list editions and translations available to the general reader. I have also included biographies of most of the major figures. The reader should thus be able to gain entry into some of the fascinating topics and make closer acquaintance with some of the outstanding figures which I have had space only to sketch.

Abbott, T.K., 1927, *Kant's Theory of Ethics,* Longmans Green & Co. Ltd., London.

Ackerknecht, E.H., 1925, *Rudolf Virchow: Doctor, Statesman. Anthropologist,* University of Wisconsin Press, Madison, Milwaukee and London.

Adelmann, H.B., 1966, *Marcello Malpighi and the Evolution of Embryology,* 5 vols., Cornell University Press, Ithaca, N.Y.

Adkins, A.W.H., 1970, *From the Many to the One,* Constable & Co. Ltd., London.

Adolph, E.F., 1961, Early concepts of physiological regulations, *Physiol. Rev.,* **41,** 737.

Agassiz, L., 1866, *The Structure of Animal Life,* Charles Scribner & Co., New York.

Alembert, J.L. d', 1751, *Preliminary Discourse to the Encyclopedia of Diderot,* Paris; trs. Schwab, R.N., Bobbs-Merrill Co. Inc., Indianapolis, New York, 1963.

Amacher, M.P., 1964, Thomas Laycock, I.M. Sechenoy and the reflex arc concept, *Bull. Hist. Med.,* **38,** 168 - 83.

Arber, A., 1964, *Goethe's Botany,* Waltham, Mass.

Aristotle: References to Aristotle's writings are made, as is customary, to Bekker's edition of the Greek text published in 1831 (*Aristotelis Opera Edidit Academia Regia Borusca,* vol. 1 - 2, Aristoteles Graece ex recognitione Immanuelis Bekkeri, Berolini, 1831). Most translations into modern European languages make use of the page number, column and line of this great edition. References are thus, for example, *Nichomachean Ethics* 1103b30: page 1103, column b (right hand of the two columns), line 30.
Categoriae, trs. Edghill, E.M.
Analytica Priora, trs. Jenkinson, A.J.
Analytica Posteriora, trs. Mure, G.R.G.
Physica, trs. Hardie, R.P. and Gaye, R.K.
De Caelo, trs. Stocks, J.L.
De Generatione et Corruptione, trs. Joachim, H.H.
Meteorologica, trs. Ross, W.D.
De Anima, trs. Smith, J.A.
De Respiratione, trs. Ross, G.R.T.
De Partibus Animalium, trs. Ogle, W.
De Motu Animalium, trs. Farquharson, A.S.L.
Historia Animalium, trs. Thompson, D'Arcy W.
De Iuventate et Senectute, trs. Ross, G.R.T.
See also Cherniss, H., 1935; Gomperz, T., 1912; Grene, M., 1963;

Heath, T., 1949; Jaeger, W., 1934; Ross, W.D., 1923; Thompson, D'Arcy W., 1923.
Avicenna: see Rahmann, F., 1952.

Bacon, F., *The Philosophical Works of Francis Bacon,* reprinted from the texts and translations, with notes and prefaces, of Ellis and Spedding, ed. Robertson, J.M., George Routledge and Sons, Ltd., London, 1905.
Baer, K.E. von, 1828, 1837, *Ueber Entwicklungsgeschichte der Thiere,* 2 vols., Konigsberg.
Bailey, C., 1928, *The Greek Atomist and Epicurus,* Oxford University Press, London.
Baker, J.R., 1948 - 1953, The Cell Theory: a restatement, history and critique, *Q. J. Micros. Sci.,* **89,** 103 - 25; **90,** 87 - 108; **93,** 151 - 90; **94,** 407 - 40.
Bastholm, E., 1950, *The History of Muscle Physiology,* Munksgaard, Kobenhavn.
Bayon, H.P., 1938, 1939, William Harvey, physician and biologist: his precursors, opponents and successors, *Annals of Sci.,* **3,** 59 - 118, 435 - 56; **4,** 65 - 109, 329 - 89.
Beare, J.I., 1906, *Greek Theories of Elementary Cognition from Alcmaeon to Aristotle,* Clarendon Press, Oxford.
Bell, A.E., 1943, The Concept of Energy, *Nature,* **151,** 519 - 23.
Bell, C., 1811, *Idea of a New Anatomy of the Brain,* London; reprinted in *J. Anat. Physiol.,* **3,** 147 - 82, 1869.
Berger, P.L. and Luckmann, T., 1967, *The Social Construction of Reality,* Penguin Books Ltd., Harmondsworth, Middx.
Bernard, C., 1865, *Introduction a l'étude de la médecine expérimentale,* Paris; facsimile reprint by Culture et Civilisation, Bruxelles, 1965; trs. Green, H.C., *Introduction to the Study of Experimental Medicine,* Dover Publications Inc., New York, 1957.
Bernard, C., 1872, *De la Physiologie Générale,* Paris; facsimile reprint by Culture et Civilisation, Bruxelles, 1965.
Bernard, C., 1878, *Leçons sur les phénoménes de la vie communs aux animaux et aux végétaux,* ed. Dastre, A., 2 vols., Paris.
Bernard, C., 1880, *Pathologie Expérimentale,* 2nd edn., Paris. See also Olmsted, J.M.D. and Olmsted, E.H., 1952.
Berzelius, J.J., 1813, *A View of the Progress and Present State of Animal Chemistry,* trs. Brunnmark, G., London.
Bichat, M.F.X., 1800, *Recherches physiologiques sur la vie et la mort,* Paris; facsimile reprint Gauthier-Villars, Paris, 1955; trs. Gold, F., *Physiological Researches on Life and Death,* London, 1815.
Bichat, M.F.X., 1801, *Anatomie Générale,* Paris; trs. Coffyn, C., *General Anatomy,* 2 vols., London, 1824.
Blake, W., *Poetry and Prose,* ed. Keynes, G., The Nonesuch Library, London, 1961.
Bonnet, C., 1769, *La Palingénésie Philosophique,* Geneva.
Borelli, G.A., 1680, 1681, *De Motu Animalium,* 2 vols., Roma. See also Hall, T.S., 1951.
Boring, E.G., 1929, *A History of Experimental Psychology,* Appleton, New York.
Borst, C.V., ed., 1970, *The Mind-Brain Identity Theory,* Macmillan & Co. Ltd., London.
Boveri, T., 1904, *Ergebnisse über die Konstitution der Chromatischen Substanz des Zellkerns,* Gustav Fischer, Jena.
Boyle, R., 1660, *New Experiments Physico-Mechanical Touching the Spring of the Air,* Oxford.
Boyle, R., 1661, *The Sceptical Chymist,* Cooke, London: facsimile reprint by Dawson & Sons Ltd., London, 1965.
Boyle, R., 1674, *Suspicions about some Hidden Qualities in the Air,* Oxford.
Boyle, R., 1688, *A Disquisition about the Final Causes of Natural Things,* London.
Brazier, M.A.B., 1959, The historical development of neurophysiology, in *Handbook of Physiology,* vol. 1 American Physiological Soc., Washington, pp. 1 - 58.
Brazier, M.A.B., 1961, *A History of the Electrical Activity of the Brain,* Pitman Medical Publishing Co. Ltd., London.
Brett, G.S., 1908, *The Philosophy of Gassendi,* Macmillan & Co. Ltd., London.
Bridges, C.B., 1916, Non-disjunction as proof of the chromosome theory of heredity, *Genetics,* **1,** 1-52; 107-63.
Broca, P-P., 1863, Siège du langage articulé, *Bull. Soc. Anthrop., Paris,* **4,** 200 - 2.

Brooks, C. McC. and Cranefield, P.F., 1959, *The Historical Development of Physiological Thought,* Hafner Publishing Co., New York.

Brown, K.C., ed., 1965, *Hobbes Studies,* Blackwell, Oxford.

Bruner, J.S. and Postgate, L., 1949, On the perception of incongruity: a paradigm, *J. Personality,* **18**, 206 - 23.

Buffon, G.L.L. de, 1749 - 1804, *Histoire Naturelle,* 44 vols., Paris.
1799 - 1805, *Histoire Naturelle,* new edition, 64 vols.
Vols. 1 - 3: *Théorie de la Terre.*
Vol. 4: *Epoques de la Nature.*
Trs. Smellie, W., *Buffon's Natural History,* London, 1791.

Burkhardt, J., 1945, *The Civilisation of the Renaissance,* trs. Middlemore, S.G.C., Phaidon Press, London.

Burnet, J., 1892, *Early Greek Philosophy,* Adam and Charles Black, London.

Burtt, E.A., 1950, *The Metaphysical Foundations of Modern Physical Science,* Routledge and Kegan Paul, London.

Caird, E., 1889, *The Critical Philosophy of Immanuel Kant,* Maclehose and Sons, Glasgow.

Cajal, C. Ramon y, 1888, Estructura del cerebelo, *Grac. méd.Catalana,* **11**, 449 - 57.

Cajal, C. Ramon y, 1909 - 1911, *Histologie du Systéme Nerveux de l'Homme et des Vertebres,* trs. Azoulay, M., 2 vols., Paris.

Candolle, A.P. de, 1820, Essai élémentaire de géographie botanique, in *Dictionnaire des Sciences Naturelles,* vol. 18, Strasbourg.

Carnot. S.N.L.. 1824, *Reflexions sur la Puissance Motrice du Feu,* Paris; facsimile reprint by Dawson and Sons, Ltd., London, 1966; trs. Thurston, R.H., *Reflections on the Motive Power of Fire,* Macmillan and Co. Ltd., London, 1890; Dover Publications Inc., New York, 1960.

Carré, M.H., 1949, *Phases of Thought in England,* Clarendon Press, Oxford.

Cassirer, E., 1945, *Rousseau, Kant, Goethe,* trs. Gutmann, J., Kristeller, P.O. and Randall, H., Princeton University Press, Princeton.

Cassirer, E., 1963, *Individual and Cosmos in Renaissance Philosophy,* trs. Domandi, M., Blackwell, Oxford.

Casson, S., 1939, *The Discovery of Man,* Hamish Hamilton, London.

Chaptal, J.A.C., 1790, *Eléments de Chimie,*Montpellier; trs. Nicholson, W., *Elements of Chemistry,* London, 1791.

Chapuis, A. and Droz, E., 1958, *Automata.* Central Book Co. Inc., New York.

Chaucer, G. *The Complete Works of Geoffrey Chaucer,* ed. Skeat, W.W., Oxford University Press, London, 1912.

Cherniss, H., 1935, *Aristotle's Criticism of Presocratic Philosophy,* Johns Hopkins University Press, Baltimore.

Childe, G., 1927, *The Dawn of European Civilisation,* Kegan Paul, London.

Clark, K., 1958, *Leonardo da Vinci,* Penguin Books, Harmondsworth, Middlesex.

Clarke, E. and O'Malley, C.D., 1968, *The Human Brain and Spinal Cord,* University of California Press, Berkeley and Los Angeles.

Clendening, L., 1960, *Source Book of Medical History,* Dover Publications Inc., New York.

Cleve, F.M., 1963, *The Giants of Pre-Sophistic Greek Philosophy,* Martinus Nighoff, The Hague.

Cohen, I.B., 1964, 'Quantum in se est': Newton's concept of inertia in relation to Descartes and Lucretius, *Roy. Soc. Notes and Records,* **19**, 132 - 55.

Coleman, W., 1964, *Georges Cuvier, Zoologist,* Harvard University Press, Cambridge, Mass.

Coleman, W., 1971, *Biology in the Nineteenth Century,* John Wiley and Sons, Inc., New York and London.

Coleridge, S.T., 1812, *Biographia Literaria,* London.

Coleridge, S.T., *Collected Letters,* 6 vols., ed Grigg, E.L., Clarendon Press, Oxford, 1956 - 1971.

Copernicus, N., 1543, *De Revolutionibus Orbium Coelestium,* Nuremberg; trs. Wallis, C.G., *On the Revolutions of the Heavenly Spheres,* Great Books of the Western World, 16, ed. Hutchins, R.M., Encyclopedia Britannica Inc., Chicago, London, Toronto,

1952; facsimile edition, Macmillan Press Ltd., London, 1973.
See also Crombie, A.C., 1952; Kesten, H., 1945; Koyré, A., 1957.

Cooke, J., 1762, *The New Theory of Generation,* London.

Correns, C., 1900, G. Mendels Regel über das Verhalten der Nachkommenschaft der Rassenbastarde, *Berichte der Deutsche Botanische Gesselleschaft,* **18,** 158 - 168; trs. Pitternick, L.K. in *Genetics,* **35,** Suppl. 33 - 41, 1950.

Crawley, A.E., 1909, *The Idea of the Soul,* Adam and Charles Black, London.

Crombie, A.C., 1952, *Augustine to Galileo,* Falcon Books, London.

Crombie, A.C., ed., 1963, *Scientific Change,* Heinemann and Co. Ltd., London.

Crombie, A.C., 1969, The influence of medieval discussions of method for the scientific revolution, in *Critical Problems in the History of Science,* ed. Claggett, M., University of Wisconsin Press, Wisconsin.

Croone, W., 1667, *De Ratione Motus Musculorum,* Amsterdam.

Croone, W., 1681, An hypothesis of the structure of a muscle and the reason of its contraction; read in the Surgeons Theatre Anno 1675, 1676, *Phil. Coll. Roy. Soc.,* **2,** 22 - 5.

Cudworth, R., 1678, *The True Intellectual System of the Universe,* ed. Birch, T., London, 1820.

Cuvier, G., 1799 - 1805, *Leçons sur l'Anatomie Comparée,* 2 vols., Bruxelles; trs. Ross, W., *Lectures on Comparative Anatomy,* 2 vols., London, 1802.

Cuvier, G., 1812, *Recherches sur les Ossemens Fossiles de Quadrupèdes,* Paris: trs. Whittaker, G.B., *Researches into Fossil Osteology,* London, 1835.

Cuvier, G., 1817, *Le Règne Animal,* 4 vols., Paris; facsimile reprint by Culture et Civilisation, Bruxelles, 1969; numerous translations.

Cuvier, G., 1825, *Discours sur les Revolutions de la Surface du Globe,* Paris; facsimile reprint by Culture et Civilisation, Bruxelles, 1969; trs. Jameson, R., *Theory of the Earth,* Edinburgh, 1827.
See also Coleman, W., 1964.

Dalton, J., 1808 - 1827, *A New System of Chemical Philosophy,* 2 vols., Manchester; facsimile reprint by Dawson and Sons Ltd., London, 1965.

Darenberg, C., see Galen.

Darlington, C.D., 1969, *The Evolution of Man and Society,* George Allen and Unwin Ltd., London.

Darlington, C.D., 1969, Genetics and Society, *Past and Present,* **43,** 3 - 37.

Darwin, C.R., 1836, *Journal of Researches,* Ward Lock and Co., London, New York, Melbourne.

Darwin, C.R., 1859, *The Origin of Species,* John Murray, London.

Darwin, C.R., 1868, *Variations of Animals and Plants under Domestication,* 2 vols., John Murray, London.

Darwin, C.R., 1871, *The Descent of Man,* 2 vols., John Murray, London.

Darwin, C.R., 1872, *Expression of Emotions in Man and Animals,* John Murray, London.

Darwin, C.R., 1888, *Darwin's Life and Letters,* 3 vols., ed. Darwin, F., John Murray, London.

Darwin, C.R., 1892, *Autobiography and Selected Letters,* ed. Darwin, F., Appleton and Co. Ltd., London; Dover Publications Inc., New York, 1958.
See also Glass, B. *et al.,* 1959; Eiseley, L., 1959; Smith, S., 1960.

Darwin, E, 1806, *The Poetical Works,* 3 vols., London.
See also King-Hele, D., 1974; Krause, E., 1879.

Daumas, M., ed., 1964, *Histoire Générale des Techniques,* Presses Universitaires de France.

Dawson, W.R., 1929, *Magician and Leech: a study of the beginnings of medicine with special reference to ancient Egypt,* Methuen, London.

Descartes, R., *Oeuvres Complètes,* ed. Cousin, V., 11 vols., Paris, 1824. *The Philosophical Works of Descartes,* ed. and trs. Haldane, E.S. and Ross, G.R.T., 2 vols., University Press, Cambridge.
See also Kemp-Smith, N., 1963; Vartanian, A., 1953.

Diderot, D., 1774 - 1780, *Eléments de Physiologie,* in *Oeuvres Complètes,* ed. Assézat et Tourneux. Paris, 1875 - 1877.

See also Vartanian, A., 1953.
Dijksterhuis, E.J., 1961, *The Mechanisation of the World Picture,* trs. Dikshoorn, C., Oxford University Press, London.
Diogenes Laertius, *Lives of Eminent Philosophers,* trs. Hicks, R.D., 2 vols., *The Loeb Classical Library,* Heinemann Ltd.,London, 1925.
Dobson, J.F., 1924, Herophilus, *Proc. Roy. Soc. Med.,* **18,** 19 - 32.
Dobson, J.F., 1926, Erasistratus, *Proc. Roy. Soc. Med.,* **20,** 825 - 32.
Dodds, E.R., 1951, *The Greeks and the Irrational,* University of California Press, Berkeley and Los Angeles.
Drake, S., 1957, *Discoveries and Opinions of Galileo,* Doubleday, New York.
Drake, S. and Drabkin, I.E., 1969, *Mechanics in Sixteenth Century Italy,* The University of Wisconsin Press, Madison, Milwaukee and London.
Driesch, H., 1891. *Die Mathematisch-mechanische Betrachtung Morphologischer Probleme der Biologie,* Jena.
Driesch, H., 1908, *The Science and Philosophy of Organism,* 2 vols., Adam and Charles Black, London.
Driesch, H., 1914, *The History and Theory of Vitalism,* trs. Ogden, C.K., Macmillan and Co. Ltd., London.
Duhem, P., 1913 - 1959, *Le Système du Monde,* 10 vols., Paris.
Dunn, L.C., 1965, *A Short History of Genetics,* McGraw-Hill Book Co., New York and London.
Durkheim, E., 1964, *Essays in Sociology and Philosophy,* ed. Wolff, K.H., Harper and Row Ltd., New York and London.
Dutrochet, R.J.H., 1824, *Recherches Anatomiques et Physiologiques sur la Structure Intime des Animaux et sur la Motilité*, Paris.

Eckermann: see Goethe, J.W.
Einstein, A. and Infield, L., 1938, *The Evolution of Modern Physics,* Simon and Schuster Ltd., New York.
Eiseley, L., 1959, *Darwin's Century,* Gollancz Ltd., London.
Eliot, T.S., 1920, *The Sacred Wood,* Faber and Faber Ltd., London.
Erasistratus: see Dobson, J.F., 1926; Sarton, G., 1952.

Farrington, B., 1961, *Greek Science,* Penguin Books, Harmondsworth, Middlesex.
Farrington, B., 1967, *The Faith of Epicurus,* Weidenfeld and Nicholson, London.
Ferrier, D., 1873, Experimental researches in cerebral physiology and pathology, *West Riding Lun. Asyl. Med. Rep.,* **3,** 30 - 96.
Ferrier, D., 1876, *The Functions of the Brain,* Smith, Elder; London.
See also Sherrington, C.S., 1928.
Fiegl, H., 1967, *The 'Mental' and the 'Physical',* University of Minnesota Press, Minneapolis.
Flourens, P., 1824, *Recherches Expérimentales sur les Propriétés et les Fonctions du Système Nerveux dans les Animaux Vertèbres,* Paris.
See also Fulton, J. and Wilson, L., 1966.
Foster, M., 1901, *History of Physiology during the Sixteenth, Seventeenth and Eighteenth Centuries,* Dover Publications Inc., New York.
Frankfort, H., Frankfort, H.A., Wilson, J.A., Jacobsen, T., 1949, *Before Philosophy: the Intellectual Adventure of Ancient Man,* Penguin Books, Harmondsworth, Middlesex.
Frazer, J, 1907 - 1915, *The Golden Bough, A Study in Magic and Religion,* 12 vols., Macmillan and Co. Ltd., London.
Freeman, K., 1946, *The Presocratic Philosophers,* Blackwell, Oxford.
French, R.K., 1969, *Robert Whytt: The Soul and Medicine,* Wellcome Institute of the History of Medicine, London.
Freud, S., 1953, *The Complete Psychological Works,* Hogarth Press, London.
Friedlander, P., 1958, *Plato,* trs. Meyerhof, H., Routledge and Kegan Paul, London.
Fulton, J. and Wilson, L., 1966, *Selected Readings in the History of Physiology,* C.C. Thomas, Springfield, Illinois.

Galen, *Opera Omnia,* ed. Kuehn, D.C.G., Leipzig, 1821 - 1833.
This is the standard edition of Galen's works in Greek and Latin, available in photo-reprints.
Oeuvres Anatomiques et Médicales de Galien, trs. and ed. Daremberg, C., Paris, 1854 - 1856. A French translation of many of the treatises.
On the Natural Faculties, trs. Brock, A.J., *The Loeb Classical Library,* Heinemann Ltd., London, 1952.
On the Usefulness of the Parts of the Body, trs. May, M.T., Cornell University Press, New York, 1968.
See also Sarton, G., 1954; Siegel, R.E., 1968.

Galilei, G., 1623, *Il Saggiatore,* Roma; trs. in Drake, S., 1957.

Galilei, G., 1632, *Dialogo . . . . sopra i Due Massimi Sistemi del Mondo, Telemaico e Copernicano,* Fiorenza; facsimile reprint by Culture et Civilisation, Bruxelles, 1966; trs. Salusbury, T., *Dialogue on the Great World Systems,* London, 1661, ed. de Santillana, G., Chicago University Press, 1953.

Galilei, G., 1638, *Discorsi . . . intorno a Due Nuove Scienze,* Leida; facsimile reprint by Culture et Civilisation, Bruxelles, 1966; trs. Crew H. and de Salvio, A., *Dialogues Concerning Two New Sciences,* Dover Books Inc., New York, 1953.
See also Crombie, A.C., 1952; Drake, S., 1957; Drake, S. and Drabkin, I.E., 1969; de Santillana, G., 1958.

Gall, F.J. and Spurzheim, J.C., 1810 - 1819, *Anatomie et Physiologie du Système Nerveux en Génerale, et du Cerveau en Particulier,* 4 vols., Paris.

Galvani, L., 1791, *De Viribus Electricitatus,* Bologna; facsimile reproduction and trs. by Foley, M.G., Burndy Library Inc., Norwalk, Connecticut, 1954.

Gassendi, P., 1649, *Animadversiones in Decimum Librum Diogenes Laertius,* Lyon.

Gassendi, P., 1658, *Syntagma Philosophicum* in *Opera Omnia,* Lyon.
See also Brett, G.S., 1908.

Ghiselin, B., ed., 1955, *The Creative Process,* University of California Press, Berkeley and Los Angeles.

Gillespie, G.C., 1951, *Genesis and Geology,* Harvard University Press, Cambridge, Mass.

Gillespie, G.C., 1960, *The Edge of Objectivity,* Princeton University Press.

Glass, B., Temkin, O., Strauss, W.L., eds., 1959, *Forerunners of Darwin,* Johns Hopkins Press, Baltimore.

Godwin, W., 1793, *Political Justice,* London.

Goethe, J.W., 1790, *Faust,* trs. Taylor, B., Worlds Classics, Oxford, 1932.

Goethe, J.W., Einwirkung der neueren Philosophie, *Naturwissenschaftliche Schriften,* ed. Beutler, E., Zurich, 1950.

Goethe, J.W., *Conversations of Goethe with Eckermann and Soret,* trs. Oxenford, K., George Bell and Sons, London, 1892.
See also Arber, A., 1964; Cassirer, E., 1945; Hume-Brown, P., 1920; Sherrington, C.S., 1949; Trevelyan, H., 1949; Vietor, K, 1950.

Golgi, C., 1908, La doctrine du neurone, *Le Prix Nobel en 1906,* Stockholm.

Gomperz, T., 1912, *Greek Thinkers,* 4 vols., John Murray, London.

Goodfield, G.J., 1960, *The Growth of Scientific Physiology,* Hutchinson and Co. Ltd., London.

Goodrich, E.S., 1930, *Studies in the Structure and Development of Vertebrates,* Macmillan and Co. Ltd., London.

Gouldner, A.W., 1965, *Enter Plato,* Basic Books Inc., New York.

Granit, R., 1965, *Charles Scott Sherrington,* Nelson, London.

Grene, M., 1963, *A Portrait of Aristotle,* Faber and Faber Ltd., London.

Grigg, E.L.: see Coleridge, S.T.

Guerlac, H., 1961, *Lavoisier, the Crucial Year, The Background and Origin of his First Experiments on Combustion, 1772,* Cornell University Press, Ithaca.

Guthrie, W.K.C., 1963, 1965, *A History of Greek Philosophy,* 2 vols., University Press, Cambridge.

Haeckel, E., 1876, *The History of Creation,* 2 vols., trs. revised by Lankester, E.R., H.S. King and Co. Ltd., London.

Hagberg, K., 1952, *Carl Linnaeus,* trs. Blair, A., Cape, London.
Haldane, E.S. and Ross, G.R.T., 1912: see Descartes, R.
Hall, M., 1833, On the reflex functions of the medulla oblongata and the medulla spinalis, *Phil. Trans. Roy. Soc.,* **123,** 635 - 65.
Hall, T.S., ed., 1951, *A Source Book in Animal Biology,* McGraw-Hill Book Co. Inc., New York.
Hall, T.S., 1969, *Ideas of Life and Matter,* 2 vols., Chicago University Press, Chicago and London.
Haller, A. von, 1751, *Primae Lineae Physiologiae,* Gottingen; trs. Cullen, W., *First Lines of Physiology,* Edinburgh, 1779.
Haller, A. von, 1757 - 1766, *Elementa Physiologiae Corporis Humani,* 8 vols., Lausanne.
Hanson, N.R., 1961, *Patterns of Discovery,* University Press, Cambridge.
Hartley, D., 1749, *Observations on Man,* London.
Harvey, W., *The Works of William Harvey translated from the Latin with a Life of the Author by Robert Willis, M.D.,* Sydenham Society, London, 1847. Some of the more important of Harvey's writings are conveniently collected in *Great Books of the Western World,* 28, *Encyclopedia Britannica Inc.,* ed. Hutchins, R.M., Chicago, London, Toronto, 1952 (trs. Robert Willis):
An Anatomical Disquisition on the Motion of the Heart and Blood in Animals (Exercitatio Anatomica de Motu Cordis et Sanguinis in Animalibus), London, 1628.
The First Disquisition to John Riolan on the Circulation of the Blood, 1649.
A Second Disquisition to John Riolan in which many objections to the Circulation of the Blood are refuted, 1649.
Anatomical Exercises in the Generation of Animals (Exercitationes de Generatione Animalium), 1653.
See also Bayon, H.P., 1938, 1939; Keynes, G., 1966; Pagel, W., 1966; Whitteridge, G., 1971.
Heath, T., 1949, *Mathematics in Aristotle,* Oxford University Press, London.
Held, R. and Hein, A., 1963, Movement-produced stimulation in the development of visually guided behaviour, *J. Comp. Physiol. Psychol.,* **56,** 872–6.
Helmholtz, H., 1847, *Ueber der Erhaltung der Kraft,* Berlin; trs. in Magie, W.F., 1935.
Helmont, J. van, 1648, *Ortus Medicinae,* Amsterdam; facsimile reprint by Culture et Civilisation, Bruxelles, 1966.
Hendril, W., 1968, *The Greek Patristic View of Nature,* University Press, Manchester.
Herophilus: see Dobson, J.F., 1924.
Hesse, M., 1961, *Forces and Fields: The Concept of Action at a Distance,* Nelson and Co. Ltd., London.
Hobbes, T., 1650, *Human Nature,* London; ed. Molesworth, W., London, 1840.
Hobbes, T., 1651, *Leviathan,* London; ed. Oakeshott, M., Blackwood, Oxford, 1946. See also Brown, K.C., 1965.
Hodges, H., 1970, *Technology in the Ancient World,* Allen Lane, The Penguin Press, London.
Holmyard, E.J., 1957, *Alchemy,* Penguin Books, Harmondsworth, Middlesex.
Homer, *The Iliad,* trs. Murray, A.T., *The Loeb Classical Library,* Heinemann Ltd., London, 1924.
Hooke, R., 1665, *Micrographia,* London; facsimile reprint by Culture et Civilisation, Bruxelles, 1966.
Hopkins, A.J., 1934, *Alchemy, Child of Greek Philosophy,* Columbia University Press, New York.
Hopstock, H., 1921, Leonardo as anatomist, in Singer, C., ed., *Studies in the History and Method of Science,* vol. 2, pp. 151 - 91, Clarendon Press, Oxford.
Hubel, D.H., 1963, The visual cortex of the brain, *Sci. Amer.,* **209,** 5, 54-62.
Hughes, A.F.W., 1959, *A History of Cytology,* Abelard-Schuman Ltd., New York.
Hume, D., 1739, *A Treatise of Human Nature,* London; in *The Philosophical Works of David Hume,* ed. Green, T.H. and Grose, T.H., 2 vols., Longmans Green and Co., London, 1874.
Hume, D., 1740, *An Abstract of a Book lately Published; Entitled, A Treatise of Human Nature,* London; University Press, Cambridge, 1938.

Hume-Brown, P., 1920, *The Life of Goethe,* 2 vols., John Murray, London.
Hutton, J., 1795, *The Theory of the Earth,* Edinburgh.
Huxley, T.H., 1858, On the theory of the vertebrate skull, *Proc. Roy. Soc. B.,* **9**, 381 - 457.
Huxley, T.H., 1869, The genealogy of animals, in *Critiques and Addresses,* Macmillan and Co., London, 1888.
Huxley, T.H., 1874, On the hypothesis that animals are automata and its history, in *Collected Essays,* vol. 1, Macmillan and Co. Ltd., London, 1898.
Huxley, T.H., 1898, *Scientific Memoirs of T.H. Huxley,* ed. Foster, M. and Lankester, E.R., 3 vols., Macmillan and Co. Ltd., London.
Huxley, T.H., 1900, *Life and Letters of T.H. Huxley,* ed. Huxley, L., 3 vols., Macmillan and Co. Ltd., London.

Iltis, H., 1932, *Life of Mendel,* trs. Paul, E. and C., George Allen and Unwin, London.

Jackson, J.H., 1931, 1932, *Selected Writings of John Hughlings Jackson,* ed. Taylor, J., 2 vols., Hodder and Stoughton, London.
Jaeger, W., 1934, *Aristotle,* trs. Robinson, E.S., Oxford University Press, London.
Jaeger, W., 1947, *Theology of the Early Greek Philosophers,* trs. Robinson, E.S., Clarendon Press, Oxford.
Jammer, M., 1957, *Concepts of Force: A Study in the Foundations of Dynamics,* Harvard University Press, Cambridge, Mass.
Jenkins, T.G., 1959, The ancient craft of coracle building, *Country Life,* **125**, 716 - 17.
John, E.R., 1972, Switchboard vs. statistical theories of learning and memory, *Science,* **177**, 850 - 64.
Jung, K., 1953, *Collected Works,* Routledge and Kegan Paul, London.

Kant, I., 1747, *Kant's Inaugural Dissertation and other writings on Space,* trs. Handyside, J., Chicago, 1929.
Kant, I., 1755, *Allgemeine Naturgeschichte und Theorie des Himmels,* Konigsberg and Leipzig; ed. and trs. Hastie, W., *Kant's Cosmogony,* Maclehose and Sons, Glasgow, 1900.
Kant, I., 1781, *Kritik der Reinen Vernunft,* Riga (second edition, 1787); trs. Meiklejohn, J.M.D., *The Critique of Pure Reason,* Encyclopedia Britannica Great Books, 42, Encyclopedia Britannica Inc., ed. Hutchins, R.M., Chicago, London, Toronto, 1952.
Kant, I., 1788, *Kritik der practischen Vernunft,* Riga; trs. Abbott, T., *The Critique of Practical Reason,* Encyclopedia Britannica Inc., ed. Hutchins R.M., Chicago, London, Toronto, 1952.
Kant, I., 1790, *Kritik der Urteilskraft,* Berlin and Leibau; trs. Meredith, J.C., *The Critique of Judgment,* Encyclopedia Britannica Inc., ed. Hutchins, R.M., Chicago, London, Toronto, 1952.
See also Abbott, T.K., 1927; Caird, E., 1889; Korner, S., 1955.
Kappers, A., Huber, G.C. and Crosbie, E.C. (1936), *The Comparative Anatomy of the Vertebrates including Man,* 2 vols., Macmillan, New York.
Kemp Smith, N., 1963, *New Studies in the Philosophy of Descartes,* Macmillan and Co. Ltd., London.
Kesten, H., 1945, *Copernicus and his World,* trs. Ashton, E. and Guterman, N., Roy Publishers, New York.
Keynes, G., 1966, *The Life of William Harvey,* Clarendon Press, Oxford.
King-Hele, D., 1974, Erasmus Darwin, master of many crafts, *Nature,* **247**, 87 - 91.
Kirk, G.S. and Raven, J.E., 1971, *The Presocratic Philosophers,* University Press, Cambridge.
Klemm, F., 1959, *A History of Western Technology,* trs. Singer, D.W., George Allen and Unwin, London.
Konorski, J., 1948, *Conditioned Reflexes and Neuron Organisation,* trs. Garry, S., University Press, Cambridge.
Konorski, J., 1967, *The Integrative Activity of the Brain,* Chicago University Press.
Korner, S., 1955, *Kant,* Penguin Books Ltd., Harmondsworth, Middlesex.
Koyré, A., 1943, Galileo and Plato, *J. Hist. of Ideas,* **4**, 400 - 28.
Koyré, A., 1957, *From the Closed World to the Infinite Universe,* The Johns Hopkins Press, Baltimore.

Kranzberg, M. and Pursell, C.W., 1967, *Technology in Western Civilisation,* 2 vols., Oxford University Press, London.

Krause, E., 1879, *The Life of Erasmus Darwin,* trs. Dallas W.S., John Murray, London.

Kristeller, P.O., 1945, The school of Salerno: its development and its contribution to the history of learning, *Bull. Hist. Med.,* **17,** 138 - 94.

Kuehn, D.C.G.: see Galen.

Kuhn, T.S., 1962, *The Structure of Scientific Revolutions,* Chicago University Press.

Lamarck, J.B.P.A. de, 1801, *Système des Animaux sans Vertèbres,* Paris; facsimile reprint by Culture et Civilisation, Bruxelles, 1969.

Lamarck, J.B.P.A. de, 1809, *Philosophie Zoologique,* 2 vols., Paris; facsimile reprint by Culture et Civilisation, Bruxelles, 1970; trs. Elliot, H., *Zoological Philosophy,* Macmillan and Co., London, 1914; Hefner Publishing Co. New York and London, 1963.

Lamarck, J.B.P.A. de, 1815 - 1822, *Histoire Naturelle des Animaux sans Vertèbres,* 7 vols., Paris; facsimile reprint by Culture et Civilisation, Bruxelles, 1969. See also Glass, B., *et al.,* 1959.

Lavoisier, A. and Laplace, P., 1780, Mémoire sur la chaleur, *Mémoires de l'Académie des Sciences,* Paris; trs. Gabriel, M.L., 'Memoir on heat' in *Great Experiments in Biology,* ed. Gabriel, M.L. and Fogel, S., Prentice-Hall, Engelwood Cliffs, N.J., 1955.

Lavoisier, A., 1789, *Traité Elémentaire de Chimie,* 2 vols., Paris; facsimile reprint by Culture et Civilisation, Bruxelles, 1965; trs. Kerr, R., *Elements of Chemistry,* Great Books of the Western World, 45, Encyclopedia Britannica Inc., ed. Hutchins, R.M., Chicago, London, Toronto, 1952.
See also Guerlac, H., 1961; Meldrum, A.V., 1930; Metzger, H., 1935.

Laycock, T., 1845, On the reflex functions of the brain, *Brit. and For. Med. Rev.,* **19,** 298 - 318.

Leeuwenhoek, A. van, 1677, Observations . . . of the carneous fibres of a muscle and the cortical and medullary part of the brain, *Phil. Trans. Roy. Soc.,* **12,** 899 - 895 (905).
See also Schierbeek, A., 1959.

Leibniz, G.W. von, 1714, *Monadology* in *Monadology and other Philosophical Writings,* trs. Latta, R., Oxford University Press, London, 1898.

Leicester, H.M. and Klickstein, H.S., ed., 1952, *A Source Book in Chemistry,* McGraw-Hill Book Co., New York.

Leonardo da Vinci
The Drawings of Leonardo da Vinci, compiled, annotated and introduced by Popham, A.E., Cape, London, 1946.
Selections from the Notebooks of Leonardo da Vinci, ed. Richter, I.A., Oxford University Press, London, 1952.
Leonardo da Vinci on the Human Body, O'Malley, C.D. and Saunders, J.B., Schumann, New York, 1962.
See also Clark, K., 1958; Hopstock, H., 1921.

Lewes, G.H., 1879, *Problems of Life and Mind,* Trubner and Co. Ltd., London.

Lewes, G.H., 1890, *A Biographical History of Philosophy,* Routledge and Sons Ltd., London.

Liddell, E.G.T., 1960, *The Discovery of Reflexes,* Clarendon Press, Oxford.

Liebig, J., 1842, *Animal Chemistry,* trs. Gregory, W., London.

Linnaeus, C., 1751, *Philosophia Botanica,* Stockholm and Amsterdam.

Linnaeus, C., 1758, *Systema Naturae,* Stockholm; facsimile reprint by British Museum (Natural History), London, 1956.

Linnaeus, C., 1766, *Clavis Medicinae,* Stockholm.
See also Hagberg, K., 1952.

Locke, J., 1690, *Essay Concerning Human Understanding,* London; ed. Yolton, J.W., 2 vols., Everyman's Library, Dent, London, 1965.

Loeb, J., 1912, *The Mechanistic Conception of Life,* ed. Fleming, D., Harvard University Press, Cambridge, Mass.

Longet, F.A., 1842, *Anatomie et Physiologie du Système Nerveux de l'Homme,* Paris.

Lovejoy, A.O., 1952, *Essays in the History of Ideas,* Johns Hopkins Press, Baltimore.

Lovejoy, A.O., 1960, *The Great Chain of Being,* Harper Torchbooks, New York.

Lucretius, *The Nature of the Universe,* trs. Latham, R., Penguin Books, Harmondsworth, Middlesex, 1951.
Lyell, C., 1830 - 1832, *Principles of Geology,* 3 vols., first edition, John Murray, London.
Lyell, C., 1881, *Life, Letters and Journals,* 2 vols., ed. Mrs. Lyell, John Murray, London. See also Wilson, L.G., 1972.

MacKay, D.M., 1967, *Freedom of Action in a Mechanistic Universe,* University Press, Cambridge.
Magendie, F., 1822, Expériences sur les fonctions des racines des nerfs rachidiens, *J. Physiol. Exp. Path.,* **2,** 276 - 9.
Magie, W.F., ed., 1935, *A Source Book in Physics,* McGraw-Hill Book Co. Inc., New York and London.
Malebranche, Le P.N., 1672, *Recherche de la Vérité*, Paris.
Malinowski, B., 1925, *Magic, Science and Religion,* The Free Press, Glencoe, Illinois.
Malthus, T., 1798, *An Essay on the Principle of Population,* London.
Mare, Walter de la, 1969, *The Complete Poems,* ed. Richard de la Mare, Faber and Faber Ltd., London.
Matteucci, C., 1840, *Essai sur les Phénoménes Electriques des Animaux,* Paris.
Maupertuis, P.L.M. de, 1756, *Système de la Nature,* Lyon.
May, M.T., 1968: see Galen.
Mayow, J., 1674, *Tractatus Quinque Medico-physici,* Oxford; trs. A.C.B. and L.D., *Medico-Physical Works,* The Alembic Club, Edinburgh.
McColley, G., 1940, Nicholas Hill and the *Philosophia Epicurea, Ann. Sci.,* **4,** 390 - 405.
Meldrum, A.V., 1930, *The Eighteenth Century Revolution in Science,* Longmans Green, Calcutta.
Melsen, A.G. van, 1960, *From Atomos to Atom, The History of the Concept 'Atom',* trs. Koren, H.J., Harper Torchbooks, New York.
Mendel, J.G., 1866, Versuch uber Pflanzenhybriden, *Verh. naturforsch. Verein Brunn,* **4,** 3 - 47; trs. Royal Horticultural Society, London, 1901, reprinted in *Experiments in Plant Hybridisation,* Oliver and Boyd, London, 1965. See also Iltis, H., 1932.
Merton, R.K., 1970, *Science, Technology and Society in Seventeenth Century England,* Howard Fertig, New York.
Mettrie, J.O. de la, 1748, *L'Homme Machine,* Leyden; original text and translation by Bussey, G.C. and Calkins, M.W., Open Court Publishing Co., Chicago, 1912. See also Vartanian, A., 1960.
Metzger, H., 1935, *La Philosophie de la Matière chez Lavoisier,* Paris.
Meyer, A.W., 1939, *The Rise of Embryology,* Oxford University Press, London.
Mill, J.S., 1873, *Autobiography,* Longmans, London.
Milton, J., *Complete Poems and Selected Prose,* ed. Visiac, E.H., The Nonesuch Press, London, 1938.
Monod, J., 1970, *Le hasard et la nécessité,* Editions du Seuil, Paris.
Monroe, A., 1781, *The Works of Alexander Monroe,* Edinburgh.
Morgan, T.H., Sturtevant, A.H., Muller, H.J. and Bridges, C.B., 1915, *The Mechanism of Mendelian Inheritance* (second edition 1926), Constable and Co., London.
Morris, D., 1967, *The Naked Ape,* Cape, London.

Needham, J., 1964 - 1974, *Science and Civilisation in China,* 4 vols. (continuing), University Press, Cambridge.
Needham, J., 1970, ed., *The Chemistry of Life,* University Press, Cambridge.
Needham, J., and Hughes, A., 1959, *A History of Embryology,* second edition, University Press, Cambridge.
Newton, I., 1687, *Principia Mathematica,* London trs. Motte, A., *Sir Isaac Newton's Mathematical Principles,* London, 1729; revised Cajori, F., University of California Press, Berkeley and Los Angeles, 1934.
Newton, I., 1704, *Opticks,* London; facsimile edition by Culture et Civilisation, Bruxelles, 1966; fourth edition, 1730, ed. Cohen, I.B., Dover Publications Inc., New York, 1952.

Newton, I., *Isaac Newton's Papers and Letters on Natural Philosophy,* ed. Cohen, I.B., University Press, Cambridge, 1958.
Newton, I., *The Correspondence of Sir Isaac Newton,* ed. Eddleston, J., 1850; facsimile edition by Frank Cass and Co. Ltd., London, 1969.
Nordenskiold, E., 1920 - 1924, *The History of Biology,* trs. Eyre, L.B., Tudor Publishing Co., New York, 1928.
Nuyens, M., 1948, *L'Evolution de la Psychologie d'Aristote,* Louvain.

Oken, L., 1809 - 1811, *Lehrbuch der Naturphilosophie,* 3 vols., Berlin; trs. Tuck, A., *Elements of Physiophilosophy,* Ray Society, London, 1874.
Olmsted, J.M.D. and Olmsted, E.H., 1952, *Claude Bernard,* Abelard Schumann, London, New York, Toronto.
O'Malley, C.D., 1964, *Andreas Vesalius of Brussels,* University of California Press, Berkeley and Los Angeles.
O'Malley, C.D. and Saunders, J.B., 1962: see Leonardo da Vinci.
Onians, R.B., 1954, *The Origins of European Thought,* University Press, Cambridge.
Oppenheimer, J.M., 1967, *Essays in the History of Embryology and Biology,* M.I.T. Press, Massachusetts and London.
Owen, R., 1848, *On the Archetype and Homologies of the Vertebrate Skeleton,* London.
Owens, J., 1963, *The Doctrine of Being in the Aristotelian Metaphysics,* Medieval Studies of Toronto Inc., Toronto.

Pagel, W., 1958, *Paracelsus: An Introduction to Philosophical Medicine in the Era of the Renaissance,* Karger, Basel and New York.
Pagel, W., 1966, *The Biological World of William Harvey,* Karger, Basel and New York.
Paley, W., 1825, *The Works of William Paley,* 5 vols., ed. Lynam, R., London.
Paracelsus (Hohenheim, T.B. de): see Pagel, W., 1958.
Paré, A., 1564, *Dix Livres de la Chirurgie,* Paris; trs. Linker, R.W. and Womack, N., *Ten Books of Surgery,* University of Georgia Press, Athens, 1969.
See also Singer, D.W., 1924.
Partington, J.R., 1959 - 1970, *A History of Chemistry,* 4 vols., Macmillan and Co. Ltd., London.
Pater, W., 1873, *Studies in the History of the Renaissance,* London.
Pavlov, I.P., 1927, *Conditioned Reflexes,* trs. and ed. Andrep, G.V., Oxford University Press, London.
Pavlov, I.P., 1955, *Selected Works,* trs. Belsky, S., ed. Gibbons, J., Foreign Languages Publishing House, Moscow.
Perrault, C., 1680, *De la Mécanique des Animaux,* Paris.
Perrier, E., 1884, *La Philosophie Zoologique avant Darwin,* ed. Alcan, F., Paris.
Perutz, M.F., 1962, *Proteins and Nucleic Acids,* Elsevier, Amsterdam, London and New York.
Planck, M., 1950, *Scientific Autobiography and other Papers,* trs. Gaynor, F., Williams and Norgate, London.
Plato, *The Dialogues of Plato translated into English,* Jowett, B., 5 vols., The Clarendon Press, Oxford.
See also Friedlander, P., 1958; Gouldner, A.W., 1965.
Playfair, J., 1802, *Illustrations to the Huttonian Theory of the Earth,* Edinburgh.
Poincaré, H., 1908, Mathematical Creation, in Ghiselin, B., 1955.
Pope, A., *The Poems of Alexander Pope,* ed. Butt, J., Methuen and Co. Ltd., London, 1963.
Popper, K.R., 1967, Quantum mechanics without 'The Observer', in *Quantum Theory and Reality,* ed. Bunge, M., Springer Verlag, Berlin, London, Heidelberg.
Porta, G.B. della, 1589, *Magiae Naturalis,* Naples; trs. Anonymous, *Natural Magick,* London, 1658; reprinted Basic Books Inc., New York, 1957.
Poynter, F.N.L., ed., 1958, *The History and Philosophy of Knowledge of the Brain and its Functions,* Blackwell, Oxford.
Priestley, J., 1774, *Experiments and Observations on Different Kinds of Air,* London; Alembic Club, Edinburgh, 1894.

Rahman, F., 1952, *Avicenna's Psychology*, Oxford University Press, London.

Rather, L.J., 1958, *Disease, Life and Death: Selected Essays by Rudolph Virchow*, Stanford.

Reymond, E. du Bois, 1848, *Untersuchungen über thierische Electricität*, Berlin; Bence-Jones, H., ed., *On Animal Electricity*, London, 1852.

Reymond, E du Bois, 1872, *Uber die Grenzen des Naturkennes*, Leipzig.

Rist, J.M., 1967, *Plotinus: The Road to Reality*, University Press, Cambridge.

Roger, J., 1963, *Les Sciences de la Vie dans la Pensée Française du XVIIIe Siècle*, Colin, Paris.

Ross, W.D., 1923, *Aristotle*, Methuen and Co., London.

Rousseau, J-J, 1755, *Discourse sur l'Origine de l'Inégalité*, Paris; Classiques Larousse, Paris, 1939.

Roux, W., 1888, Contributions to the developmental mechanics of the embryo, *Virchows Archiv*, **114**, 113 - 53.

Russell, B., 1946, *A History of Western Philosophy*, George Allen and Unwin, London.

Russell, B., 1969, *The ABC of Relativity*, George Allen and Unwin, London.

Russell, E.S., 1916, *Form and Function: A Contribution to the History of Animal Morphology*, John Murray, London.

Saint-Hilaire, G..1818, *Philosophie Anatomique*, 2 vols., Paris; facsimile reprint by Culture et Civilisation, Bruxelles, 1968.

Sambursky, S., 1956, *The Physical World of the Greeks*, Routledge and Kegan Paul, London.

Sambursky, S., 1959, *Physics of the Stoics*, Routledge and Kegan Paul, London.

Santillana, G. de, 1958, *The Crime of Galileo*, Heinemann Ltd., London.

Santillana, G. de and Dechend, H. von, 1969, *Hamlet's Mill: An Essay in Myth and the Frame of Time*, Gambit Inc., Boston.

Sarton, G., 1952, *A History of Science*, 2 vols., Harvard University Press, Cambridge, Mass.

Sarton, G., 1954, *Galen of Pergamon*, University of Kansas Press.

Sartre, J-P., 1957, *Being and Nothingness*, trs. Barnes, H.E., Methuen and Co. Ltd., London.

Scheler, M., 1926, *Die Wissenformen und die Gesselleschaft*, Leipzig.

Schierbeek, A., 1959, *Measuring the Invisible World*, Abelard-Schumann, New York.

Schiller, J., 1969, Physiology's struggle for independence in the first half of the nineteenth century, *Hist. Sci.*, **8**, 64 - 89.

Schleiden, M.J., 1838, Beitrage zur Phytogenesis, *Arch. fur Anat. und Physiol.*, trs. Smith, H., Contributions to phytogenesis, Sydenham Society, London, 1847.

Schofield, R.E., 1970, *Mechanism and Materialism*, Princeton University Press.

Schopenhauer, A., 1836, *On the Will in Nature* in *Schopenhauer Selections*, ed. Parker, D.H., Charles Scribner's Sons Ltd., London, 1928.

Schwann, T., 1839, *Mikroskopische Untersuchungen über die Ubereinstimmung in der Struktur dem Wachstein der Thiere und Pflanzen*, trs. Smith, H., *Microscopical Researches into the accordance in Structure and Growth of Animals and Plants*, Sydenham Society, London, 1847.

Sears, P.B., 1963, Utopia and the living landscape, *Daedalus*, **94**, 474 - 86.

Sechenov, I.M., 1863, *Physiologische Studien über die Hemmungsmechanismen für die Reflexthatigkeit im Gehirne des Frosches*, Berlin.

Sechenov, I.M., 1962, *Selected Physiological and Psychological Works*, trs. Belsky, S., ed. Gibbons, G., Foreign Languages Publishing House, Moscow.

Seneca, *Naturales Quaestiones*, trs. Corcoran, T.H., The Loeb Classical Library, Heinemann Ltd., London.

Serres, E., 1842, *Précis d'Anatomie Transcendente*, Paris.

Shelley, P.B., *Selected Poetry and Prose*, The Nonesuch Press, London, 1951.

Sherrington, C.S., 1906, *The Integrative Action of the Nervous System*, University Press, Cambridge.

Sherrington, C.S., 1928, Obituary notice of Sir David Ferrier, *Proc. Roy. Soc. B., 103*, viii - xvi.

Sherrington, C.S., 1946, *The Endeavour of Jean Fernel*, University Press, Cambridge.

Sherrington, C.S., 1949, *Goethe on Nature and Science*, University Press, Cambridge.

Sherrington, C.S., 1955, *Man on his Nature,* Penguin Books, Harmondsworth, Middlesex. See also Granit, R., 1965.

Siegel, R.E., 1968, *Galen's System of Physiology and Medicine,* Karger, Basel and New York.

Singer, C., ed., 1921, *Studies in the History and Method of Science,* 2 vols., Clarendon Press, Oxford.

Singer, C., 1952, *Vesalius on the Human Brain,* Oxford University Press, London.

Singer, C., 1957, *A Short History of Anatomy,* Constable and Co., London.

Singer, C., Holmyard, E.J., Hall, A.R. and Williams, T.I., 1954 - 1958, *A History of Technology,* 5 vols., Clarendon Press, Oxford.

Singer, D.W., 1924, *Selections from the Works of Ambroise Paré,* Classics of Medicine, London.

Smith, C.U.M., 1970, *The Brain, Towards an Understanding,* Faber and Faber Ltd., London.

Smith, S., 1960, The origin of 'The Origin' as discerned from Charles Darwin's notebooks and his annotations in the books he read between 1837 and 1842, *Adv. of Sci.,* **16,** 391 - 401.

Spemann, H., 1938, *Embryonic Development and Induction,* Yale University Press.

Spencer, H., 1864, *The Principles of Biology,* Williams and Norgate, London.

Storey, R.L., 1973, *Chronology of the Medieval World,* Barrie and Jenkins, London.

Strauss, L., 1965, On the spirit of Hobbes' political philosophy, in Brown K.C., 1965.

Streeter, E.C., 1916, The role of certain Florentines in the history of anatomy, artistic and practical, *Johns Hopkins Hosp. Bull,* **27,** 113 - 18.

Sutton, W.S., 1903, The chromosomes in heredity, *Biol. Bull.,* **4,** 231 - 51.

Temkin, O., 1949, Metaphors in human biology, in *Science and Civilisation* ed. Stauffer, R.C., Madison, Wisconsin.

Thompson, D'Arcy W., 1923, Natural Science: Aristotle, in *The Legacy of Greece,* ed. Livingstone, R.W., Clarendon Press, Oxford.

Thorndike, L., 1929 - 1956, *A History of Magic and Experimental Science,* 8 vols., Macmillan and Co. Ltd., London.

Tillyard, E.M.W., 1948, *The Elizabethan World Picture,* Chatto and Windus, London.

Toulmin, S. and Goodfield, J., 1965, *The Discovery of Time,* Hutchinson and Co., London.

Trevelyan, H., 1949, Goethe the Thinker, in *Essays on Goethe,* ed. Rose, E., Cassell and Co. Ltd., London.

Turnbull, C.M., 1965, *Wayward Servants: The Two Worlds of African Pygmies,* Eyre and Spottiswood, London.

Underwood, E.A., ed., 1953, *Science, Medicine and History,* 2 vols., Oxford University Press, London.

Underwood, E.A., 1963, The early teaching of anatomy at Padua, *Ann. Sci.,* **19,** 1 - 26.

Vartanian, A., 1953, *Diderot and Descartes,* Princeton University Press.

Vartanian, A., 1960, *La Mettrie's 'L'Homme Machine',* Princeton University Press.

Vasari, G., 1530, *Lives of the Artists,* Florence; trs. Bull, G., Penguin Books, Harmondsworth, Middlesex.

Vaucanson: see Chapuis, A. and Droz, E., 1958.

Verbeke, G., 1945, *L'Evolution de la Doctrine de Pneuma du Stoicisme à S. Augustine,* Louvain.

Verworn, M., 1893–1894, Modern physiology, *Monist,* **4,** 361.

Vesalius, A., 1543, *De Humani Corporis Fabrica,* Basileae; facsimile reprint by Culture et Civilisation, Bruxelles, 1964.
See also O'Malley, C.D., 1964; Singer, C., 1952.

Vietor, K., 1950, *Goethe the Thinker,* Harvard University Press, Cambridge, Mass.

Vinci, Leonardo da: see Leonardo.

Virchow, R., 1855, Cellular-Pathologie, *Virchows Archiv.,* **8,** 1.

Virchow, R., 1858, *Die Cellular-Pathologie in ihrer Begrundung auf physiologische und pathologische Gewebelehre,* Berlin; trs. Chance, F., *Cellular Pathology,* London, 1860.

See also Ackerknecht, E.H., 1925; Rather, L.J., 1958.
Voltaire, F.M., 1738, *Eléments de la Philosophie de Newton,* Amsterdam; trs. Hanna, J., *The Elements of Sir Isaac Newton's Philosophy,* London, 1738; facsimile reprint by Frank Cass and Co. Ltd., London, 1967.

Wallace, A.R.: see Williams-Ellis, A., 1966.
Walter, G.O., 1969, Typesetting, *Scient. Am.* **220,** (5), 60 - 9.
Watson, J.D., 1965, *Molecular Biology of the Gene,* Benjamin, New York.
Watson, J.D. and Crick, F.H.C., 1953, Molecular structure of nucleic acids: a structure for deoxyribose nucleic acid, *Nature,* **171,** 737 - 8.
Weismann, A., 1904, *The Evolution Theory,* 2 vols., trs. Thomson, J.A., and M.A., Edward Arnold Ltd., London.
Wheeler, L.R., 1939, *Vitalism, its History and Development,* H.F. and G. Witherby, London.
White, L., 1967, The historical roots of the ecological crisis, *Science,* **155,** 1203 - 7.
White, L., 1967, Technology in the middle ages, in Kranzberg, M. and Pursell, C.W., vol. 1., 1967.
Whitteridge, G., 1971, *William Harvey and the Circulation of the Blood,* Macdonald, London.
Whytt, R., 1751, *Essay on the Vital and other Involuntary Motions of Animals,* Edinburgh.
Wightman, W.P.D., 1951, *The Growth of Scientific Ideas,* Oliver and Boyd, Edinburgh.
Willey, B., 1934, *The Seventeenth Century Background: Studies in the Thought of the Age in relation to its Poetry and Religion,* Chatto and Windus, London.
Willey, B., 1965, *The Eighteenth Century Background: Studies on the Idea of Nature in the Thought of the Period,* Chatto and Windus, London.
Williams-Ellis, A., 1966, *Darwin's Moon: A Biography of Alfred Russell Wallace,* Blackie, London and Glasgow.
Willier, B.H. and Oppenheimer, J.M., 1964, *Foundations of Experimental Embryology,* Prentice-Hall, New York.
Willis, T., 1672, *De Anima Brutorum,* London; trs. Pordage, S., *Dr. Willis'Practice of Physic,* London, 1684.
Wilson, J.W., 1969, Biology attains maturity, in Clagett, M. ed., *Critical Problems in the History of Science,* University of Wisconsin Press, Madison, Milwaukee & London.
Wilson, L.G., 1959, Erasistratus, Galen and the pneuma, *Bull. Hist. Med.,* **33,** 293 - 314.
Wilson, L.G., 1961, William Croone's theory of muscular contraction , *Notes and Records of Roy. Soc.,* **16,** 158 - 78.
Wilson, L.G., 1972, *Charles Lyell, The Years to 1841: The Revolution in Geology,* Yale University Press, New Haven, London
Wittgenstein, L., 1922, *Tractatus Logico-Philosophicus,* ed. Pears, D.F. and McGuinness, B.F., Routledge, 1961.
Wittgenstein, L., 1963, *Philosophical Investigations,* trs. Anscombe, G.E.M., Blackwell, Oxford.
Wolff, C.F., 1759, *Theoria Generationis,* Halle.
Wolff, C.F., 1768, *De Formatione Intestorum,* republished as *Ueber die Bildung des Darmkanals in bebruteten Hunchen,* Halle, 1812.
Wood, W.B. and Edgar, R.S., 1967, Building a bacterial virus, *Scient. Am.,* **217,** 1, 60 - 75.
Woodger, J.H., 1967, *Biological Principles, A Critical Study,* Routledge and Kegan Paul, London.
Wordsworth, W., *The Poetical Works of William Wordsworth,* ed. Hutchinson, T., revised de Selincourt, E., Oxford University Press, London, 1936.

Xenophon, *Memorabilia,* trs. Marchant, E.C., The Loeb Classical Library, Heinemann Ltd., London, 1923.

Young, R.M., 1969, Malthus and the evolutionists: the common context of biological and social theory, *Past and Present,* **43,** 109 - 45.
Young, R.M., 1970, *Mind, Brain and Adaptation in the Nineteenth Century,* University Press, Cambridge.

Zeller, E., 1892, *The Stoics, Epicureans and Sceptics,* trs. Reichel, O.J., Longmans Green Ltd., London.
Zirkle, C., 1935, The inheritance of acquired characteristics and the provisional hypothesis of pangenesis, *Amer. Nat.,* **69**, 417 - 45.
Zirkle, C., 1951, The knowledge of heredity before 1900, in *Genetics in the Twentieth Century,* ed. Dunn, L.C., Macmillan, New York.

INDEX *

*NOTE: page numbers followed by a small n, e.g. 153n, refer the reader to the footnote reference on that page.